Comet for Data Science

Enhance your ability to manage and optimize the life cycle of your data science project

Angelica Lo Duca

BIRMINGHAM—MUMBAI

Comet for Data Science

Group Product Manager: Gebin George

Publishing Product Manager: Dinesh Chaudhary

Senior Editor: David Sugarman

Technical Editor: Devanshi Ayare

Copy Editor: Safis Editing

Project Coordinator: Farheen Fathima

Proofreader: Safis Editing

Indexer: Sejal Dsilva

Production Designer: Shyam Sundar Korumilli

Marketing Coordinators: Shifa Ansari and Abeer Riyaz Dawe

First published: August 2022

Production reference: 1280722

Published by Packt Publishing Ltd.
Livery Place
35 Livery Street
Birmingham
B3 2PB, UK.

ISBN 978-1-80181-443-0

www.packt.com

Jesus Christ: yesterday, today, forever.

- Angelica Lo Duca

Foreword

We founded Comet in 2017 to provide machine learning developers with tools and empower them to create business value with artificial intelligence. Since then, tens of thousands of community practitioners worldwide have adopted Comet to manage and optimize models across the complete MLOps life cycle.

We work with organizations across industries, including Affirm, Ancestry, Etsy, The RealReal, Uber, and Zappos, who use Comet to increase productivity, accelerate model development, and achieve value with AI.

We were thrilled to have Dr. Angelica Lo Duca join our community when she shared her work on Heartbeat, Comet's editorially independent publication. As she learned more about our MLOps platform, she shared those learnings with readers through clear, concise language and detailed code snippets and model visualizations.

A data journalist, Dr. Lo Duca guides the reader through the machine learning development journey with Comet, from exploratory data analysis through model deployment, showing the process of creating business value. In reading this book, you will come to understand Dr. Lo Duca's strategic use of Comet and her perspectives on how our platform plays a key role in the modern MLOps stack.

Writing a book is a huge undertaking, and while Dr. Lo Duca wrote and produced this book independently, we have supported her by reviewing the content. We hope it is useful as you explore what is possible when you use Comet to manage and optimize machine learning and deep learning models across the complete development life cycle.

Sign up for your free Comet account and join our Slack community. Simply visit Comet's website (www.comet.com) to find us. Also, I personally invite you to share your work with our community! We look forward to learning about the amazing discoveries and business value you create with Comet.

Happy reading,

Gideon Mendels
CEO and Founder
Comet

Contributors

About the author

Angelica Lo Duca is a researcher at the Institute of Informatics and Telematics of the National Research Council, Italy. She is also an external professor of data journalism at the University of Pisa. Her research includes data science, data journalism, and web applications. She used to work on network security, semantic web, linked data, and blockchain. She has published more than 40 scientific papers at national and international conferences and journals and has participated in many international projects and events, including as a member of the Program Committee. She is also part of the editorial team of the HighTech And Innovation Journal. She owns a personal blog, where she publishes articles on her research interests.

First of all, I would like to thank Packt's fantastic editorial team. Thank you for your support, commitment, competence, and availability throughout this long journey toward the publication of this book.

Secondly, I would like to thank Comet's wonderful team, who contributed to defining the topics of each chapter of this book. Without their contribution, it would not have been possible to carry out this project. A special thanks to Emilie Lewis and Dhruv Nair from the Comet team for their patience and dedication.

Next, a big thank you to my husband, my father, my sister, Nando, and Silvana, for their patience and for supporting me during the good times and the most difficult ones.

Finally, a big thank you to my children, who pave the way toward a future with hope through their eyes and their smiles.

About the reviewers

Devanshu Tayal is a data science enthusiast with experience in travel and banking. With a master's degree from BITS Pilani, he has also studied mechanical engineering at IK Gujral Punjab Technical University. In his spare time, Devanshu enjoys researching new applications for data science, playing music, and playing badminton. Diversity and inclusion, Python, algorithms, data structures, machine learning, natural language processing, Tableau, Power BI, data visualization, and AI are some of his other interests.

Emil Bogomolov is a machine learning lead at Youpi Inc. He is engaged in creating new ways of collaboration using video. Previously, he was a research engineer in the computer vision group at the Skolkovo Institute of Science and Technology. He is the co-author of papers published at international conferences, such as VISAPP, WACV, and CVPR, and an educational courses author on data analysis at online schools. Emil is also a frequent speaker at technology conferences and author of tech articles on machine learning and AI.

Table of Contents

3

Model Evaluation in Comet

Section 2 – A Deep Dive into Comet

4

Workspaces, Projects, Experiments, and Models

5

Building a Narrative in Comet

6

Integrating Comet into DevOps

7

Extending the GitLab DevOps Platform with Comet

Section 3 – Examples and Use Cases

8

Comet for Machine Learning

9

Comet for Natural Language Processing

10

Comet for Deep Learning

11

Comet for Time Series Analysis

Preface

A recent survey of machine learning professionals (`https://www.comet.com/site/about-us/news-and-events/press-releases/comet-releases-new-survey-highlighting-ais-latest-challenges-too-much-friction-too-little-ml/`) concluded that about 40%–60% of interviewed professionals abandoned their data science projects because they were not able to manage the full life cycle process of their data science projects. I'm a data science researcher, and before encountering Comet, I belonged to that 40%–60% of professionals who abandon their data science projects. In fact, during my working experience, I have abandoned many projects without concluding them because of the nature of research, where you test an idea and, if it does not work, you drop it.

Almost a year ago, I discovered Comet, a platform for model tracking and monitoring, and some wonderful people from its team, who opened my mind to the many features provided by Comet. I began to study it, with the hope of keeping my projects organized and moving them from early stages to production. I realized that I was able to conclude all the projects I implemented in Comet because of the simplicity of the platform.

Comet for Data Science is the result of my studies and tests, as well as the countless biweekly meetings with the Comet team. The book aims at helping you to learn how to manage a data science project workflow, from its early stages up to project deployment and reporting. In a single sentence, *Comet for Data Science* is written to help you to conclude your data science projects successfully.

By picking this book, you will look at the general concepts of data science from a Comet perspective, with the hope that you will increase your productivity. The book will take you through the journey of building a data science project and integrating it into Comet, including exploratory data analysis, model building and evaluation, report building, and, finally, moving the model to production. Throughout the book, you will implement many practical examples that you can use to better understand the described concepts, as well as starting points for your projects.

I hope that this book will add something to your knowledge, and – why not? – help you to become a better data scientist!

Happy reading!

Who this book is for

This book is for data scientists and data analysts who want to learn how to manage and optimize a complete data science project life cycle using Comet and other DevOps platforms. This book is also useful for those who aim at increasing their productivity by means of a practical tool for model tracking and monitoring. Prior programming knowledge of Python is assumed.

What this book covers

Chapter 1, An Overview of Comet, is a general introduction to Comet, an experimentation platform, which allows you to manage and optimize machine learning projects, from their early stages to their final deployment. First, you will learn what Comet is and who its target users are. Then, you will get familiar with the Comet basic concepts, including projects, experiments, workspaces, and panels. Finally, you will build two basic use cases in Comet.

Chapter 2, Exploratory Data Analysis in Comet, guides you to use Comet to perform **Exploratory Data Analysis (EDA)**. First, you will be introduced to the main steps to perform EDA, including problem setting, data preparation, preliminary data analysis, and preliminary results. Then, you will review the main two techniques used to perform EDA: visual and non-visual EDA. Finally, you will learn how to use Comet for EDA through a practical example.

Chapter 3, Model Evaluation in Comet, guides you to use Comet to perform model evaluation. First, you will be introduced to the main concepts to evaluate the performance of a model, such as data splitting, how to choose metrics for evaluation, and the basic concepts behind error analysis. Then, you will see the main model evaluation techniques for different data science tasks (classification, regression, and clustering). Finally, you will learn how to use Comet for model evaluation, through a practical example.

Chapter 4, Workspaces, Projects, Experiments, and Models, deepens some concepts regarding Comet. First, you will see some advanced concepts on workspaces, projects, and experiments, as well as how to perform parameter optimization in Comet. Then, you will learn how to implement a Comet experiment using R or Java as the main programming language. Finally, you will extend the basic examples implemented in *Chapter 1, An Overview of Comet*.

Chapter 5, Building a Narrative in Comet, describes some strategies to build a good report in Comet. First, you will review the basic concepts and techniques to build a narrative from data, including an overview of the DIKW pyramid, and how to turn your data into a story. Then, you will learn how to build a narrative in Comet through two practical examples.

Chapter 6, Integrating Comet into DevOps, provides you with practical concepts and examples on DevOps and MLOps and how to integrate them into Comet. First, you will review the basic concepts and best practices related to DevOps and MLOps. Then, you will learn how to integrate Comet into the DevOps/MLOps paradigm through the concept of the REST API. Next, you will analyze Docker and Kubernetes, two of the most common frameworks for DevOps. Finally, you will learn how to integrate Comet in Docker and Kubernetes through two practical examples.

Chapter 7, Extending the GitLab DevOps Platform with Comet, describes the concept of **Continuous Integration (CI)** and **Continuous Delivery (CD)**, how to implement it using GitLab, and how to integrate Comet in a CI/CD workflow. First, you will review the basic concepts of CI/CD, including the CI/CD workflow and the concept of a source control system. Then, you will see the GitLab basic concepts, including an overview of its architecture, how versioning works, and the basic GitLab commands.

Then, you will configure GitLab to work with Comet. Finally, you will see a practical example that will help you to get familiar with the described concepts.

Chapter 8, Comet for Machine Learning, provides you with an overview of the **Machine Learning (ML)** concepts, with a focus on the `scikit-learn` library, and how to integrate them in Comet. First, you will review the basic ML concepts, including a classification of the main ML systems, the main ML models, and their main challenges. Then, you will review the scikit-learn package, with a focus on preprocessing, dimensionality reduction, model selection, supervised learning, and unsupervised learning. Finally, you will learn how to integrate Comet with scikit-learn through a practical example.

Chapter 9, Comet for Natural Language Processing, illustrates the main concepts behind **Natural Language Processing (NLP)**, with a focus on the Spark NLP library, and how to integrate the main concepts in Comet. First, you will review the basic NLP workflow and also learn how to classify the main NLP systems and what their main challenges are. Then, you will review the Spark NLP library, including the concepts of annotation and pipeline. Finally, you will learn how to integrate Comet with Spark NLP through a practical example.

Chapter 10, Comet for Deep Learning, describes the main concepts behind **Deep Learning (DL)**, with a focus on the TensorFlow library, and how to integrate them in Comet. First, you will review the basic concepts behind neural networks, their difference with respect to DL networks, and how to classify DL networks. Then, you will review the TensorFlow library, with a focus on how to load a dataset, as well as how to build a train a model. Finally, you will learn how to integrate Comet with TensorFlow through a practical example.

Chapter 11, Comet for Time Series Analysis, reviews the main concepts related to **Time Series Analysis (TSA)**, with a focus on the Prophet library, and how to integrate them into Comet. First, you will review the basic concepts behind TSA, including the concept of stationarity, the time series components, and how to check the presence of breakpoints in a time series. Then, you will be introduced to the Prophet library, with a focus on how to build a prediction model. Finally, you will learn how to integrate Comet with Prophet through a practical example.

To get the most out of this book

You should have a basic knowledge of data science, as well as its general objectives. In addition, you should be familiar with the Python language, and, in particular with the `pandas` and `matplotlib` libraries.

You will need a version of Python installed on your computer – Python 3.8, if possible. Almost all code examples have been tested using macOS Monterey 12.0.1, with the exception of software in *Chapter 10, Comet for Deep Learning*, which has been tested on Google Colab. However, code examples should work with future version releases too.

Software/hardware covered in the book	Operating system requirements
Python 3.8	Windows, macOS, or Linux
Python libraries: 1. `comet-ml==3.23.0` 2. `findspark==1.4.2` 3. `gradio==3.2.2` 4. `matplotlib==3.4.3` 5. `numpy==1.19.5` 6. `pandas==1.3.4` 7. `pandas-profiling==3.1.0` 8. `pyspark==3.2.1` 9. `scikit-learn==1.0` 10. `seaborn==0.11.2` 11. `shap==0.40.0` 12. `spark-nlp==3.4.4` 13. `sweetviz==2.1.3`	Windows, macOS, or Linux.
1. `tensorflow==2.8.2`	Windows, macOS, or Linux If you are using macOS, please make sure that the chip is not Apple M1. To overcome this problem, you can use Google Colab.
Java SE Development Kit 17.0.2 (optional for *Chapter 4, Workspaces, Projects, Experiments, and Models*)	Windows, macOS, or Linux
Java libraries (optional for *Chapter 4, Workspaces, Projects, Experiments, and Models*): 1. `comet-java-sdk-1.1.10` `weka 3.8.6`	Windows, macOS, or Linux
R software (optional)	Windows, macOS, or Linux
R libraries (optional): 1. `caret` 2. `cometr`	Windows, macOS, or Linux
Docker	Windows, macOS, or Linux
Kubernetes	Windows, macOS, or Linux
`git`	Windows, macOS, or Linux
Java 8 (required for *Chapter 9, Comet for Natural Language Processing*)	Windows, macOS, or Linux
Apache Spark 3.1.2	Windows, macOS, or Linux

You should notice that the code examples described in *Chapter 4, Workspaces, Projects, Experiments, and Models*, require a different version of Java with respect to those described in *Chapter 9, Comet for Natural Language Processing*.

In addition, to make Comet work, you need to sign up to the Comet platform (`https://www.comet.com/signup`) and create an account.

If you are using the digital version of this book, we advise you to type the code yourself or access the code from the book's GitHub repository (a link is available in the next section). Doing so will help you avoid any potential errors related to the copying and pasting of code.

Download the example code files

You can download the example code files for this book from GitHub at `https://github.com/PacktPublishing/Comet-for-Data-Science`. If there's an update to the code, it will be updated in the GitHub repository.

We also have other code bundles from our rich catalog of books and videos available at `https://github.com/PacktPublishing/`. Check them out!

Download the color images

We also provide a PDF file that has color images of the screenshots and diagrams used in this book. You can download it here: `https://packt.link/sJpZu`.

Conventions used

There are a number of text conventions used throughout this book.

`Code in text`: Indicates code words in text, database table names, folder names, filenames, file extensions, pathnames, dummy URLs, user input, and Twitter handles. Here is an example: "We note that some columns, such as country and `reservation_status_date`, have a high cardinality."

A block of code is set as follows:

```
import pandas as pd
import matplotlib.pyplot as plt
import seaborn as sns
from datetime import datetime
```

Any command-line input or output is written as follows:

```
pip install pandas-profiling
```

Bold: Indicates a new term, an important word, or words that you see onscreen. For instance, words in menus or dialog boxes appear in **bold**. Here is an example: "Regarding the **Variables** section, we can distinguish between categorical and numerical data."

> **Tips or Important Notes**
> Appear like this.

Get in touch

Feedback from our readers is always welcome.

General feedback: If you have questions about any aspect of this book, email us at `customercare@ packtpub.com` and mention the book title in the subject of your message.

Errata: Although we have taken every care to ensure the accuracy of our content, mistakes do happen. If you have found a mistake in this book, we would be grateful if you would report this to us. Please visit `www.packtpub.com/support/errata` and fill in the form.

Piracy: If you come across any illegal copies of our works in any form on the internet, we would be grateful if you would provide us with the location address or website name. Please contact us at `copyright@packt.com` with a link to the material.

If you are interested in becoming an author: If there is a topic that you have expertise in and you are interested in either writing or contributing to a book, please visit `authors.packtpub.com`.

Share Your Thoughts

Once you've read *Comet for Data Science*, we'd love to hear your thoughts! Scan the QR code below to go straight to the Amazon review page for this book and share your feedback.

`https://packt.link/r/1-801-81443-0`

Your review is important to us and the tech community and will help us make sure we're delivering excellent quality content.

Section 1 – Getting Started with Comet

This section will introduce the basic concepts behind Comet, including the main Comet components: workspaces, projects, experiments, and panels (*Chapter 1, An Overview of Comet*). In addition, you will learn how to use Comet to perform exploratory data analysis (*Chapter 2, Exploratory Data Analysis in Comet*) and model evaluation in Comet (*Chapter 3, Model Evaluation in Comet*). Throughout this section, we do not cover model building because we will deal with this aspect in a later section, with more details and examples.

To get more familiar with the described concepts, you will be guided to implement some basic use cases, through step-by-step and commented examples in Python.

The main focus of this section is to get you ready to work with all the main features provided by Comet.

This section includes the following chapters:

- *Chapter 1, An Overview of Comet*
- *Chapter 2, Exploratory Data Analysis in Comet*
- *Chapter 3, Model Evaluation in Comet*

An Overview of Comet

Data science is a set of strategies, algorithms, and best practices that we exploit to extract insights and trends from data. A typical data science project life cycle involves different steps, including problem understanding, data collection and cleaning, data modeling, model evaluation, and model deployment and monitoring. Although every step requires some specific skills and capabilities, all the steps are strictly connected to each other and, usually, they are organized as a pipeline, where the output of a module corresponds to the input of the next one.

In the past, data scientists built complete pipelines manually, which required much attention: a little error in a single step of the pipeline affected the following steps. This manual management led to an extension of the time to market for complete data science projects.

Over the last few years, thanks to the improvements introduced in the fields of artificial intelligence and cloud computing, many online platforms have been deployed, for the management and monitoring of the different steps of a data science project life cycle. All these platforms allow us to shorten and facilitate the time to market of data science projects by providing well-integrated tools and mechanisms.

Among the most popular platforms for managing (almost) the entire life cycle of a data science project, there is **Comet**. Comet is an experimentation platform that provides an easy interface with the most popular data science programming languages, including Python, Java, JavaScript, and R software. This book provides concepts and extensive examples of how to use Comet in Python. However, we will give some guidelines on how to exploit Comet with other programming languages in *Chapter 4, Workspaces, Projects, Experiments, and Models*.

The main objective of this chapter is to provide you with a quick-start guide to implementing your first simple experiments. You will learn the basic concepts behind the Comet platform, including accessing the platform for the first time, the main Comet dashboard, and two practical examples, which will help you to get familiar with the Comet environment. We will also introduce the Comet terminology, including the concepts of workspaces, projects, experiments, and panels. In this chapter, we will also provide an overview of Comet, by focusing on the following topics:

- Motivation, purpose, and first access to the Comet platform
- Getting started with workspaces, projects, experiments, and panels

- First use case – tracking images in Comet
- Second use case – simple linear regression

Before moving on to how to get started with Comet, let's have a look at the technical requirements to run the experiments in this chapter.

Technical requirements

The examples illustrated in this book use Python 3.8. You can download it from the official website at `https://www.python.org/downloads/` and choose version 3.8.

The examples described in this chapter use the following Python packages:

- `comet-ml 3.23.0`
- `matplotlib 3.4.3`
- `numpy 1.19.5`
- `pandas 1.3.4`
- `scikit-learn 1.0`

comet-ml

`comet-ml` is the main package to interact with Comet in Python. You can follow the official procedure to install the package, as explained at this link: `https://www.comet.ml/docs/quick-start/`.

Alternatively, you can install the package with `pip` in the command line, as follows:

```
pip install comet-ml==3.23.0
```

matplotlib

`matplotlib` is a very popular package for data visualization in Python. You can install it by following the official documentation, found at this link: `https://matplotlib.org/stable/users/getting_started/index.html`.

In `pip`, you can easily install `matplotlib`, as follows:

```
pip install matplotlib== 3.4.3
```

numpy

numpy is a package that provides useful functions on arrays and linear algebra. You can follow the official procedure, found at `https://numpy.org/install/`, to install `numpy`, or you can simply install it through `pip`, as follows:

```
pip install numpy==1.19.5
```

pandas

pandas is a very popular package for loading, cleaning, exploring, and managing datasets. You can install it by following the official procedure as explained at this link: `https://pandas.pydata.org/getting_started.html`.

Alternatively, you can install the `pandas` package through `pip`, as follows:

```
pip install pandas==1.3.4
```

scikit-learn

scikit-learn is a Python package for machine learning. It provides different machine learning algorithms, as well as functions and methods for data wrangling and model evaluation. You can install `scikit-learn` by following the official documentation, as explained at this link: `https://scikit-learn.org/stable/install.html`.

Alternatively, you can install `scikit-learn` through `pip`, as follows:

```
pip install scikit-learn==1.0
```

Now that we have installed all the required libraries, we can move on to how to get started with Comet, starting from the beginning. We will cover the motivation, purpose, and first access to the Comet platform.

Motivation, purpose, and first access to the Comet platform

Comet is a cloud-based and self-hosted platform that provides many tools and features to track, compare, describe, and optimize data science experiments and models, from the beginning up to the final monitoring of a data science project life cycle.

In this section, we will describe the following:

- **Motivation** – why and when to use Comet
- **Purpose** – what Comet can be used for and what it is not suitable for
- **First access to the Comet platform** – a quick-start guide to access the Comet platform

Now, we can start learning more about Comet, starting with the motivation.

Motivation

Typically, a data science project life cycle involves the following steps:

1. **Understanding the problem** – Define the problem to be investigated and understand *which* types of data are needed. This step is crucial, since a misinterpretation of data may produce the wrong results.

2. **Data collection** – All the strategies used to collect and extract data related to the defined problem. If data is already provided by a company or stakeholder, it could also be useful to search for other data that could help to better model the problem.

3. **Data wrangling** – All the algorithms and strategies used to clean and filter data. The use of **Exploratory Data Analysis (EDA)** techniques could be used to get an idea of data shape.

4. **Feature engineering** – The set of techniques used to extract from data the input features that will be used to model the problem.

5. **Data modeling** – All the algorithms implemented to model data, in order to extract predictions and future trends. Typically, data modeling includes machine learning, deep learning, text analytics, and time series analysis techniques.

6. **Model evaluation** – The set of strategies used to measure and test the performance of the implemented model. Depending on the defined problem, different metrics should be calculated.

7. **Model deployment** – When the model reaches good performance and passes all the tests, it can be moved to production. Model deployment includes all the techniques used to make the model ready to be used with real and unseen data.

8. **Model monitoring** – A model could become obsolete; thus, it should be monitored to check whether there is performance degradation. If this is the case, the model should be updated with fresh data.

We can use Comet to organize, track, save, and make secure almost all the steps of a data science project life cycle, as shown in the following figure. The steps where Comet can be used are highlighted in green rectangles:

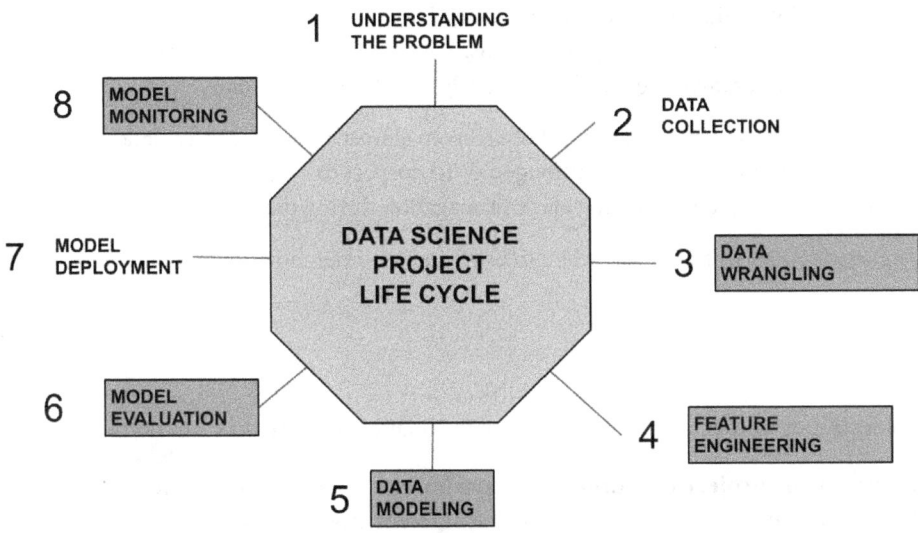

Figure 1.1 – The steps in a data science project life cycle, highlighting
where Comet is involved in green rectangles

The steps involved include the following:

- **Data wrangling** – thanks to the integration with some popular libraries for data visualization, such as the `matplotlib`, `plotly`, and `PIL` Python libraries, we can build **panels** in Comet to perform EDA, which can be used as a preliminary step for data wrangling. We will describe the concept of a panel in more detail in the next sections and chapters of this book.

- **Feature engineering** – Comet provides an easy way to track different experiments, which can be compared to select the best input feature sets.

- **Data modeling** – Comet can be used to debug your models, as well as performing hyperparameter tuning, thanks to the concept of **Optimizer**. We will illustrate how to work with Comet Optimizer in the next chapters of this book.

- **Model evaluation** – Comet provides different tools to evaluate a model, including panels, evaluation metrics extracted from each experiment, and the possibility to compare different experiments.

- **Model monitoring** – Once a model has been deployed, you can continue to track it in Comet with the previously described tools. Comet also provides an external service, named **Model Production Monitoring (MPM)**, that permits us to monitor the performance of a model in real time. The MPM service is not included in the Comet free plan.

We cannot exploit Comet directly to deploy a model. However, we can easily integrate Comet with GitLab, one of the most famous DevOps platforms. We will discuss the integration between Comet and GitLab in *Chapter 7, Extending the GitLab DevOps Platform with Comet*.

To summarize, Comet provides a single point of access to almost all the steps in a data science project, thanks to the different tools and features provided. With respect to a traditional and manual pipeline, Comet permits automating and reducing error propagation during the whole data science process.

Now that you are familiar with why and when to use Comet, we can move on to looking at the purpose of Comet.

Purpose

The main objective of Comet is to provide users with a platform where they can do the following:

- **Organize your project into different experiments** – This is useful when you want to try different strategies or algorithms or produce different models.

- **Track, reproduce, and store experiments** – Comet assigns to each experiment a unique identifier; thus, you can track every single change in your code without worrying about recording the changes you make. In fact, Comet also stores the code used to run each experiment.

- **Share your projects and experiments with other collaborators** – You can invite other members of your team to read or modify your experiments, thus making it easy to extract insights from data or to choose the best model for a given problem.

Now that you have learned about the purpose of Comet, we can illustrate how to access the Comet platform for the first time.

First access to the Comet platform

Using Comet requires the creation of an account on the platform. The Comet platform is available at this link: `https://www.comet.ml/`. Comet provides different plans that depend on your needs. In the free version, you can have access to almost all the features, but you cannot share your projects with your collaborators.

If you are an academic, you can create a premium Comet account for free, by following the procedure for academics: `https://www.comet.ml/signup?plan=academic`. In this case, you must provide your academic account.

You can create a free account simply by clicking on the **Create a Free Account** button and following the procedure.

Getting started with workspaces, projects, experiments, and panels

Comet has a nested architecture, as shown in the following figure:

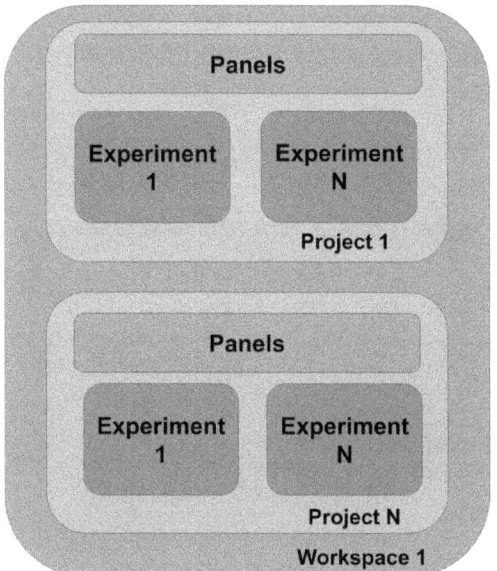 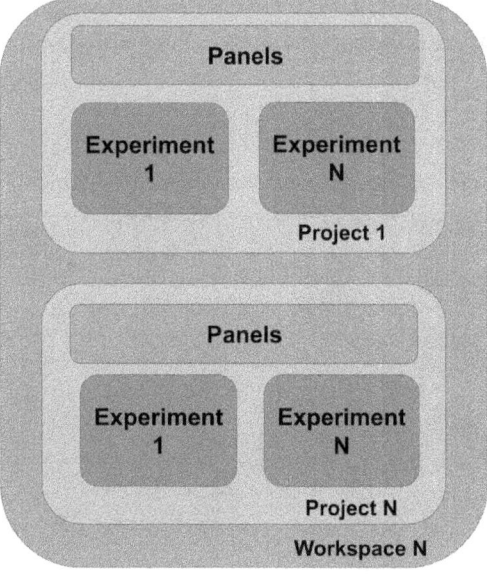

Figure 1.2 – The modular architecture in Comet

The main components of the Comet architecture include the following:

- **Workspaces**
- **Projects**
- **Experiments**
- **Panels**

Let's analyze each component separately, starting with the first one – workspaces.

Workspaces

A Comet workspace is a collection of projects, private or shared with your collaborators. When you create your account, Comet creates a default workspace, named with your username. However, you can create as many workspaces as you want and share them with different collaborators.

To create a new workspace, we can perform the following operations:

1. Click on your username on the left side of the page on the dashboard.
2. Select **Switch Workspace | View all Workspaces**.
3. Click on the **Create a workspace** button.
4. Enter the workspace name, which must be unique across the whole platform. We suggest prepending your username to the workspace name, thus avoiding conflicting names.
5. Click the **Save** button.

The following figure shows an account with two workspaces:

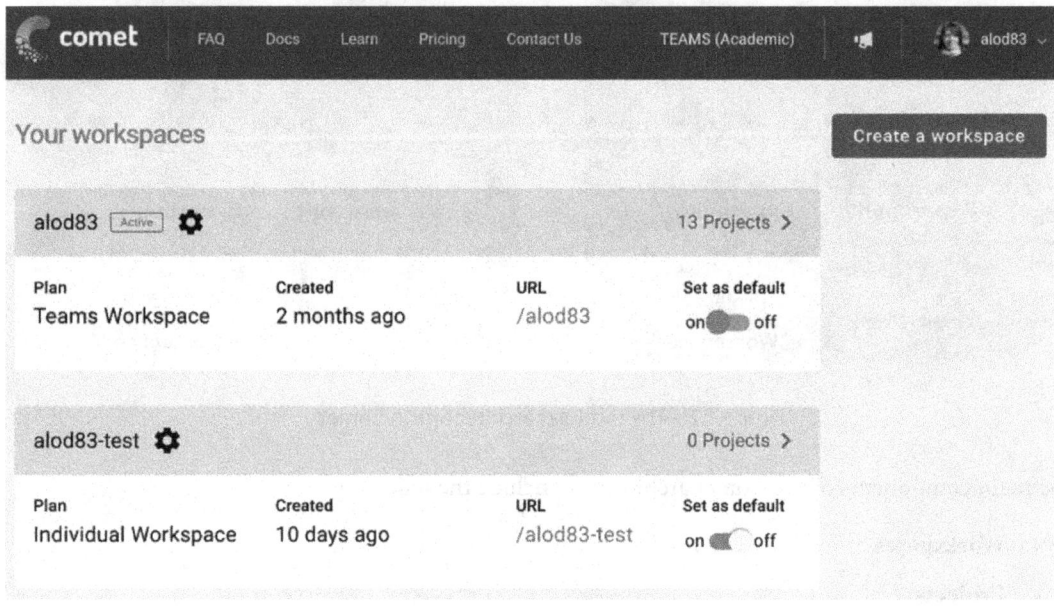

Figure 1.3 – An account with two workspaces

In the example, there are two workspaces, **alod83**, which is the default and corresponds to the username, and **alod83-test**.

Now that we have learned how to create a workspace, we can move on to the next feature: **projects**.

Projects

A Comet project is a collection of experiments. A project can be either private or public. A private project is visible only by the owner and their collaborator, while a public project can be seen by everyone.

In Comet, you can create as many projects as you want. A project is identified by a name and can belong only to a workspace, and you cannot move a project from one workspace to another.

To create a project, we perform the following operations:

1. Launch the browser and go to `https://www.comet.ml/`.

2. If you have not already created an account, click on the top-right **Create a free account** button and follow the procedure. If you already have an account, jump directly to *step 3*.

3. Log in to the platform by clicking the **Login** button located in the top-right part of the screen.

4. Select a **Workspace**.

5. Click on the **new project** button in the top-right corner of the screen.

6. A new window opens where you can enter the project information, as shown in the following figure:

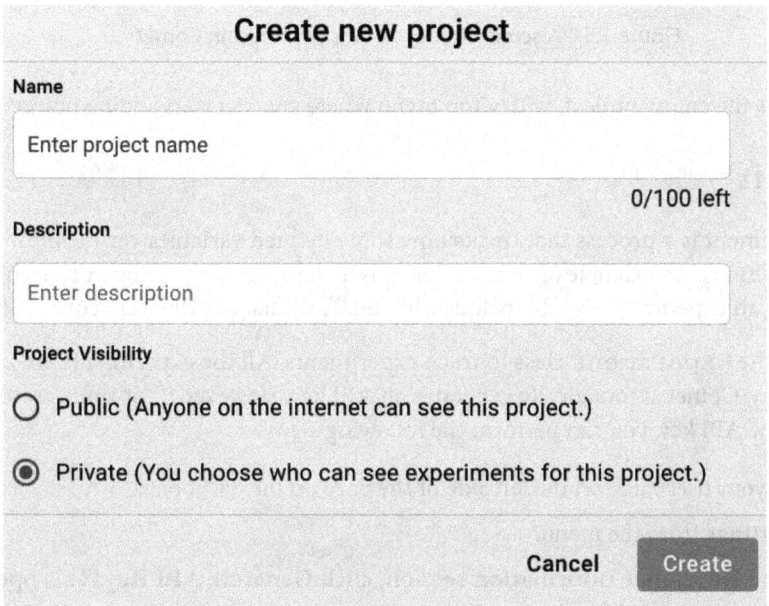

Figure 1.4 – The Create new project window

The window contains three sections: **Name**, where you write the project name, **Description**, a text box where you can provide a summary of the project, and **Project Visibility**, where you can decide to make the project either private or public. Enter the required information.

7. Click on the **Create** button.

The following figure shows the just-created Comet dashboard:

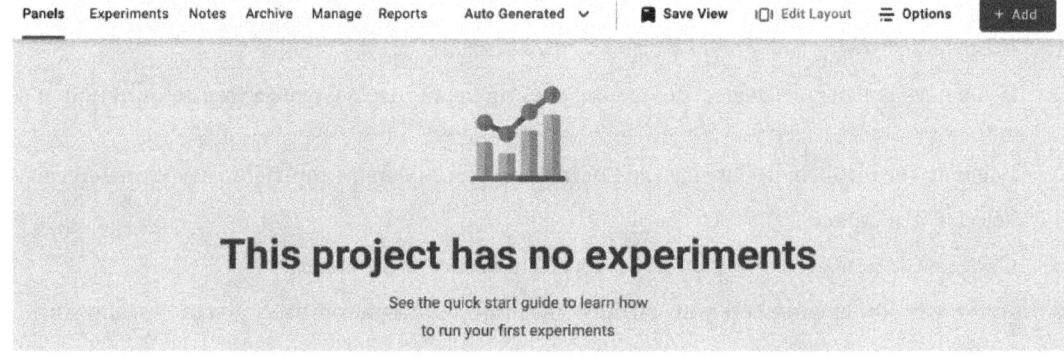

Figure 1.5 – A screenshot of an empty project in Comet

The figure shows the empty project, with a top menu where you can start adding objects.

Experiments

A Comet experiment is a process that tracks how some defined variables vary while changing the underlying conditions. An example of an experiment is hyperparameter tuning in a machine learning project. Usually, an experiment should include different runs that test different conditions.

Comet defines the `Experiment` class to track experiments. All the experiments are identified by the same API key. Comet automatically generates an API key. However, if for some reason you need to generate a new API key, you can perform the following steps:

1. Click on your username on the left side of the page on the dashboard.
2. Select **Settings** from the menu.
3. Under the **Developer Information** section, click **Generate API Key**. This operation will override the current API key.
4. Click **Continue** on the pop-up window. You can copy the value of the new API key by clicking on the **API Key** button.

We can add an experiment to a project by taking the following steps:

1. Click on the **Add** button, located in the top-right corner.
2. Select **New Experiment** from the menu.

3. Copy the following code (*Ctrl* + *C* on Windows/Linux environments or *Cmd* + *C* on macOS) and then click on the **Done** button:

```
# import comet-ml at the top of your file
from comet-ml import Experiment
# Create an experiment with your api key
experiment = Experiment(
    api_key="YOUR API KEY",
    project_name="YOUR PROJECT NAME",
    workspace="YOUR WORKSPACE",
)
```

The previous code creates an experiment in Comet. Comet automatically sets the `api_key`, `project_name`, and `workspace` variables. We can paste the previous code after the code we have already written to load and clean the dataset.

In general, storing an API key in code is not secure, so Comet defines two alternatives to store the API key. Firstly, in a Unix-like console, create an environment variable called COMET_API_KEY, as follows:

```
export COMET_API_KEY="YOUR-API-KEY"
```

Then, you can access it from your code:

```
import os
API_KEY = os.environ['COMET_API_KEY']
```

The previous code uses the `os` package to extract the `COMET_API_KEY` environment variable.

As a second alternative, we can define a `.comet.config` file and put it in our home directory or our current directory and store the API key, as follows:

```
[comet]
api_key=YOUR-API-KEY
project_name=YOUR-PROJECT-NAME,
workspace=YOUR-WORKSPACE
```

Note that the variable values should not be quoted or double-quoted. In this case, when we create an experiment, we can do so as follows:

```
experiment = Experiment()
```

We can use the `Experiment` class to track DataFrames, models, metrics, figures, images, metrics, and artifacts. For each element to be tracked, the `Experiment` class provides a specific method. Every tracking function starts with `log_`. Thus, we have `log_metric()`, `log_model()`, `log_image()`, and so on.

To summarize it in one sentence, experiments are the bridge between your code and the Comet platform.

Experiments can also be used as context managers, for example, to split the training and test phases into two different contexts. We will discuss these aspects in more detail in *Chapter 3, Model Evaluation in Comet*.

Now that you are familiar with Comet experiments, we can describe the next element – panels.

Panels

A Comet **panel** is a visual display of data. It can be either static or interactive. You can choose whether to make a panel private, share it with your collaborators, or make it public. While charts refer to a single experiment, panels can display data across different experiments, so we can define a panel only within the Comet platform and not in our external code.

Comet provides different ways to build a panel:

- Use a panel template.
- Write a custom panel in Python.
- Write a custom panel in JavaScript.

Let's analyze the three ways separately, starting with the first one: *Use a panel template*.

From the main page of our project, we can perform the following operations:

1. Click on the **Add** button located in the top-right corner of the page.
2. Select **New Panel**.
3. Choose the panel and select **Add** in the new window, as shown in the following figure:

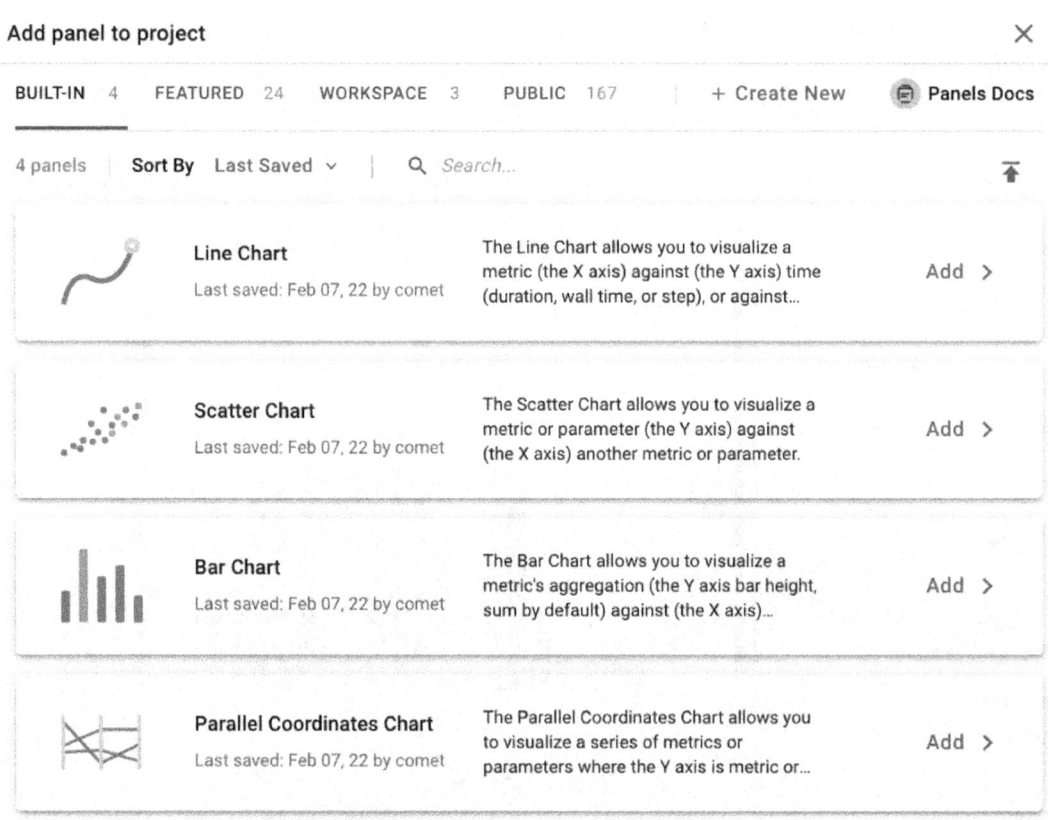

Figure 1.6 – The main window to add a panel in Comet

The figure shows the different types of available panels, divided into four main categories, identified by the top menu:

- **BUILT-IN** – Comet provides some basic panels.
- **FEATURED** – The most popular panels implemented by the Comet community and made available to everyone.
- **WORKSPACE** – Your private panels, if any.
- **PUBLIC** – All the public panels implemented by Comet users.

We can configure a built-in panel by setting some parameters in the Comet interface, as shown in the following figure:

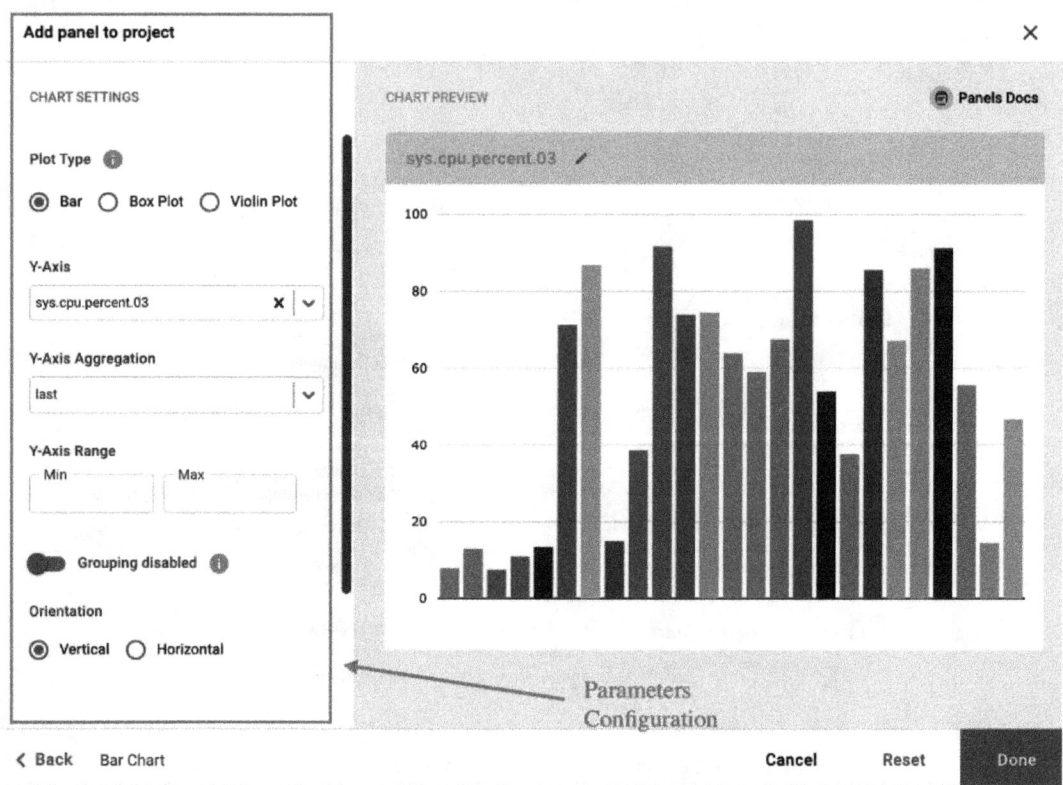

Figure 1.7 – Parameter configuration in a built-in Comet panel

The figure shows the form for parameter configuration on the left and a preview of the panel on the right.

While we can configure built-in panels through the Comet interface, we need to write some code to use the other categories of panels – featured, workspace, and public. In these cases, we select the type of panel, and then we configure the associated parameters, in the form of key-value pairs, as shown in the following figure:

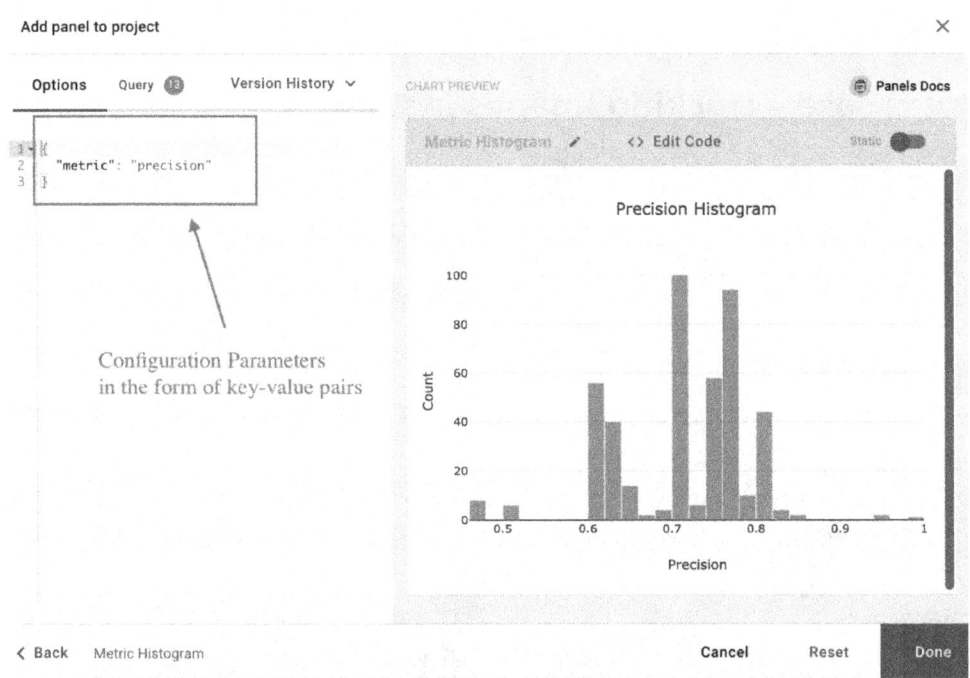

Figure 1.8 – Parameters configuration in a featured, workspace, or public panel

On the left, the figure shows a text box, where we can insert the configuration parameters in the form of a Python dictionary:

```
{"key1" : "value1",
 "key2" : "value2",
 "keyN" : "valueN"
}
```

Note the double quotes for the strings. The key-value pairs depend on the specific panel.

The previous figure also shows a preview of the panel on the right, with the possibility to access the code that generates the panel directly, through the **Edit Code** button.

So far, we have described how to build a panel through an existing template. Comet also provides the possibility to write custom panels in Python or JavaScript. We will describe this advanced feature in the next chapter, on EDA in Comet.

Now that you are familiar with the Comet basic concepts, we can implement two examples that use Comet:

- Tracking images in Comet
- Building a panel in Comet

Let's start with the first use case – tracking images in Comet.

First use case – tracking images in Comet

For the first use case, we describe how to build a panel showing a time series related to some Italian performance indicators in Comet. The example uses the `matplotlib` library to build the graphs.

You can download the full code of this example from the official GitHub repository of the book, available at the following link: `https://github.com/PacktPublishing/Comet-for-Data-Science/tree/main/01`.

The dataset used is provided by the World Bank under the CC 4.0 license, and we can download it from the following link: `https://api.worldbank.org/v2/en/country/ITA?downloadformat=csv`.

The dataset contains more than 1,000 time series indicators regarding the performance of Italy's economy. Among them, we will focus on the 52 indicators related to **Gross Domestic Product** (GDP), as shown in the following table:

Indicator Name	Indicator Code
Agriculture, forestry, and fishing, value added (% of GDP)	NV.AGR.TOTL.ZS
Broad money (% of GDP)	FM.LBL.BMNY.GD.ZS
Central government debt, total (% of GDP)	GC.DOD.TOTL.GD.ZS
Claims on other sectors of the domestic economy (% of GDP)	FS.AST.DOMO.GD.ZS
Coal rents (% of GDP)	NY.GDP.COAL.RT.ZS
Current account balance (% of GDP)	BN.CAB.XOKA.GD.ZS
Current health expenditure (% of GDP)	SH.XPD.CHEX.GD.ZS
Domestic credit provided by financial sector (% of GDP)	FS.AST.DOMS.GD.ZS
Domestic credit to private sector (% of GDP)	FS.AST.PRVT.GD.ZS
Domestic credit to private sector by banks (% of GDP)	FD.AST.PRVT.GD.ZS
Domestic general government health expenditure (% of GDP)	SH.XPD.GHED.GD.ZS
Expense (% of GDP)	GC.XPN.TOTL.GD.ZS
Exports of goods and services (% of GDP)	NE.EXP.GNFS.ZS
External balance on goods and services (% of GDP)	NE.RSB.GNFS.ZS
Final consumption expenditure (% of GDP)	NE.CON.TOTL.ZS
Foreign direct investment, net inflows (% of GDP)	BX.KLT.DINV.WD.GD.ZS
Foreign direct investment, net outflows (% of GDP)	BM.KLT.DINV.WD.GD.ZS
Forest rents (% of GDP)	NY.GDP.FRST.RT.ZS
General government final consumption expenditure (% of GDP)	NE.CON.GOVT.ZS

Government expenditure on education, total (% of GDP)	SE.XPD.TOTL.GD.ZS
Government expenditure per student, primary (% of GDP per capita)	SE.XPD.PRIM.PC.ZS
Government expenditure per student, secondary (% of GDP per capita)	SE.XPD.SECO.PC.ZS
Government expenditure per student, tertiary (% of GDP per capita)	SE.XPD.TERT.PC.ZS
Gross capital formation (% of GDP)	NE.GDI.TOTL.ZS
Gross domestic savings (% of GDP)	NY.GDS.TOTL.ZS
Gross fixed capital formation (% of GDP)	NE.GDI.FTOT.ZS
Gross fixed capital formation, private sector (% of GDP)	NE.GDI.FPRV.ZS
Gross national expenditure (% of GDP)	NE.DAB.TOTL.ZS
Gross savings (% of GDP)	NY.GNS.ICTR.ZS
Households' and NPISHs' final consumption expenditure (% of GDP)	NE.CON.PRVT.ZS
Imports of goods and services (% of GDP)	NE.IMP.GNFS.ZS
Industry (including construction), value added (% of GDP)	NV.IND.TOTL.ZS
Manufacturing, value added (% of GDP)	NV.IND.MANF.ZS
Market capitalization of listed domestic companies (% of GDP)	CM.MKT.LCAP.GD.ZS
Merchandise trade (% of GDP)	TG.VAL.TOTL.GD.ZS
Military expenditure (% of GDP)	MS.MIL.XPND.GD.ZS
Mineral rents (% of GDP)	NY.GDP.MINR.RT.ZS
Natural gas rents (% of GDP)	NY.GDP.NGAS.RT.ZS
Net acquisition of financial assets (% of GDP)	GC.AST.TOTL.GD.ZS
Net incurrence of liabilities, total (% of GDP)	GC.LBL.TOTL.GD.ZS
Net investment in nonfinancial assets (% of GDP)	GC.NFN.TOTL.GD.ZS
Net lending (+) / net borrowing (-) (% of GDP)	GC.NLD.TOTL.GD.ZS
Oil rents (% of GDP)	NY.GDP.PETR.RT.ZS
Personal remittances, received (% of GDP)	BX.TRF.PWKR.DT.GD.ZS
Research and development expenditure (% of GDP)	GB.XPD.RSDV.GD.ZS
Revenue, excluding grants (% of GDP)	GC.REV.XGRT.GD.ZS
Services, value added (% of GDP)	NV.SRV.TOTL.ZS
Stocks traded, total value (% of GDP)	CM.MKT.TRAD.GD.ZS
Tax revenue (% of GDP)	GC.TAX.TOTL.GD.ZS
Total natural resources rents (% of GDP)	NY.GDP.TOTL.RT.ZS
Trade (% of GDP)	NE.TRD.GNFS.ZS
Trade in services (% of GDP)	BG.GSR.NFSV.GD.ZS

Figure 1.9 – Time series indicators related to GDP used to build the dashboard in Comet

The table shows the indicator name and its associated code.

We implement the Comet dashboard through the following steps:

1. Download the dataset.
2. Clean the dataset.
3. Build the visualizations.
4. Integrate the graphs in Comet.
5. Build a panel.

So, let's move on to building your first use case in Comet, starting with the first steps, downloading and cleaning the dataset.

Downloading the dataset

As already said at the beginning of this section, we can download the dataset from this link: `https://api.worldbank.org/v2/en/country/ITA?downloadformat=csv`. Before we can use it, we must remove the first two lines from the file, because they are simply a header section. We can perform the following steps:

1. Download the dataset from the previous link.
2. Unzip the downloaded folder.
3. Enter the unzipped directory and open the file named `API_ITA_DS2_en_csv_v2_3472313.csv` with a text editor.
4. Select and remove the first four lines, as shown in the following figure:

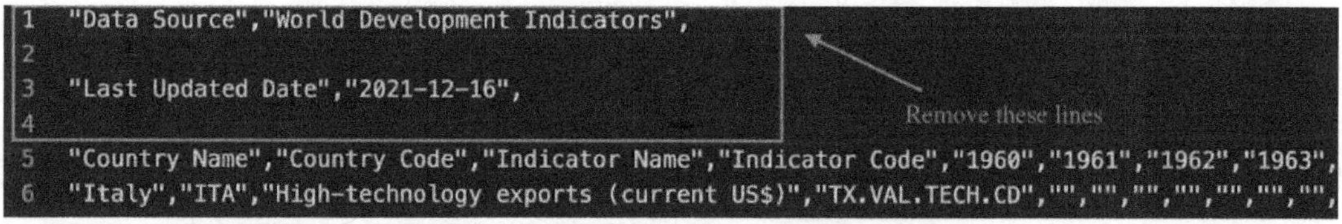

Figure 1.10 – Lines to be removed from the API_ITA_DS2_en_csv_v2_3472313.csv file

5. Save and close the file. Now, we are ready to open and prepare the dataset for the next steps. Firstly, we load the dataset as a `pandas` DataFrame:

```
import pandas as pd
df = pd.read_csv('API_ITA_DS2_en_csv_v2_3472313.csv')
```

We import the `pandas` library, and then we read the CSV file through the `read_csv()` method provided by `pandas`.

The following figure shows an extract of the table:

Country Name	Country Code	Indicator Name	Indicator Code	1960	1961	1962	1963	...	2017	2018	2019	2020
Italy	ITA	Age dependency ratio, old (% of working-age population)	SP.POP.DPND.OL	14,56	14,78	14,98	15,18		35,14	35,59	36,06	36,57
Italy	ITA	Population ages 65-69, female (% of female population)	SP.POP.6569.FE.5Y	3,90	3,93	3,98	4,05		6,05	6,00	5,96	5,97
Italy	ITA	Population ages 25-29, female (% of female population)	SP.POP.2529.FE.5Y	7,36	7,34	7,37	7,40		5,06	5,02	4,98	4,95
Italy	ITA	Life expectancy at birth, total (years)	SP.DYN.LE00.IN	69,12	69,76	69,15	69,25		82,95	83,35	83,20	
Italy	ITA	Labor force participation rate, female (% of female population ages 15+) (national estimate)	SL.TLF.CACT.FE.NE.ZS		24,62				40,94	41,10	41,26	39,84
Italy	ITA	Merchandise exports to low- and middle-income economies in Europe & Central Asia (% of total merchandise exports)	TX.VAL.MRCH.R2.ZS	1,71	1,68	1,07	1,12		6,23	5,92	5,83	

Figure 1.11 – An extract of the API_ITA_DS2_en_csv_v2_3472313.csv dataset

The dataset contains 66 columns, including the following:

- **Country Name** – Set to `Italy` for all the records.
- **Country Code** – Set to `ITA` for all the records.
- **Indicator Name** – Text describing the name of the indicator.
- **Indicator Code** – String specifying the unique code associated with the indicator.
- *Columns from 1960 to 2020* – The specific value for a given indicator in that specific year.
- *Empty column* – An empty column. The `pandas` DataFrame names this column `Unnamed: 65`.

Now that we have downloaded and loaded the dataset as a `pandas` DataFrame, we can perform dataset cleaning.

Dataset cleaning

The dataset presents the following problems:

- Some columns are not necessary for our purpose.
- We need only indicators related to GDP.
- The dataset contains the years' names in the header.
- Some rows contain missing values.

We can solve the previous problems by cleaning the dataset with the following operations:

- Drop unnecessary columns.
- Filter only GDP-based indicators.
- Transpose the dataset to obtain years as a single column.
- Deal with missing values.

Now, we can start the data cleaning process from the first step – drop unnecessary columns. We should remove the following columns:

- `Country Name`
- `Country Code`
- `Indicator Code`
- `Unnamed: 65`

We can perform this operation with a single line of code, as follows:

```
df.drop(['Country Name', 'Country Code', 'Indicator
Code','Unnamed: 65'], axis=1, inplace = True)
```

The previous code exploits the `drop()` method of the `pandas` DataFrame. As the first parameter, we pass the list of columns to be dropped. The second parameter, `(axis = 1)`, specifies that we want to drop columns, and the last parameter, `(inplace = True)`, specifies that all the changes must be stored in the original dataset.

Now that we have removed unnecessary columns, we can filter only GDP-based indicators. All the GDP-based indicators contain the following text: `(% of GDP)`. So, we can select only these indicators by searching all the records where the `Indicator Name` column contains that text, as shown in the following piece of code:

```
df = df[df['Indicator Name'].str.contains('(% of GDP)')]
```

We can use the operation contained in square brackets (`df['Indicator Name'].str.contains('(% of GDP)')`) to extract only the rows that match our criteria, in this case, the rows that contain the string `(% of GDP)`. We have exploited the `str` attribute of the DataFrame to extract all the strings of the `Indicator Name` column, and then we have matched each of them with the string `(% of GDP)`. The `contains()` method returns `True` only if there is a match; otherwise, it returns `False`.

Now we have only the interesting metrics. So, we can move on to the next step: transposing the dataset to obtain years as a single column. We can perform this operation as follows:

```
df = df.transpose()
```

```
df.columns = df.iloc[0]
df = df[1:]
```

The `transpose()` method exchanges rows and columns. In addition, in the transposed DataFrame, we want to rename the columns to the indicator name. We can achieve this through the last two lines of code.

Now that we have transposed the dataset, we can proceed with missing values management. We could adopt different strategies, such as interpolation or average values. However, to keep the example simple, we decide to simply drop rows from 1960 to 1969 that do not contain values for almost all the analyzed indicators. In addition, we drop the indicators that contain less than 30 no-null values.

We can perform these operations through the following line of code:

```
df.dropna(thresh=30, axis=1, inplace = True)
df = df.iloc[10:]
```

Firstly, we drop all the columns (each representing a different indicator) with less than 30 non-null values. Then, we drop all the rows from 1960 to 1969.

Now, the dataset is cleaned and ready for further analysis. The following figure shows an extract of the final dataset:

	Trade (% of GDP)	Government expenditure on education, total (% of GDP)	Gross capital formation (% of GDP)	...	Net lending (+) / net borrowing (-) (% of GDP)	Stocks traded, total value (% of GDP)
1970	30,09885	3,36681	26,00610			
1971	30,39609	2,89877	23,83427			
1972	31,82281		22,93538			
1973	34,06467		26,35899		-4,05517	
1974	40,80289	3,83514	29,77102		-3,68272	
1975	37,40045	3,71161	23,77739		-7,97296	0,71894
1976	41,19215	3,91952	26,59778		-5,48234	0,55892
1977	41,80956		24,44196		-4,54745	0,33618
1978	41,26819	3,78036	23,88776		-5,60589	0,62465
1979	43,71042	4,16373	24,42132		-7,16932	0,87907
1980	42,97266		26,76769		-8,69110	1,79648
1981	45,28204		24,69333			2,51916
1982	43,50829		23,73429			0,65235

Figure 1.12 – The final dataset, after the cleaning operations

The figure shows that we have grouped years from different columns into a single column, as well as splitting indicators from one column into different columns. Before the dropping operation, the dataset before had 61 rows and 52 columns. After dropping, the resulting dataset has 51 rows and 39 columns.

We can save the final dataset, as follows:

```
df.to_csv('API_IT_GDP.csv')
```

We have exploited the to_csv() method provided by the pandas DataFrame.

Now that we have cleaned the dataset, we can move on to the next step – building the visualizations.

Building the visualizations

For each indicator, we build a separate graph that represents its trendline over time. We exploit the matplotlib library.

Firstly, we define a simple function that plots an indicator and saves the figure in a file:

```python
import matplotlib.pyplot as plt
import numpy as np
def plot_indicator(ts, indicator):
    xmin = np.min(ts.index)
    xmax = np.max(ts.index)
    plt.xticks(np.arange(xmin, xmax+1, 1),rotation=45)
    plt.xlim(xmin,xmax)
    plt.title(indicator)
    plt.grid()
    plt.figure(figsize=(15,6))
    plt.plot(ts)
    fig_name = indicator.replace('/', "") + '.png'
    plt.savefig(fig_name)
    return fig_name
```

The plot_indicator() function receives a time series and the indicator name as input. Firstly, we set the range of the x axis through the xticks() method. Then, we set the title through the title() method. We also activate the grid through the grid() method, as well as setting the figure size, through the figure() method. Finally, we plot the time series (plt.plot(ts)) and save it to our local filesystem. If the indicator name contains a /, we replace it with an empty value. The function returns the figure name.

The previous code generates a figure like the following one:

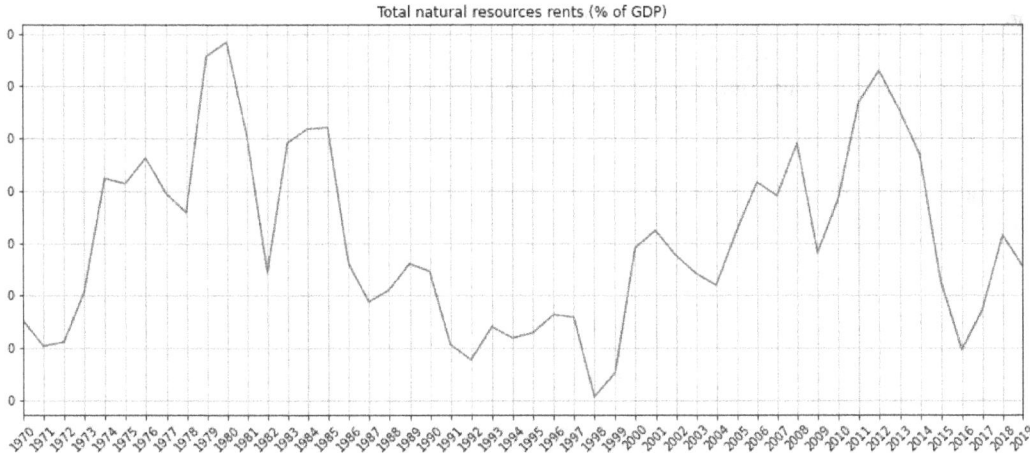

Figure 1.13 – An example of a figure generated by the plot_indicator() function

Now that we have set up the code to create the figures, we can move on to the last step: integrating the graphs in Comet.

Integrating the graphs in Comet

Now, we can create a project and an experiment by following the procedure described in the *Getting started with workspaces, projects, experiments, and panels* section. We copy the generated code and paste it into our script:

```
from comet-ml import Experiment
experiment = Experiment()
```

We have just created an experiment. We have stored the API key and the other parameters in the .comet.config file located in our working directory.

Now we are ready to plot the figures. We use the log_image() method provided by the Comet experiment to store every produced figure in Comet:

```
for indicator in df.columns:
    ts = df[indicator]
    ts.dropna(inplace=True)
    ts.index = ts.index.astype(int)
    fig = plot_indicator(ts,indicator)
    experiment.log_image(fig,name=indicator, image_
format='png')
```

We iterate over the list of indicators, identified by the DataFrame columns (`df.columns`). Then, we extract the associated time series from the current indicator. After, we save the figure through the `plot_indicator()` function, previously defined. Finally, we load the image in Comet.

The code is complete, so we run it.

When the running phase is complete, the results of the experiments are available in Comet. To see the produced graphs, we perform the following steps:

1. Open your Comet project.

2. Select **Experiments** from the top menu.

3. A table with all the experiments that have run appears. In our case, there is only one experiment.

4. Click on the experiment name, in the first cell on the left. Since we have not set a specific name for the experiment, Comet will add it for us. An example of a name given by Comet is *straight_contract_1272*.

5. Comet shows a dashboard with all the experiment details. Click on the **Graphics** section from the left menu. Comet shows all your produced graphs, as shown in the following figure:

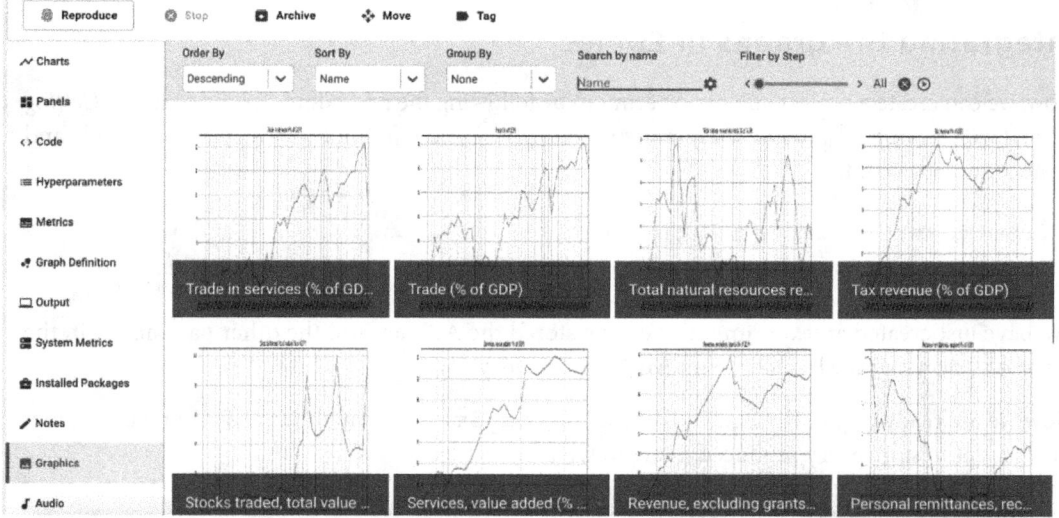

Figure 1.14 – A screenshot of the Graphics section

The figure shows all the indicators in descending order. Alternatively, you can display figures in ascending order and you can sort and group them.

Comet also stores the original images under the **Assets & Artifacts** section of the left menu, as shown in the following figure:

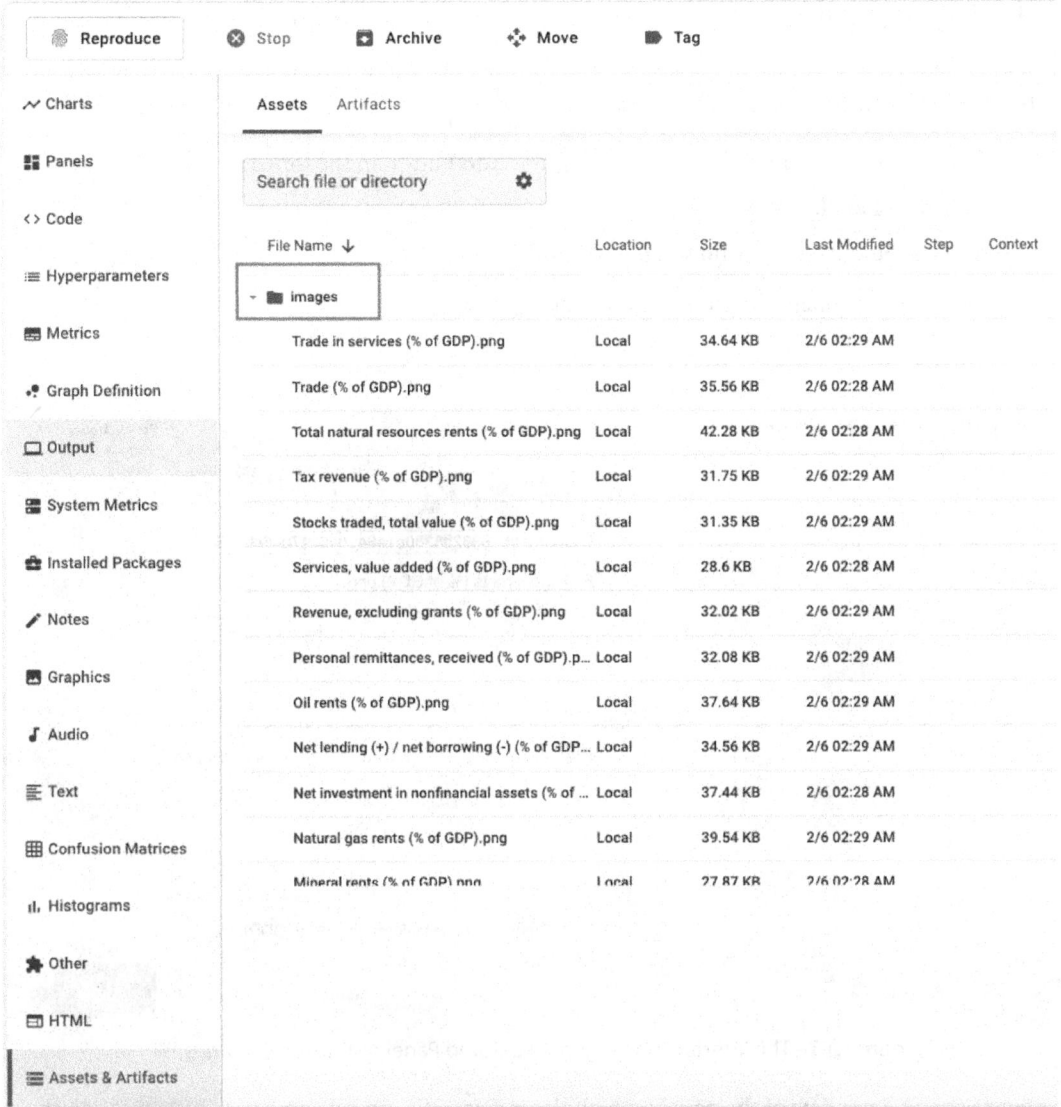

Figure 1.15 – A screenshot of the Assets & Artifacts section

The figure shows that all the images are stored in the **images** directory. The **Assets & Artifacts** directory also contains two directories, named **notebooks** and **source-code**.

Finally, the **Code** section shows the code that has generated the experiment. This is very useful when running different experiments.

Now that you have learned how to track images in Comet, we can build a panel with the created images.

Building a panel

To build the panel that shows all the tracked images, you can use the **Show Images** panel, which is available on the **Featured Panels** tab of the Comet panels. To add this panel, we perform the following steps:

1. From the Comet main dashboard, click on the **Add** button in the top-right corner.
2. Select **New Panel**.
3. Select the **Featured** tab | **Show Images** | **Add**.

    A new window opens, as shown in the following figure:

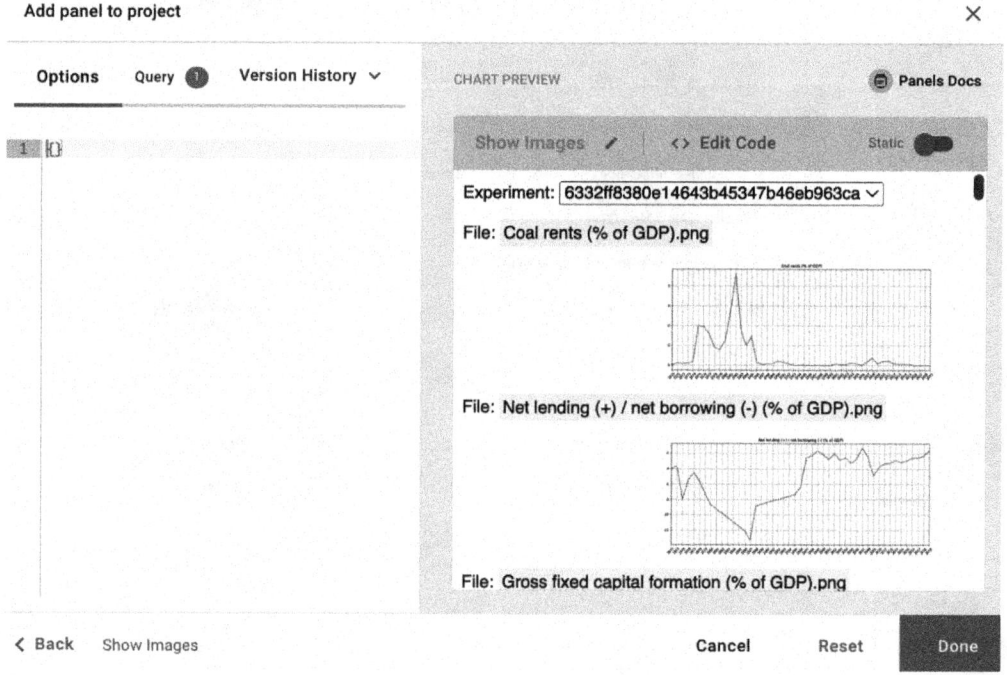

Figure 1.16 – The interface to build the Featured Panel called Show Images

The figure shows a preview of the panel with all the images. We can set some variables, through the key-value pairs, that contain only one image, by specifying the following parameters in the left text box:

```
{   "imageFileName": "Gross fixed capital formation (% of GDP).
png" }
```

The "imageFileName" parameter allows us to specify exactly the image to be plotted. The following figure shows a preview of the resulting panel, after specifying the image filename:

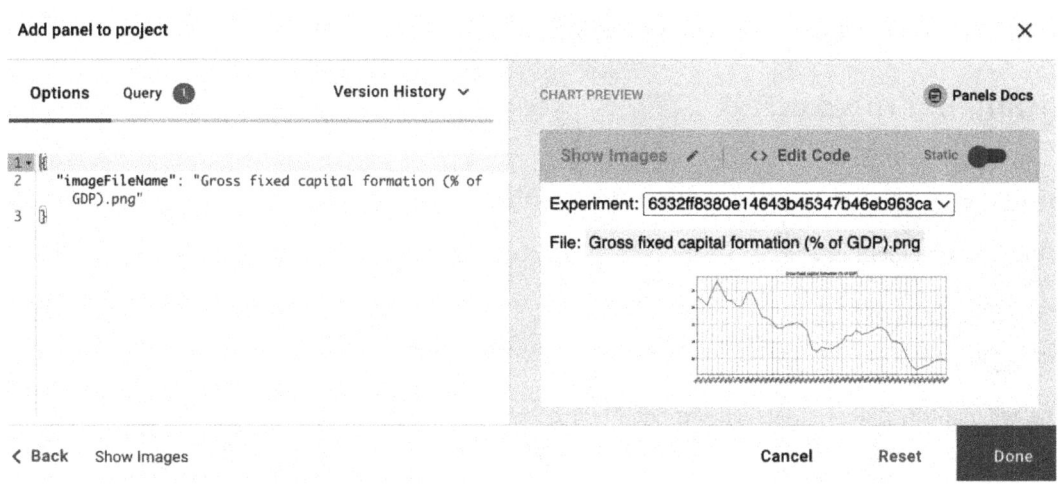

Figure 1.17 – A preview of the Show Images panel after specifying key-value pairs

The figure shows how we can manipulate the output of the panel on the basis of the key-value variables. The variables accepted by a panel depend on the specific panel. We will investigate this aspect more deeply in the next chapter.

Now that we have learned how to get started with Comet, we can move on to a more complex example that implements a simple linear regression model.

Second use case – simple linear regression

The objective of this example is to show how to log a metric in Comet. In detail, we set up an experiment that calculates the different values of **Mean Squared Error** (**MSE**) produced by fitting a linear regression model with different training sets. Every training set is derived from the same original dataset, by specifying a different seed.

You can download the full code of this example from the official GitHub repository of the book, available at the following link: `https://github.com/PacktPublishing/Comet-for-Data-Science/tree/main/01`.

We use the `scikit-learn` Python package to implement a linear regression model. For this use case, we will use the diabetes dataset, provided by `scikit-learn`.

We organize the experiment in three steps:

- Initialize the context.
- Define, fit, and evaluate the model.
- Show results in Comet.

Let's start with the first step: initializing the context.

Initializing the context

Firstly, we create a `.comet.config` file in our working directory, as explained in the *Getting started with workspaces, projects, experiments, and panels*, in the *Experiment* section.

As the first statement of our code, we create a Comet **experiment**, as follows:

```
from comet-ml import Experiment
experiment = Experiment()
```

Now, we load the diabetes dataset, provided by `scikit-learn`, as follows:

```
from sklearn.datasets import load_diabetes
  diabetes = load_diabetes()
X = diabetes.data
y = diabetes.target
```

We load the dataset and stored `data` and `target` in X and y variables, respectively.

Our context is now ready, so we can move on to the second part of our experiment: defining, fitting, and evaluating the model.

Defining, fitting, and evaluating the model

Let's start with the imports:

1. Firstly, we import all the libraries and functions that we will use:

    ```
    import numpy as np
    from sklearn.model_selection import train_test_split
    from sklearn.linear_model import LinearRegression
    from sklearn.metrics import mean_squared_error
    ```

    We import NumPy, which we will use to build the array of different seeds, and other `scikit-learn` classes and methods used for the modeling phase.

2. We set the seeds to test, as follows:

    ```
    n = 100
    seed_list = np.arange(0, n+1, step = 5)
    ```

    We define 20 different seeds, ranging from 0 to 100, with a step of 5. Thus, we have 0, 5, 10, 15 ... 95, 100.

3. For each seed, we extract a different training and test set, which we use to train the same model, calculate the MSE, and log it in Comet, using the following code:

```
for seed in seed_list:
    X_train, X_test, y_train, y_test = train_test_
split(X, y, test_size=0.20, random_state=seed)
    model = LinearRegression()
    model.fit(X_train,y_train)
    y_pred = model.predict(X_test)
    mse = mean_squared_error(y_test,y_pred)
    experiment.log_metric("MSE", mse, step=seed)
```

In the previous code, firstly we split the dataset into training and test sets, using the current seed. We reserve 20% of the data for the test set and the remaining 80% for the training set. Then, we build the linear regression model, we fit it (`model.fit()`), and we predict the next values for the test set (`model.predict()`). We also calculate MSE through the `mean_squared_error()` function. Finally, we log the metric in Comet through the `log_metric()` method. The `log_metric()` method receives the metric name as the first parameter and the metric value as the second parameter. We also specify the step for the metric that corresponds to the seed, in our case.

Now, we can launch the code and see the results in Comet.

Showing results in Comet

To see the results in Comet, we perform the following steps:

1. Open your Comet project.

2. Select **Experiments** from the top menu.

3. The dashboard shows a list of all the experiments. In our case, there is just one experiment.

4. Click on the experiment name, in the first cell on the left. Since we have not set the experiment name, Comet has set the experiment name for us. Comet shows a dashboard with all the experiment details.

In our experiment, we can see some results in three different sections: **Charts**, **Metrics**, and **System Metrics**.

Under the **Charts** section, Comet shows all the graphs produced by our code. In our case, there is only one graph referring to MSE, as shown in the following figure:

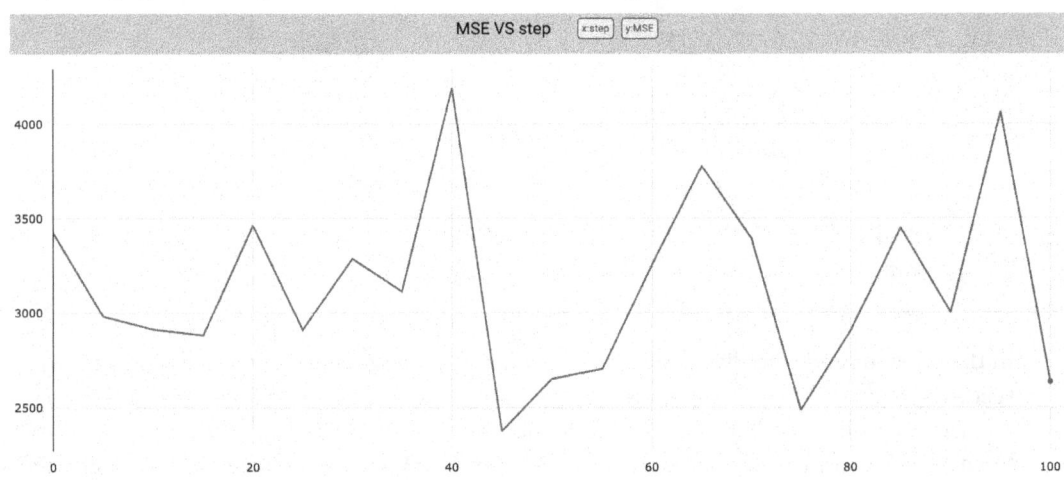

Figure 1.18 – The value of MSE for different seeds provided as input to train_test_split()

The figure shows how the value of MSE depends on the seed value provided as input to the `train_test_split()` function. The produced graph is interactive, so we can view every single value in the trend line. We can download all the graphs as `.jpeg` or `.svg` files. In addition, we can download as a `.json` file of the data that has generated the graph. The following piece of code shows the generated JSON:

```
[{"x": [0,5,10,15,20,25,30,35,40,45,50,55,60,65,70,75,80,85,90
,95],
"y":[3424.3166882137334,2981.5854714667616,2911.8279516891607,2
880.7211416534115,3461.6357411723743,2909.472185919797,3287.490
246176432,3115.203798301772,4189.681600195272,2374.331082443185
6,2650.9384531241985,2702.2483323059314,3257.2142019316807,3776
.092087838954,3393.8576719100192,2485.7719017765257,2904.061086
5479025,3449.620077951196,3000.56755060663,4065.6795384526854],
"type":"scattergl",
"name":"MSE"}]
```

The previous code shows that there are two variables, x and y, and that the type of graph is `scattergl`.

Under the **Metrics** section, Comet shows a table with all the logged metrics, as shown in the following figure:

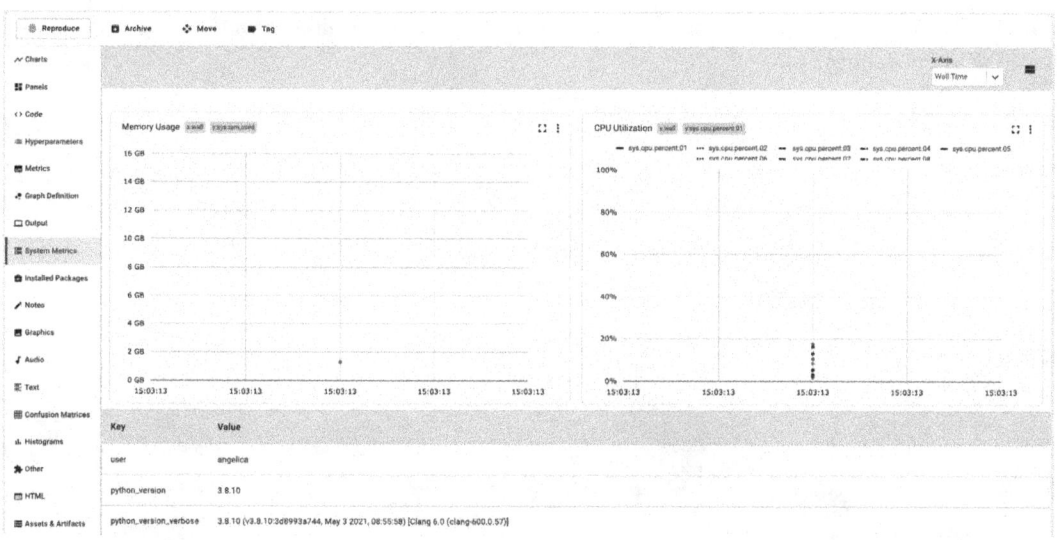

Figure 1.19 – The Metrics section in Comet

For each metric, Comet shows the last, minimum, and maximum values, as well as the step that determined those values.

Finally, under the **System Metrics** section, Comet shows some metrics about the system conditions, including memory usage and CPU utilization, as well as the Python version and the type of operating system, as shown in the following figure:

Figure 1.20 – System Metrics in Comet

The **System Metrics** section shows two graphs, one for memory usage (on the left in the figure) and the other for the CPU utilization (on the right in the figure). Under the graph, the **System Metrics** section also shows a table with other useful information regarding the machine that generated the experiment.

Summary

We just completed our first steps in the journey of getting started with Comet!

Throughout this chapter, we described the motivation and purpose of the Comet platform, as well as how to get started with it. We showed that Comet can be used in almost all the steps of a data science project.

We also illustrated the concepts of workspaces, projects, experiments, and panels, and we set up the environment to work with Comet. Finally, we implemented two practical examples to get familiar with Comet.

Now that you have learned the basic concepts and we have set up the environment, we can move on to more specific and complex features provided by Comet.

In the next chapter, we will review some concepts about EDA and how to perform it in Comet.

Further reading

- Grus, J. (2019). *Data Science from Scratch: First Principles with Python*. O'Reilly Media.
- Prevos, P. (2019). *Principles of Strategic Data Science: Creating value from data, big and small*. Packt Publishing Ltd.

2
Exploratory Data Analysis in Comet

To successfully carry out a data science project, we must first try to understand the data and ask ourselves the right questions. **Exploratory Data Analysis (EDA)** is precisely this preliminary phase that allows you to extract important information from data, and understand which questions data can and cannot answer. Therefore, a data science project should always include the EDA phase.

There are several tools for carrying out EDA, some of which require specific programming skills, such as the many visual libraries provided by Python and JavaScript, as well as others that do not, such as Tableau and Weka.

As already seen in the previous chapter, Comet is an experimentation platform that can be used in almost all phases of a data science project life cycle. In this chapter, we will see how to use Comet to perform EDA. Comet provides different features we can use to perform EDA, including panels, reports, and metric logs. We will also describe each of these three features by providing general concepts and practical examples.

In this chapter, we will adopt the following terminology:

- **Dataset** or **Data** – All the data we want to analyze. We will work with a tabular dataset containing rows and columns.
- **Feature** – A subset of the columns, typically used as input to a machine learning model.
- **Target** – A subset of the columns, typically used as the output of a machine learning model.
- **Record** – A row that contains features and targets.
- **Variable** – A column.

In this chapter, we will focus on the following topics:

- Introducing EDA
- Exploring EDA techniques
- Using Comet for EDA

You can download the full code of the examples described in this chapter from the official GitHub repository of the book, available at the following link: `https://github.com/PacktPublishing/Comet-for-Data-Science/tree/main/02`.

Before moving on to how to use Comet for EDA, let's install all the Python packages needed to run the code and the experiments contained in this chapter.

Technical requirements

We will run all the experiments and code in this chapter using Python 3.8. You can download it from the official website: `https://www.python.org/downloads/` – make sure to choose the 3.8 version.

The examples described in this chapter use the following Python packages:

- `comet-ml 3.23.0`
- `matplotlib 3.4.3`
- `numpy 1.19.5`
- `pandas 1.3.4`
- `scikit-learn 1.0`
- `pandas-profiling 3.1.0`
- `seaborn 0.11.2`
- `sweetviz 2.1.3`

We have already described the first five packages and how to install them in *Chapter 1, An Overview of Comet*. So please refer to that chapter for further details on installation. In this section, we describe the last two packages: `pandas-profiling` and `seaborn`.

pandas Profiling

`pandas-profiling` is a Python package that generates reports, both visual and quantitative, on pandas DataFrames. The official documentation of this package is available at the following link: `https://pandas-profiling.ydata.ai/docs/master/index.html`. You can install the Pandas Profiling package by running the following command in a terminal:

```
pip install pandas-profiling
```

seaborn

seaborn is a useful package for data visualization. It is fully compatible with `Matplotlib`. You can install it by running the following command in a terminal:

```
pip install seaborn
```

For more details, you can refer to the `seaborn` official documentation, available at the following link: `https://seaborn.pydata.org/installing.html`.

sweetviz

sweetviz is a Python package that generates automatic reports for EDA, starting from a Pandas DataFrame. You can install it by running the following command:

```
pip istall sweetviz
```

For more details, you can refer to the `sweetviz` official documentation, available at the following link: `https://github.com/fbdesignpro/sweetviz`.

Now that you have installed all the software needed in this chapter, let's move on to how to use Comet for EDA, starting from reviewing some basic concepts on EDA.

Introducing EDA

Exploratory Data Analysis (EDA) is one of the preliminary steps in a data science project life cycle. It enables us to understand our data in order to extract meaningful information from it. Through EDA, we can understand the underlying structure in the data.

We can think about the EDA phase as a small data science project, in which the real data analysis part (model definition and evaluation) is missing. Therefore, a typical EDA process is composed of the steps shown in the following figure:

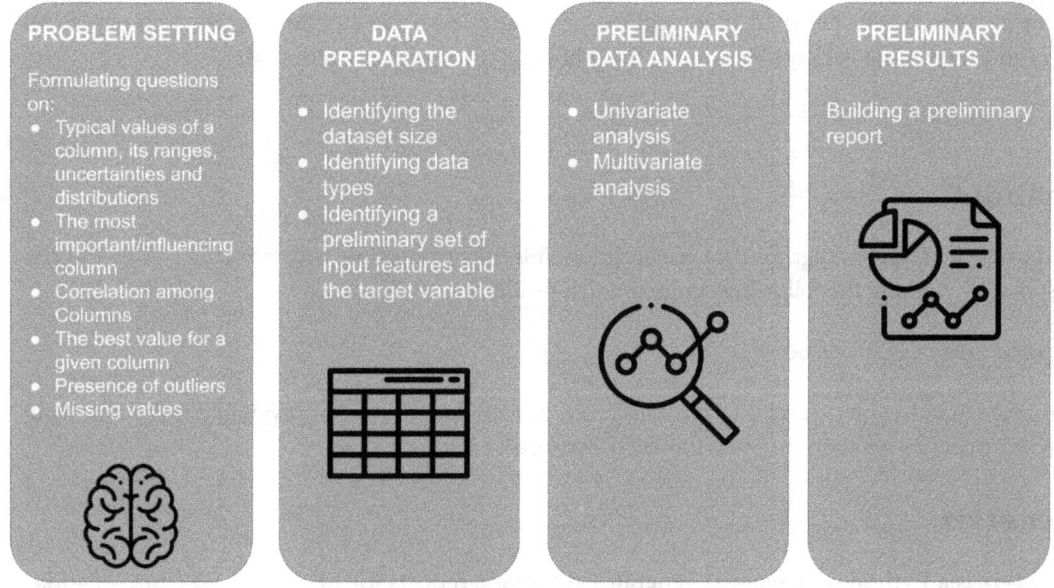

Figure 2.1 – The main steps of an EDA process

The previous figure shows that an EDA process is composed of the following steps:

- **Problem setting**
- **Data preparation**
- **Preliminary data analysis**
- **Preliminary results**

Let's investigate each step separately, starting from the first step – problem setting.

Problem setting

Problem setting is the capability to define which kind of questions our dataset can answer. The problem-setting phase is strictly related to the first step of the data science project life cycle – problem understanding – because it permits us to understand whether our dataset can answer questions in line with the main objectives of the project.

Some typical questions include, but are not limited to, the following:

- What are the typical values of a column, its ranges, uncertainties, and distributions?

- Which are the most important/influential columns?

- Can we establish a correlation among columns?

- Which is the best value for a given column?

- Does a specific column present one or more outliers?

- For a specific column, are there missing values?

Not all the questions may be relevant for a given dataset, so we should select only those that fit the specific case. When we have defined the target questions, we should order them in decreasing order of importance, thus always maintaining the focus on the most important questions first.

When we have formulated all the possible questions, we must prepare our data for further analysis. So, let's move to the next step, which is data preparation.

Data preparation

Data preparation involves all the techniques for preparing the dataset or the datasets for the next step. In this phase, we define the relevant datasets and delete the others, as well as clean and transform the selected datasets. In other words, in this phase we structure our data to be consumed in the preliminary data analysis.

This phase includes the following steps:

- **Identifying the dataset size** – In this phase, we should determine whether the number of records is potentially sufficient for our problem. Depending on the number of samples we could adopt different strategies in the next steps. For example, a little number of samples may require additional data collection, while, if we have a big dataset, we could think about the use of more sophisticated platforms for further analysis.

- **Identifying data types** – There are the following data types:

  - **Numerical** – A column that can be quantified. A numerical column can be either discrete or continuous. A discrete column is countable while a continuous column is measurable.

  - **Categorical** – A column that can assume only a limited number of values.

  - **Ordinal** – A numeric column that can be sorted.

- **Identifying a preliminary set of input features and the target variable** (or the target variables) – Depending on the problem to solve, we should define which columns of the dataset will be used as features and which as target(s).

Once we have structured our data, typically in a tabular form, we can move to the next step, which is preliminary data analysis.

Preliminary data analysis

Preliminary data analysis aims at dissecting the data to discover hidden patterns, relationships between the data and any recurring trends, extract important variables, detect outliers and anomalies, and so on. Through our preliminary analysis, we can provide an initial answer to the questions we asked previously.

We can perform two types of preliminary data analysis:

- **Univariate analysis** – When we focus only on a single variable at a time. Usually, we calculate statistics about each column, such as the minimum, maximum, and average value, as well as data distribution, the most frequent values, and so on.

- **Multivariate analysis** – When we focus on multiple variables at a time. Usually, we calculate the correlation among the variables.

In both cases, we can use **hypothesis testing** to verify whether data satisfies a certain hypothesis.

We can perform preliminary data analysis through different techniques, both visual and non-visual. We will describe them in more detail in the next section of this chapter.

When we have completed the preliminary data analysis phase, we can move on to the last step: preliminary results.

Preliminary results

In this phase, we draw the first conclusions about our data, determining which items of data are relevant to our questions and discarding the rest. To confirm our decisions, we can select some graphs or statistics that we made in the previous phase. If something is not clear or if we still have some doubts about the answers provided, we can go back to the previous steps and try to answer the questions.

Usually, the output of this phase is a **preliminary report**, which motivates our choices and includes some statistics and graphs.

We may be tempted to confuse the whole EDA process with building a summary. Actually, a summary is a fairly passive operation that tries to reduce the numerical data. EDA, on the other hand, is an active process that tries to understand the message contained in the data.

Now that we have described an overview of the EDA process, we can investigate the main EDA techniques in more detail.

Exploring EDA techniques

We can perform EDA through different techniques. In this chapter, we focus on two techniques:

- **Non-visual EDA** – We calculate some statistics or metrics to extract insights from data.
- **Visual EDA** – We use graphs to extract insights from data.

You will see the main concepts behind the two techniques through a practical example in Python.

This section is organized as follows:

- Load and prepare the dataset.
- Non-visual EDA.
- Visual EDA.

Let's start from the first step: loading and preparing the dataset.

Loading and preparing the dataset

Let's consider the Hotel Booking dataset available at `https://www.kaggle.com/jessemostipak/hotel-booking-demand?select=hotel_bookings.csv` under the CC-BY 4.0 license. Let's proceed as follows:

1. Firstly, we import all the Python packages we will use in this example:

```
import pandas as pd
import matplotlib.pyplot as plt
import seaborn as sns
from datetime import datetime
```

We will use `matplotlib` and `seaborn` for visual EDA and `datetime` to make some simple transformations to the data.

2. Then, we load the dataset as a `pandas` DataFrame:

```
df = pd.read_csv('hotel_bookings.csv')
```

The dataset contains 119,390 rows and 32 columns. The following figure shows a sample of the dataset:

Column Name	Column Value
hotel	Resort Hotel
is_canceled	0
lead_time	342
arrival_date_year	2015
arrival_date_month	July
arrival_date_week_number	27
arrival_date_day_of_month	1
stays_in_weekend_nights	0
stays_in_week_nights	0
adults	2
children	0
babies	0
meal	BB
country	PRT
market_segment	Direct
distribution_channel	Direct
is_repeated_guest	0
previous_cancellations	0
previous_bookings_not_canceled	0
reserved_room_type	C
assigned_room_type	C
booking_changes	3
deposit_type	No Deposit
agent	NULL
company	NULL
days_in_waiting_list	0
customer_type	Transient
adr	0
required_car_parking_spaces	0
total_of_special_requests	0
reservation_status	Check-Out
reservation_status_date	2015-07-01

Figure 2.2 – An example record in the Hotel Booking dataset

The previous figure shows that there is more than one column used to represent the date.

Now, we perform the following preliminary operations on the dataset:

- Extract the month number associated with each record.

- Extract the date of each record.

- Extract the season of each record.

3. We want to extract the month number associated with each record since the `arrival_date_month` column is provided as a string. We build a new column that extracts the number associated with the month by defining the following function:

```
def get_month(x):
    month_name = datetime.strptime(x, "%B")
    return month_name.month
```

The previous function simply uses the `strptime()` function provided by the `datetime` package, and returns the month.

4. We apply the previous function to the `arrival_date_month` column to build a new column in the dataset:

```
df['arrival_date_month_number'] = df['arrival_date_
month'].apply(lambda x: get_month(x))
```

We have used the `apply()` method of the DataFrame to operate on every single value of the column.

5. Now, we build a new column that contains the full date:

```
df['arrival_date'] = df[['arrival_date_year','arrival_
date_month_number','arrival_date_day_of_month']].
apply(lambda x: '-'.join(x.dropna().astype(str)), axis=1)
df['arrival_date'] = pd.to_datetime(df['arrival_date'])
```

We have selected all columns containing the arrival date and have merged them through the `join()` method. Then, we converted the built date to a `datetime` object through the `to_datetime()` function provided by the `pandas` package.

6. Finally, we extract the season from the date. We define the following function that extracts the season:

```
def get_season(date):
    md = date.month * 100 + date.day
    if ((md > 320) and (md < 621)):
```

```
        return 'spring'
elif ((md > 620) and (md < 923)):
        return 'summer'
elif ((md > 922) and (md < 1223)):
        return 'fall'
else:
        return 'winter'
```

The previous function converts the date into a number and calculates in which range this number falls.

7. We apply the previous function to the arrival date column:

```
df['arrival_season'] = df['arrival_date'].apply(lambda x:
get_season(x))
```

Similar to the previous case, we used the `apply()` method to extract the season of each record.

Now that we have prepared our dataset, we are ready to perform EDA. Let's start with non-visual EDA.

Non-visual EDA

Non-visual EDA calculates some statistics on the given data, and then presents the output in a numeric or tabular format. We can perform non-visual EDA both for univariate and multivariate analysis. In the case of univariate analysis, we can calculate descriptive statistics (for numerical data), missing values, negative and distinct values, and memory size. In the case of multivariate analysis, usually we calculate the correlation among columns.

To illustrate the most common techniques for non-visual EDA, you can use different packages that automatically build a report, such as `pandas_profiling` and `sweetviz`. To build the report in `pandas_profiling`, we can run the following code:

```
from pandas_profiling import ProfileReport
profile = ProfileReport(df, title="Hotel Booking EDA")
profile.to_file("hotel_bookings_eda.html")
```

Firstly, we created a `ProfileReport()` object by passing the pandas DataFrame as input. Then, we saved the produced report as an external HTML file. The produced report is organized into the following sections:

- Overview

- Variables

- Interactions

- Correlations

- Missing values

- Sample

- Duplicate rows

The **Overview** part contains a summary of the datasets, as shown in the following figure:

Overview

Overview	Alerts 59	Reproduction

Dataset statistics		Variable types	
Number of variables	35	Categorical	19
Number of observations	119390	Numeric	15
Missing cells	129425	DateTime	1
Missing cells (%)	3.1%		
Duplicate rows	17		
Duplicate rows (%)	< 0.1%		
Total size in memory	31.9 MiB		
Average record size in memory	280.0 B		

Figure 2.3 – The first part of the report produced by the pandas-profiling package

The report contains information on the whole dataset, including the number of variables, observations, missing cells, duplicate rows, total size in memory, and the variable types.

In addition, the summary contains information on the cardinality and correlation between the different columns, as shown in the following figure:

Figure 2.4 – A summary of the correlations between the different columns in the dataset

We note that some columns, such as `country` and `reservation_status_date`, have a high cardinality.

Regarding the **Variables** section, we can distinguish between categorical and numerical data. For categorical data, the report provides many details, as shown in the following figure:

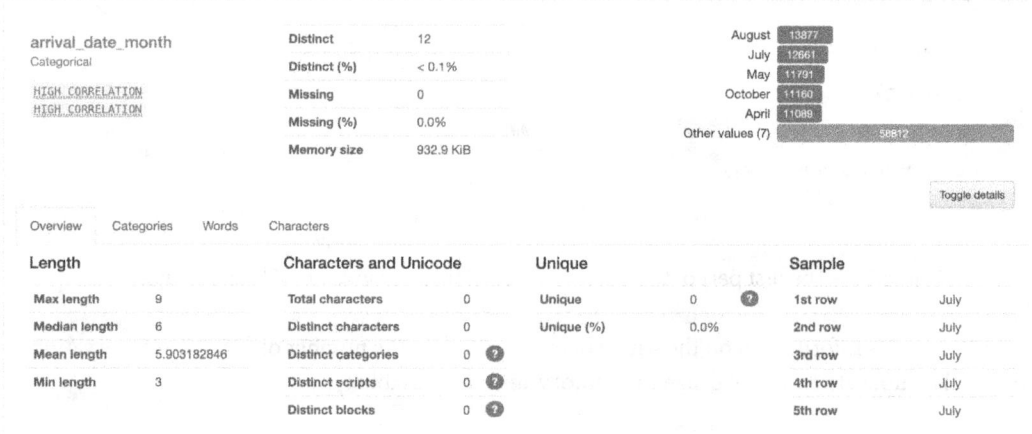

Figure 2.5 – Information provided for categorical data

For example, the report shows details on the distinct and missing values, as well as the minimum, median, and maximum length of labels.

For numerical variables, the report provides additional information, as shown in the following figure:

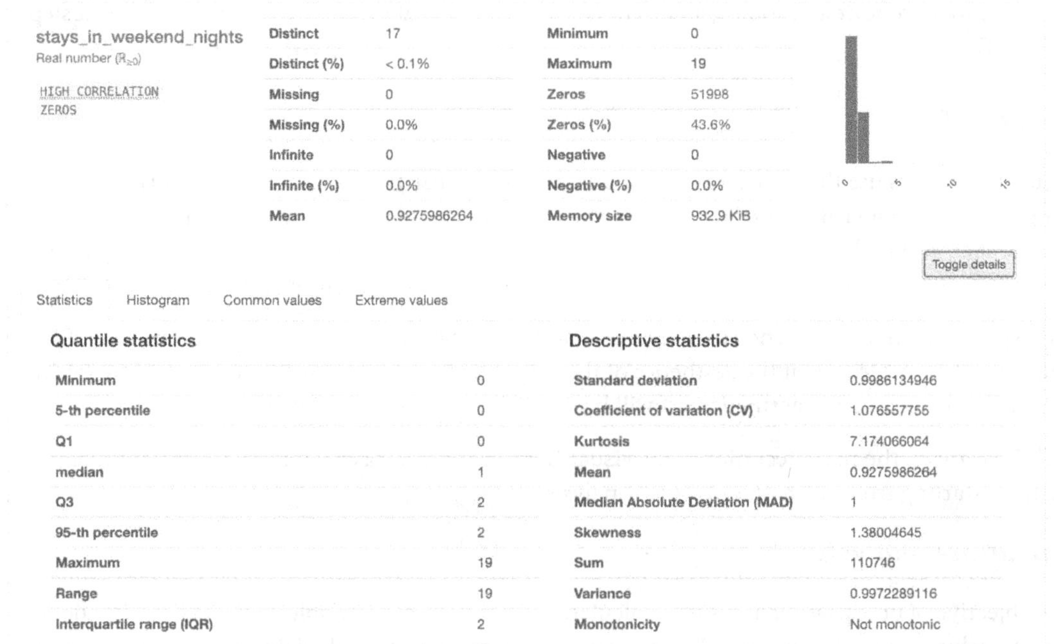

stays_in_weekend_nights	Distinct	17	Minimum	0	
Real number ($\mathbb{R}_{\geq 0}$)	Distinct (%)	< 0.1%	Maximum	19	
HIGH CORRELATION	Missing	0	Zeros	51998	
ZEROS	Missing (%)	0.0%	Zeros (%)	43.6%	
	Infinite	0	Negative	0	
	Infinite (%)	0.0%	Negative (%)	0.0%	
	Mean	0.9275986264	Memory size	932.9 KiB	

Toggle details

Statistics   Histogram   Common values   Extreme values

Quantile statistics		Descriptive statistics	
Minimum	0	Standard deviation	0.9986134946
5-th percentile	0	Coefficient of variation (CV)	1.076557755
Q1	0	Kurtosis	7.174066064
median	1	Mean	0.9275986264
Q3	2	Median Absolute Deviation (MAD)	1
95-th percentile	2	Skewness	1.38004645
Maximum	19	Sum	110746
Range	19	Variance	0.9972289116
Interquartile range (IQR)	2	Monotonicity	Not monotonic

Figure 2.6 – Information provided for numerical data

For example, the report shows various statistics on the data, including the minimum, maximum, percentiles, standard deviation, mean, skewness, and so on. Similar to the categorical data, the report also shows distinct and missing values.

The other sections of the report show information in the form of graphs; thus, we can consider them as part of visual EDA.

As an alternative to pandas-profiling, you can use sweetviz. To build a report in sweetviz, you can run the following code:

```
import sweetviz as sv
report = sv.analyze(df)
report.show_html('report.html')
```

We use the `analyze()` function provided by the `sweetviz` library, and then we save the produced report as an HTML file.

The `sweetviz` library generates similar results to those already described for `pandas-profiling`.

Now that we have reviewed some general concepts on non-visual EDA, we can move to the next step: visual EDA.

Visual EDA

Visual techniques use the ability of the human eye to recognize trends, patterns, and correlations by simply looking at a visual representation of the data. We can plot our data using different types of graphs, such as line charts, bar charts, scatter plots, area plots, table charts, histograms, lollipop charts, maps, and much more. We can use visual techniques for both univariate and multivariate analysis.

The type of graph used depends on the nature of the question we want to answer. During the EDA phase, we do not care about the aesthetics of the graph, because we are only interested in answering the questions we ask. The aesthetic part will be studied in the narrative data phase.

Let's investigate the main techniques for visual EDA in the two cases, univariate and multivariate analysis, starting from the first: univariate analysis.

Univariate analysis

The objective of univariate analysis is to calculate the behavior of a variable, including the frequency of each value and its distribution. If the variable is categorical, we calculate the frequency of each value; if the variable is numerical, we calculate its distribution and other statistics, such as mean, median, and so on.

We can consider two types of univariate analysis, one for categorical variables, and the other for numerical variables. Regarding categorical variables, we can plot the following graphs:

- **Countplot** – This counts the frequency of a variable and shows it as a bar chart. For example, we can build the countplot graph of the **arrival date month** through the following code:

```
colors = sns.color_palette('mako_r')
sns.countplot(df['arrival_date_month'], palette=colors)
plt.show()
```

In the previous example, we firstly set the color palette to `mako_r` through the `color_palette()` function provided by `seaborn`. We note that the `seaborn` library provides a function named `countplot()` that automatically builds the countplot of the variable passed as input. The following figure shows the produced graph:

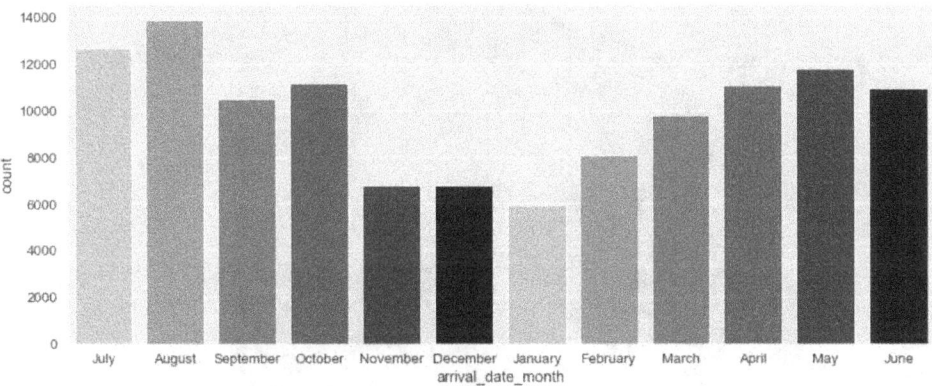

Figure 2.7 – A countplot graph for the arrival month date column of the dataset

We note that August is the most frequent category.

- **Pie chart** – This is very similar to the countplot graph, but it also shows the percentage of each category. We can build a pie chart of the **arrival month date** column as follows:

```
values = df['arrival_date_month'].value_counts()
values.plot(kind='pie', colors = colors, fontsize=17,
autopct='%.2f')
plt.legend(labels=values.index, loc="best")
plt.show()
```

Note that we used the `plot()` function provided by pandas. The following figure shows the resulting graph:

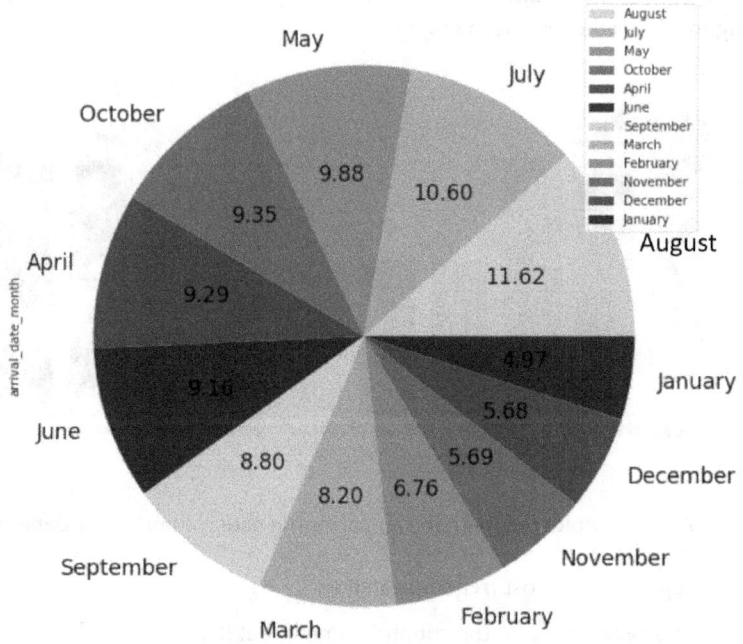

Figure 2.8 – A pie chart graph for the arrival month date column of the dataset

In the previous graph, we also have information on percentages displayed.

We have just reviewed the most common graphs for univariate categorical analysis. Now we can illustrate the most common graphs for univariate numerical analysis. This type of graph includes the following:

- **Histogram** – A distribution plot, usually used to represent continuous variables. It splits all the possible values into bins and calculates how many values fall in each bin. We can build a histogram of the `stays_in_week_nights` variable as follows:

```
plt.hist(df['stays_in_week_nights'], bins=50,
color='#40B7AD')
plt.show()
```

In the previous code, we used the `hist()` function of the `matplotlib` library and we set the number of bins to `50`. The following figure shows the resulting plot:

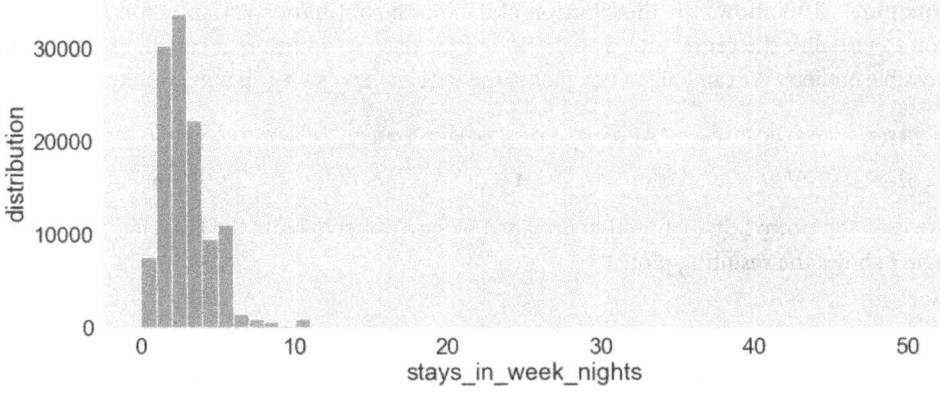

Figure 2.9 – A histogram for the stays_in_week_nights column of the dataset

The previous figure shows that almost all the values are concentrated in the first 10 bins.

- **Dist plot** – Similar to the histogram plot, but it also shows the **Kernel Density Estimate** (**KDE**). We can build a dist plot of the stays_in_week_nights variable as follows:

```
plt.figure(figsize=(15,6))
sns.distplot(df['stays_in_week_nights'], color='#40B7AD')
plt.show()
```

We used the distplot() function of the seaborn library. The following figure shows the resulting plot:

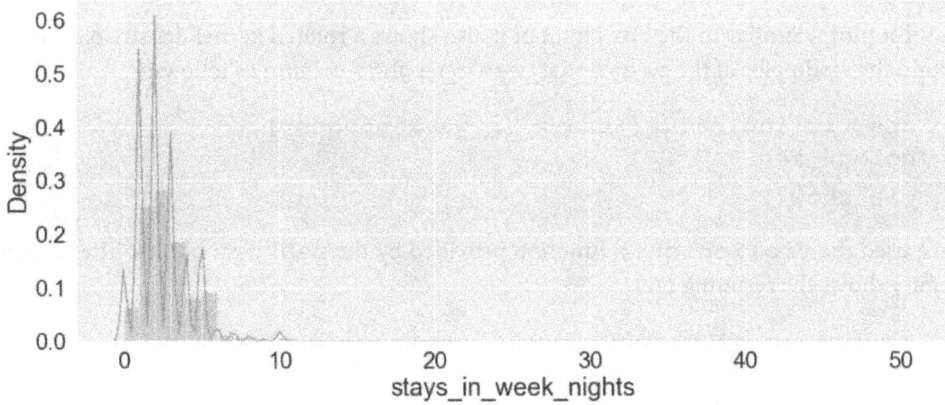

Figure 2.10 – A dist plot for the stays_in_week_nights column of the dataset

The previous figure shows the KDE as a continuous line.

- **Box plot** – This shows the distribution of data for a continuous variable. A box plot allows you to visualize the center and distribution of the data. In addition, it can be used to identify possible outliers. We can build a box plot of the stays_in_week_nights column as follows:

```
sns.boxplot(df['stays_in_week_nights'], color='#40B7AD')
plt.show()
```

We used the boxplot() function provided by seaborn to build the box plot. The following figure shows the resulting plot:

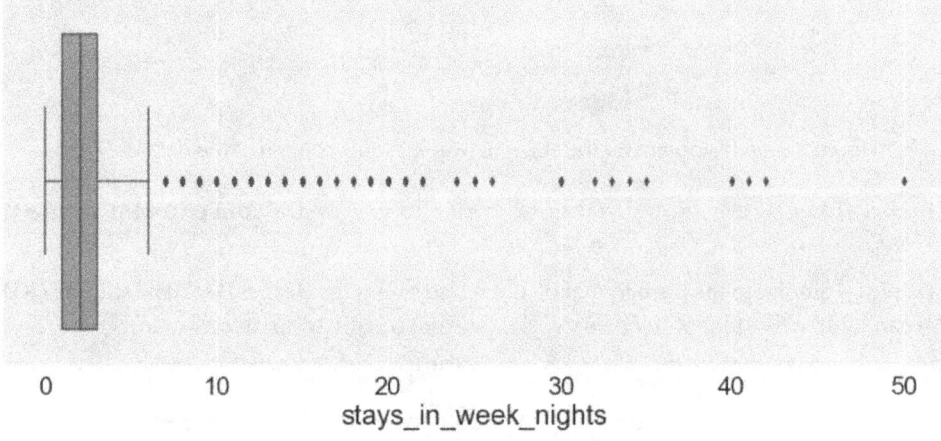

Figure 2.11 – A box plot for the stays_in_week_nights column of the dataset

The previous figure shows the average value (in color), as well as other information, such as the presence of outliers (near the value 50).

- **Violin plot** – Similar to the box plot, but it also shows a rotated kernel density plot. We can build the violin plot of the stays_in_week_nights column as follows:

```
sns.violinplot(df['stays_in_week_nights'],
color='#40B7AD')
plt.show()
```

We used the violinplot() function provided by the seaborn library. The following figure shows the resulting plot:

Figure 2.12 – A violin plot for the stays_in_week_nights column of the dataset

The previous graph is a combination of the box plot and dist plot.

We can use all the previously described graphs for univariate analysis. Let's now look at which types of visualizations we can use for multivariate analysis.

Multivariate analysis

Multivariate analysis considers multiple variables at a time. In this chapter, we focus on two variables at a time (**bivariate analysis**), but we can calculate multivariate analysis also for more variables, through techniques for dimensionality reduction or by using the style options provided by the graphs.

There are three types of bivariate analysis:

- **Numerical to numerical**
- **Categorical to categorical**
- **Numerical to categorical**

Let's analyze the graphs provided by each type separately, starting from the first: numerical to numerical.

Numerical-to-numerical analysis

In numerical-to-numerical bivariate analysis, we can plot a **scatter plot** that shows the relationship between two variables. The following piece of code shows how to build a scatter plot in seaborn:

```
sns.scatterplot(df['adults'], df['stays_in_week_
nights'],color='#40B7AD')
plt.show()
```

We used the `scatterplot()` function to represent the **adults** column on the *x* axis and the **stays_in_week_nights** on the *y* axis. The following figure shows the resulting figure:

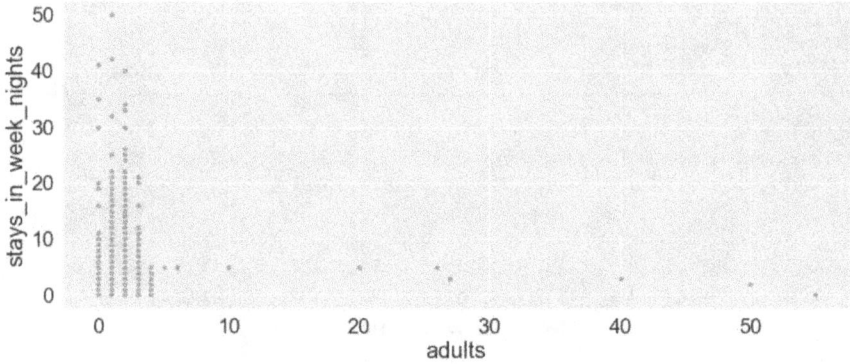

Figure 2.13 – A scatter plot for adults versus stays_in_week_nights

The previous figure shows the correlation between the **stays_in_week_nights** and **adults** variables.

Categorical-to-categorical analysis

In the categorical-to-categorical bivariate analysis, the most common graph is the **heatmap**, which plots the correlation between two variables through colors. We can plot a heatmap as follows:

```
sns.heatmap(pd.crosstab(df['customer_type'], df['arrival_date_
month']), cmap='mako_r')
plt.show()
```

The previous piece of code plots the heatmap of two categorical variables, `customer_type` and `arrival_date_month`. In addition, it uses the `crosstab()` function provided by the `pandas` library to build the matrix to be plotted. The following figure shows the resulting plot:

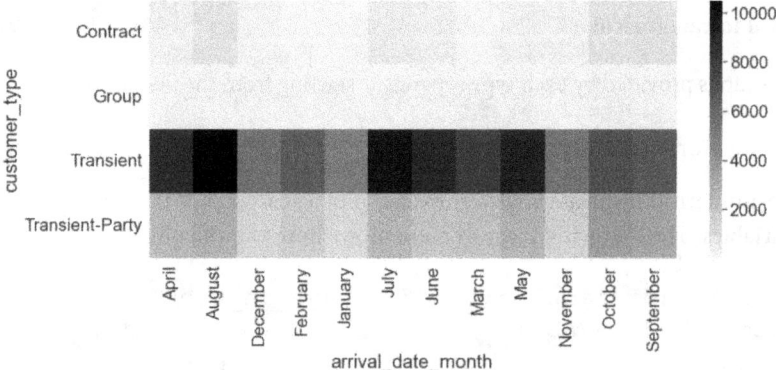

Figure 2.14 – A heatmap for customer_type and arrival_date_month

The previous figure shows the relationship between the category labels in terms of different gradients. The stronger the relationship, the darker the color.

Number-to-category analysis

In the number-to-category bivariate analysis, the most common graph is the **bar plot**, which draws categorical variables as rectangular bars with lengths proportional to the numerical variables they are compared with. We can plot a bar plot as follows:

```
sns.barplot(df['adults'], df['arrival_date_
month'],color='#40B7AD')
plt.show()
```

In the previous example, we used the `barplot()` function provided by `seaborn` to plot the `arrival_date_month` column with respect to the number of adults. The following figure shows the resulting plot:

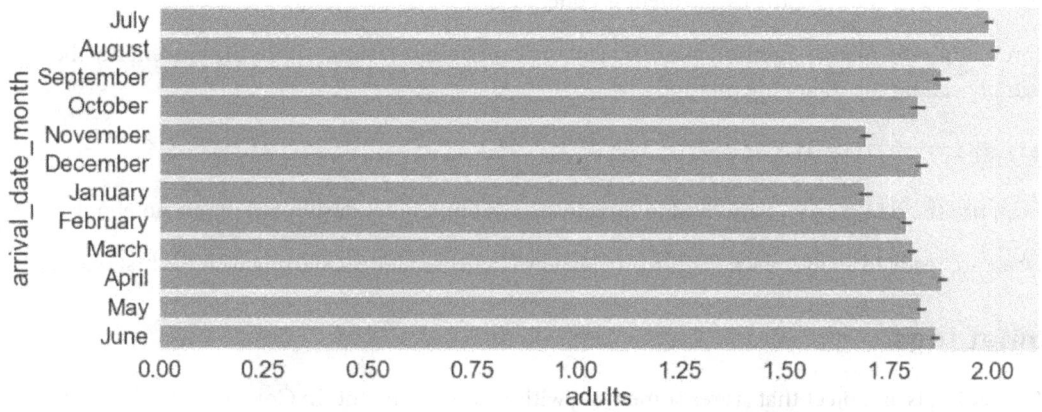

Figure 2.15 – A bar chart of the arrival_date_month column versus the number of adults

There are two additional types of graphs, geographical maps, and trend lines, that we can use for EDA. We can use geographical maps to study the distribution of a variable over a map and we can use trend lines to understand the behavior of a variable over time.

So far, we have reviewed the general concepts behind EDA. Now it is time to apply them to Comet. So, let's move to the next section: Comet for EDA.

Using Comet for EDA

Comet provides the following features to deal with EDA:

- **Log** – A mechanism to store assets, metrics, and objects in general in Comet

- **Panel** – A visual representation of one or more logged objects

- **Report** – A combination of panels

From a logical point of view, firstly we log all the needed objects, then we build all the designed panels, and, finally, we build a report. We can adopt this strategy in all the phases of the data science project life cycle, such as EDA and model evaluation. In this chapter, we focus on how to adopt this strategy during the EDA phase.

Before describing the single features, separately, here is a practical tip that permits you to integrate Comet with notebook documents. Usually, we use notebook documents (or simply *notebooks*) to perform EDA because we can use them to show both code and text, to run temporary code, and so on. Comet can also be integrated with Jupyter notebooks, by simply adding one line of code at the end of the experiment:

```
experiment.end()
```

After the end() method, the experiment is concluded.

Before calling the end() method, we can view the experiment results within the current notebook, by simply calling the following method:

```
experiment.display()
```

We can use the display() method to directly access our Comet dashboard in our notebook.

Now we are ready to analyze each feature provided by Comet separately, starting with the first one: logs.

Comet logs

A **Comet log** is an object that stores something within an experiment. In Comet we can log metrics, models, figures, images, and much more. Comet defines two types of logs:

- **Values** – This type of log includes metrics and other parameters that are available under the Comet dashboard. To log one or more values, we can use the following methods of the Experiment class:

  - log_metrics() – Logs a dictionary of key-value pairs that conceptually are metrics, such as precision, recall, and accuracy. We can access the logged values in the **Metrics** menu of the Comet dashboard.

  - log_metric() – Logs a single key-value pair. Similar to the previous method, we can access the logged value in the **Metrics** menu of the Comet dashboard.

  - log_parameters() – Logs a dictionary of key-value pairs that conceptually are hyperparameters. We can access the logged values in the **Hyperparameters** menu of the Comet dashboard.

- log_parameter() – Similar to log_parameters(), but it logs a single key-value pair.

- log_others() – Logs a dictionary of key-value pairs. These values could be used to keep track of some used parameters. We can access the logged values in the **Others** menu of the Comet dashboard.

- log_other() – Similar to log_others(), but it logs a single key-value pair.

- log_html_url() – Logs an HTML URI, which can be accessed in the **HTML** menu of the Comet dashboard.

- log_text() – Logs a text, which can be accessed in the **Text** menu of the Comet dashboard.

- **Files** – This type of log includes datasets, figures, models, and files in general that are available in the **Assets & Artifacts** menu. We can log different types of files in Comet. Referring only to EDS, we can use the following methods provided by the Experiment class:

- log_dataframe_profile() – Logs a summary of a pandas DataFrame, produced by the pandas profiling library

- log_histogram_3d() – Logs a 3D histogram chart

- log_html() – Logs an HTML file

- log_image() – Logs an image

- log_table() – Logs a CSV file or a pandas DataFrame

In this section, we have described an overview of the most important methods provided by the Experiment class for logging. For more details on the parameters they receive as input, you can refer to the Comet official documentation, available at this link: https://www.comet.ml/docs/python-sdk/Experiment/.

We have already described how to log an image and a metric in Comet in *Chapter 1, An Overview of Comet*. In this section, we describe how to log a pandas-profiling report and different values for the same metric.

We will use the hotel_booking.csv dataset we described in the previous section:

1. Firstly, we import all the packages needed for this experiment:

    ```
    import pandas as pd
    from comet_ml import Experiment
    ```

We will use two packages: pandas to transform the original dataset into a DataFrame, and comet_ml to build the Comet Experiment. Make sure to create a new project from the main dashboard, as illustrated in *Chapter 1, Overview of Comet*.

2. Then, we create a new Experiment, as follows:

    ```
    experiment = Experiment()
    ```

We have created the `Experiment` object. Note that to make the previous code run, we should configure the `.comet.config` file, as explained in *Chapter 1, Overview of Comet*.

3. Let's suppose that we have already loaded the `hotel_bookings.csv` file as a `pandas` DataFrame, as described in the previous section, and we have stored it in a variable named `df`. We can log the associated `pandas` profile report in Comet as follows:

```
experiment.log_dataframe_profile(df, "hotel_bookings")
```

We have used the `log_dataframe_profile()` method, which generates an asset containing all the required statistics. The logging process may require some time. Some progressive bars, such as those shown in the following figure, show when the logging process is complete:

Figure 2.16 – The progressive bar produced by the log_dataframe_profile() method

The previous figure shows that there are three progressive bars: **Summarize dataset**, **Generate report structure**, and **Render HTML**. When all the bars are complete, the result is available as a Comet asset.

4. Now, we make sure that Comet has logged all the outputs. We access the **Assets & Artifacts** menu item from the Experiment dashboard in Comet. In the **Assets** tab, there is the **dataframes** directory, which contains our original dataset and the logged profile summary, as shown in the following figure:

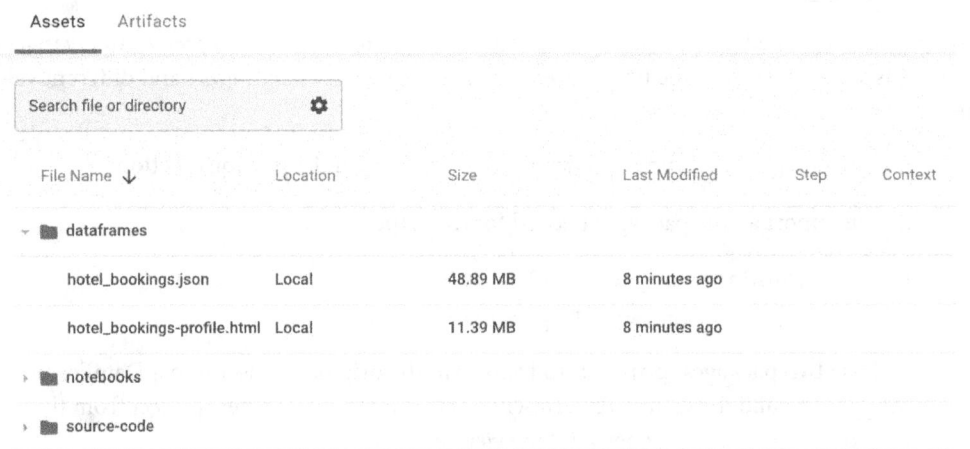

Figure 2.17 – The Assets tab, with a focus on the dataframes directory

The original dataset (`hotel_bookings.json`) is provided as a JSON object, while the summary is an HTML file. We can either download or view both of the files by simply clicking on the respective button near the file. Now that you have logged the pandas profiling object, you can see how to log multiple values for a single metric:

1. Let's suppose that we want to log at what time families check in at some hotels. For simplicity, we consider as a family a record with at least two adults, as follows:

    ```
    families = df[df['adults'] > 1]
    families.set_index('arrival_date', inplace=True)
    ```

    We assume that the DataFrame `df` contains the `hotel_bookings` dataset.

2. Now we group all the families by `arrival_date`, and we count the number of families for each date:

    ```
    ts_families = families['adults'].groupby('arrival_date').
    count()
    ```

    We used the `groupby()` method to group the dataset by families and the `count()` method to count the number of records in each group.

3. Now, we can log the `ts_families` values through the `log_metric()` method of the `Experiment` class, as follows:

    ```
    import time
    import datetime
    for i in ts_families.index:
        index = time.mktime(i.timetuple())
        experiment.log_metric('ts_families', ts_families[i],
    step=index)
    ```

    We used the step parameter of the `log_metric()` method to log different values of the same metric. In addition, we converted the `arrival_date` to timestamp, since `log_metric()` does not support dates.

4. We can see the logged metric in the Comet dashboard, under the **Metrics** menu item, as shown in the following figure.

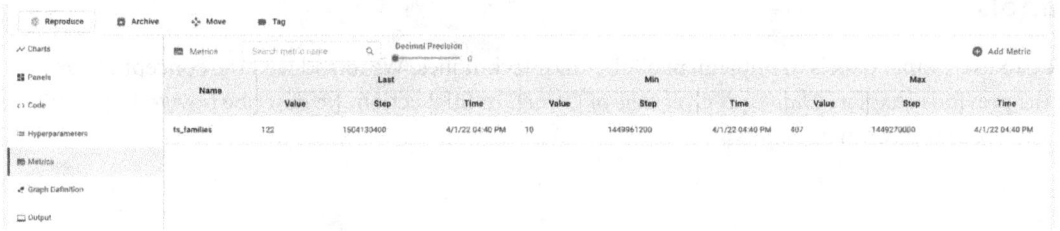

Figure 2.18 – The logged metric in Comet

The Comet dashboard shows the last, minimum, and maximum values for the `ts_families` variable.

As an alternative to the `pandas_profiling` library, you can use `sweetviz`, which is fully integrated with Comet. To log a `sweetviz` report, you can simply run the following code:

```
report.log_comet(experiment)
```

The `log_comet()` function receives as input an `experiment` object. As a result, the report is saved in Comet under the HTML menu item, as shown in the following figure:

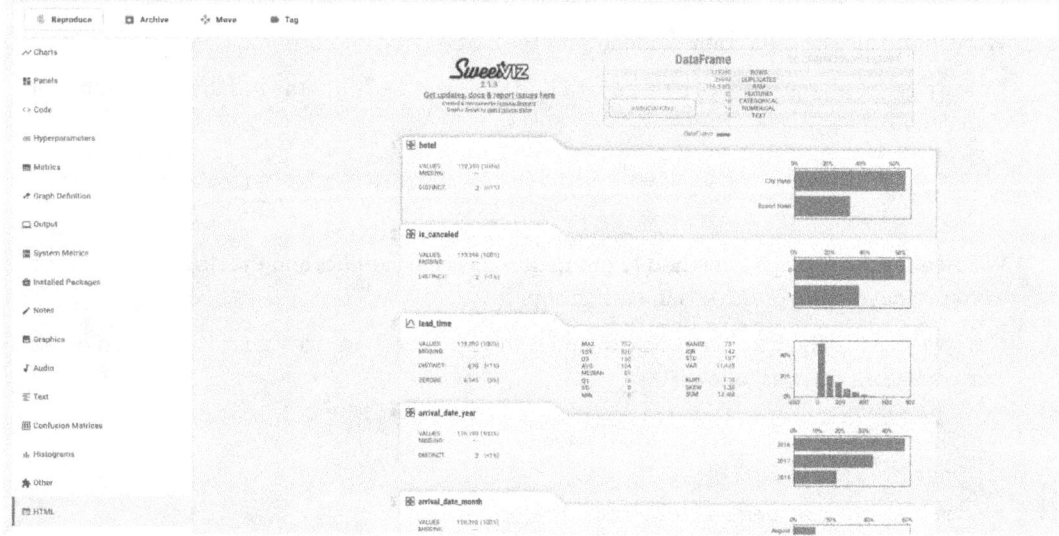

Figure 2.19 – The output of the log_comet() method provided by the sweetviz library

The HTML file generated by the `log_comet()` method is interactive, so you can browse it to explore every column of the dataset.

Now that you are familiar with Comet logs, we can move to the next feature provided by Comet for EDA: panels.

Panels

We can use Comet panels to implement all the EDA techniques. We introduced the concept of panels in the previous chapter: *Chapter 1, Overview of Comet*. In this section, we describe how to implement custom panels in Comet.

We can create custom panels in Comet either in Python or JavaScript. Although the focus of this book is mainly on the Python language, at the end of this section we will also provide some general concepts to build custom panels in JavaScript. Comet provides a practical Python/JavaScript SDK to build custom panels directly from the online platform. At the moment, we cannot implement our custom panels offline.

Building a custom panel involves the following four steps:

- Accessing the SDK
- Retrieving the environment variables
- Building the panel content
- Showing the panel in Comet

Let's investigate all the steps separately, starting from the first step: environment setup.

Accessing the SDK

The SDK is the place where we write the code to build our custom panel. To access the Comet SDK, perform the following operations:

1. From the main dashboard, click **Add → New Panel → Create New**.

2. A new window opens with the online SDK, as shown in the following figure:

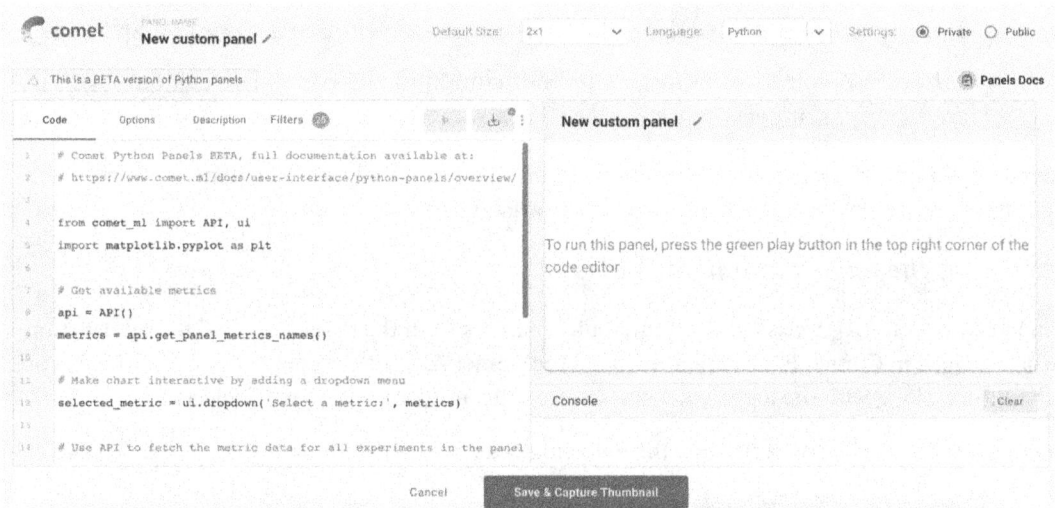

Figure 2.20 – The online SDK in Comet to build custom panels

At the top of the window, there are some options that permit us to configure the panel name, its size, the preferred language (Python or JavaScript), and the panel visibility (private or public). The main part of the window is divided into two parts: on the left, there is the editor, where we can write the code to produce the panel, while on the right, there is a preview of the panel. To see the preview, we should click the run button (the green triangle in the editor). We can save the panel by clicking on the button near the run button. In the bottom part of the window, there are two buttons used to save or cancel changes.

The editor part of the window is composed of the following four tabs:

- **Code** – The main code editor.

- **Options** – An option is a runtime value provided as input to the panel. For example, we can build generic panels that can be used in multiple projects. For each project, we need to change only the current values, but the type of panel remains constant. We can define options as key-value pairs, as shown in the following piece of code:

```
{
"key 1" : "value 1",
"key 2" : "value 2",
...
"key N" : "value N"
}
```

- **Description** – A textual description of the panel, including a name and an image.

- **Filters** – A selection of targets to be included in the panel, such as only certain experiments or certain metrics.

Now that you have learned how to set up the environment to build a custom panel in Comet, let's move to the next step – retrieve the environment variables.

Retrieving the environment variables

Comet provides the API class to easily and quickly access from the Comet SDK all the information we have logged in Comet. For example, through the Comet API, we can select all the experiments or just a subset, along with the logged metrics, assets, artifacts, workspaces, and projects.

To import the API class, we can write the following code:

```
from comet_ml import API
api = API()
```

The `API` class permits you to read all information saved in the workspaces, projects, and experiments, but it does not permit you to log new values. When the `api` object has been created, we can call many methods. Among them, the most important are the following:

- `get()` – Returns all the workspaces, projects or experiments associated with the current user. The type of object returned by this method depends on the level of granularity passed as input parameters. For example, `get(MY_WORKSPACE)` returns all the projects contained in the `MY_WORKSPACE` workspace, and `get(MY_WORKSPACE, MY_PROJECT)` returns all the experiments contained in the `MY_PROJECT` project, itself contained in the `MY_WORKSPACE` workspace.

- `get_experiment(MY_WORKSPACE, MY_PROJECT, MY_EXPERIMENT_KEY)` – Returns a specific experiment. The experiment is returned as an object of the `APIExperiment` class, which differs from the `Experiment` class because it is used within a Comet panel. There are three ways to retrieve an experiment key:

  - Through the `get_Panel_experiment_keys()` method

  - By setting the experiment key manually when creating the experiment, as follows:

    ```
    experiment = Experiment(experiment_key = MY_EXPERIMENT_
    KEY)
    ```

  - Copying the value returned by the Comet output when running the experiment, as shown in the following figure:

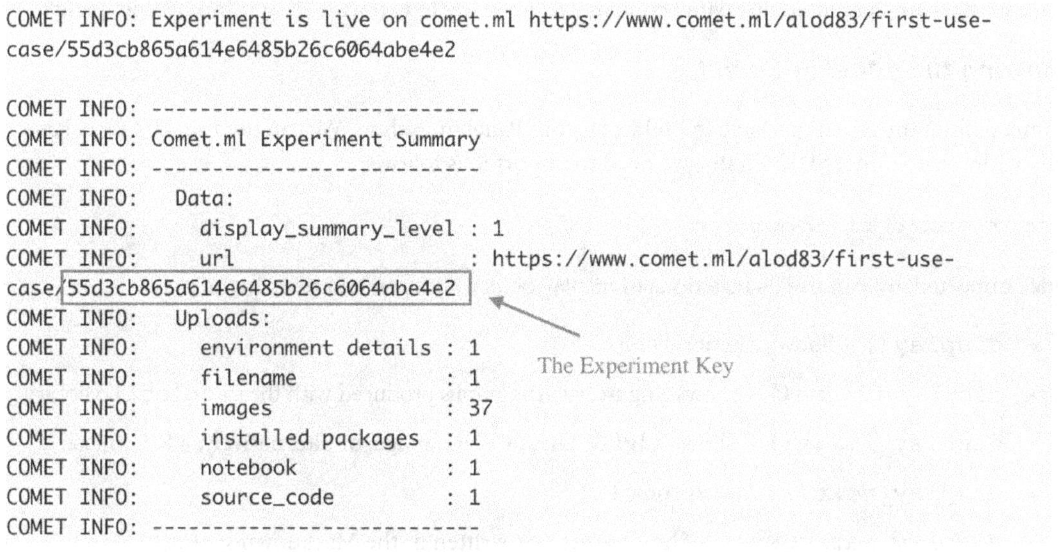

```
COMET INFO: Experiment is live on comet.ml https://www.comet.ml/alod83/first-use-
case/55d3cb865a614e6485b26c6064abe4e2

COMET INFO: ---------------------------
COMET INFO: Comet.ml Experiment Summary
COMET INFO: ---------------------------
COMET INFO:    Data:
COMET INFO:        display_summary_level : 1
COMET INFO:        url                   : https://www.comet.ml/alod83/first-use-
case/55d3cb865a614e6485b26c6064abe4e2
COMET INFO:    Uploads:
COMET INFO:        environment details : 1
COMET INFO:        filename            : 1         The Experiment Key
COMET INFO:        images              : 37
COMET INFO:        installed packages  : 1
COMET INFO:        notebook            : 1
COMET INFO:        source_code         : 1
COMET INFO: ---------------------------
```

Figure 2.21 – The experiment key returned when running the experiment

In the previous figure, the experiment key is 55d3cb865a614e6485b26c6064abe4e2.

- get_panel_options() – Returns the options provided as input in the **Option** tab of the SDK. This method returns options as a Python dictionary.

- get_experiment_*() – A collection of methods to access all the information logged in an experiment. The * must be replaced by one of the following words: code, curves, graphs, HTML, metric, or images. So, for example, to access the experiment graphs, we can use the get_experiment_graphs() method, while to access the experiment code, we can call the get_experiment_code() method.

For a complete list of methods provided by the API class, you can refer to the Comet official documentation, available at this link: https://www.comet.ml/docs/user-interface/python-Panels/API/.

Now that you have learned how to access all the environment variables within a Comet panel, let's analyze how to build the content of a panel.

Building the panel content

We can use all the environment variables to build our graphs, tables, and whatever else. Comet panels are integrated with the Matplotlib and Plotly libraries, as well as with the Python Image library. Thus, we can implement our charts as we usually do, directly in the Comet SDK.

We will describe some practical examples in the next sections of this chapter.

Once we have implemented our panel content, we are ready to export it as a custom Comet panel.

Showing the panel in Comet

Comet defines the ui subpackage to build a custom Panel in Python. We can use the ui subpackage only within the Comet SDK. Firstly, we need to import it as follows:

```
from comet_ml import ui
```

Once imported, we can use its functions to display objects, figures, and texts in the panel:

- display() – Shows a generic object
- display_figure() – Shows a figure, such as graphs produced with the matplotlib library
- display_image() – Shows a JPEG, GIF, or PNG image, or alternatively, a PIL image
- display_text() – Shows some text
- display_markdown() – Shows some text written in the Markdown syntax

- `dropdown()` – Creates a drop-down menu with the list of items passed as an argument, returns the selected item, and triggers a change to run the code
- `add_css()`/`set_css()` – Adds/sets an additional CSS style to display

We can use and combine the previous functions to build our custom panels.

Now that you have learned how to build a custom panel in Python, we will quickly describe how to build a custom panel in JavaScript.

Custom panels in JavaScript

The Comet SDK permits you to also build your custom panels in JavaScript. To create a custom panel in JavaScript, select JavaScript as the language in the panel window. In addition to the tabs described in the previous section, the tab menu of the JavaScript editor contains these additional tabs:

- **HTML** – For HTML containers.
- **CSS** – For styles.
- **Resources** – To import new libraries or CSS stylesheets. Any additional resources need to be provided as a URL.

The JavaScript SDK defines the `Comet.Panel` class as the starting point to build a panel. The `Comet.Panel` class defines the following methods:

- `draw()`/`drawOne()` – Draws one or more experiments
- `print()` – Prints some text in the panel
- `getOption()` – Gets an option
- `clear()` – Clears all of the printed objects
- `select()` – Creates a HTML select widget

For more details on the methods defined by the `Comet.Panel` class, you can refer to the Comet official documentation, available at this link: `https://www.comet.ml/docs/javascript-sdk/getting-started/`.

The `Comet.Panel` class also defines the interface to the Comet workspaces, projects, and experiments through the `this.api` variable, which is an interface to the JavaScript API class. Similar to the API class defined in Python, the JavaScript API class provides methods to access metrics, images, code, graphs, and so on. We can use the `experiment*()` method, where * must be replaced by one of the following keywords: `Code`, `HTML`, `Images`, `Graph`, or `Metric`. For example, to get a given metric, we should use the `experimentMetric()` method. For more details on the JavaScript API, you can refer to the official Comet documentation, available at this link: `https://www.comet.ml/docs/javascript-sdk/api/`.

To build a custom panel, we need to extend the Comet.Panel class, as follows:

```
class MyPanel extends Comet.Panel {
    ...
}
```

If we want to access the options defined in the **Option** tab, we can use the this.options variable. Typically, we define some default values of this variable in the setup() method, and then, at runtime, the JavaScript SDK will substitute them with the actual values set in the **Option** tab.

Within a JavaScript panel, we can use all the JavaScript libraries that we prefer, including Plotly, Highcharts, d3.js, Google Charts, and much more.

Now that you have learned how to build custom panels in Comet, we can build a practical example using the hotel_bookings.csv dataset and the ts_families logged metric from the previous section.

An example of a custom panel

The log_metric() method produces the following default chart:

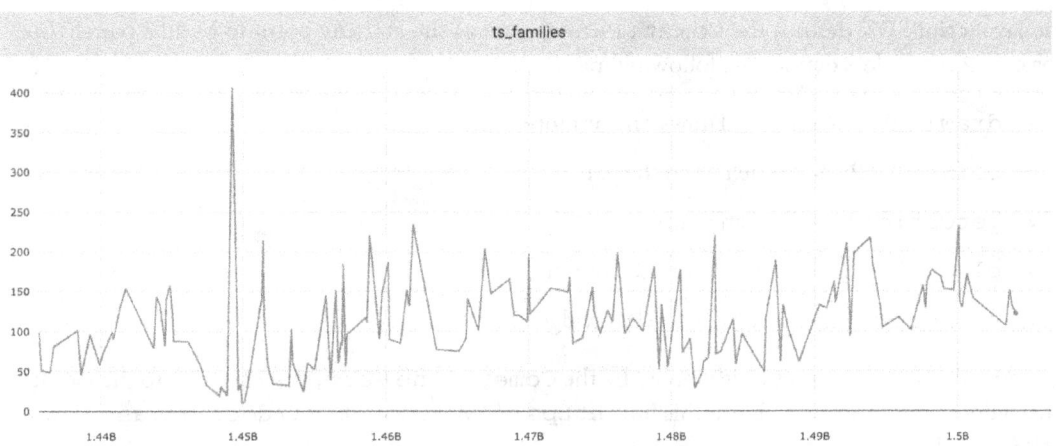

Figure 2.22 – The default chart produced by the log_metric() method

We note that the *x* axis does not correspond to dates, but to timestamps. So, in this example, we build a custom panel that converts timestamps into dates:

1. Firstly, we access the Comet online SDK and we create a new panel. We make sure that the selected language is Python. Then, in the SDK editor, we start writing our code. We import all the required libraries, as follows:

```
from comet_ml import API, ui
import matplotlib.pyplot as plt
from datetime import datetime
```

To access the Comet objects, we will use the API, while to display objects in the panel, we will use the `ui` package.

2. Now, we retrieve the logged metric as follows:

```
api = API()
experiment_keys = api.get_Panel_experiment_keys()
metric = api.get_metrics_for_chart(experiment_keys, ['ts_
families'])
```

We used the `get_metrics_for_chart()` method provided by the API to retrieve the specific `ts_families` metric.

3. After that, we build the graph, as follows:

```
for experiment_key in metrics:
    for metric in metrics[experiment_key]["metrics"]:
        cdate = [datetime.fromtimestamp(x) for x in
metric['steps']]
        plt.figure(figsize=(15,6))
        plt.grid()
        plt.xticks(rotation=45)
        plt.ylabel('Number of travelling families')
        plt.xlim(cdate[0], cdate[len(cdate)-1])
        plt.plot(cdate, metric['values'])
```

We loop over all the experiments (in our case, there is just one experiment) and all the possible metrics (just one in our case) and after converting the timestamp contained in the `metric['steps']` variable to a date, we plot the graph by calling the `display()` function:

```
ui.display(plt)
```

4.   We click on the run button, and the Comet SDK produces the following graph:

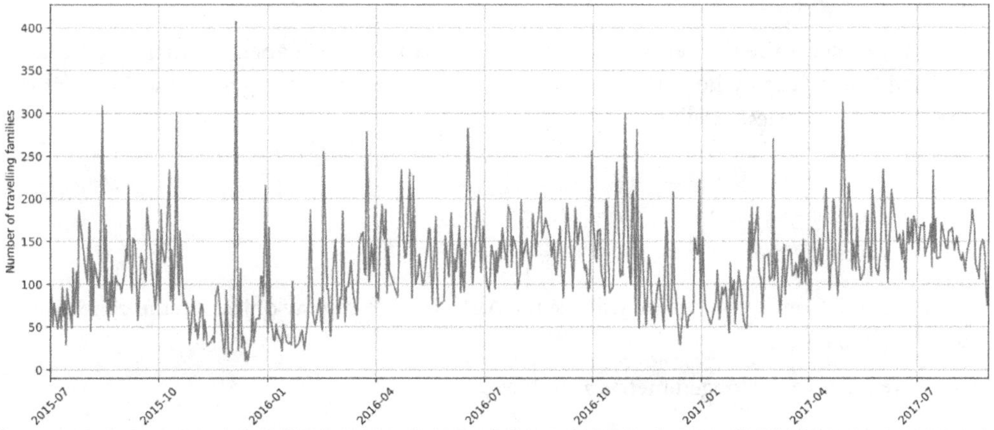

Figure 2.23 – The output of the custom panel

We have now converted timestamps to dates. Now we can add the produced custom panel to our project from the Comet dashboard simply by clicking **Add** | **New Panel** | **Workspace** | **Number of Travelling Families** | **Add**.

Now that you are familiar with custom panels, we can move on to the next feature provided by Comet for EDA: Comet Report.

Comet Report

A Comet Report is an interactive document that can contain text, panels, and experiments. A Comet Report is associated with a single project and we can create a new Report in Comet simply by clicking **Add → New Report** from the main dashboard. The following figure shows an example of a newly created empty report:

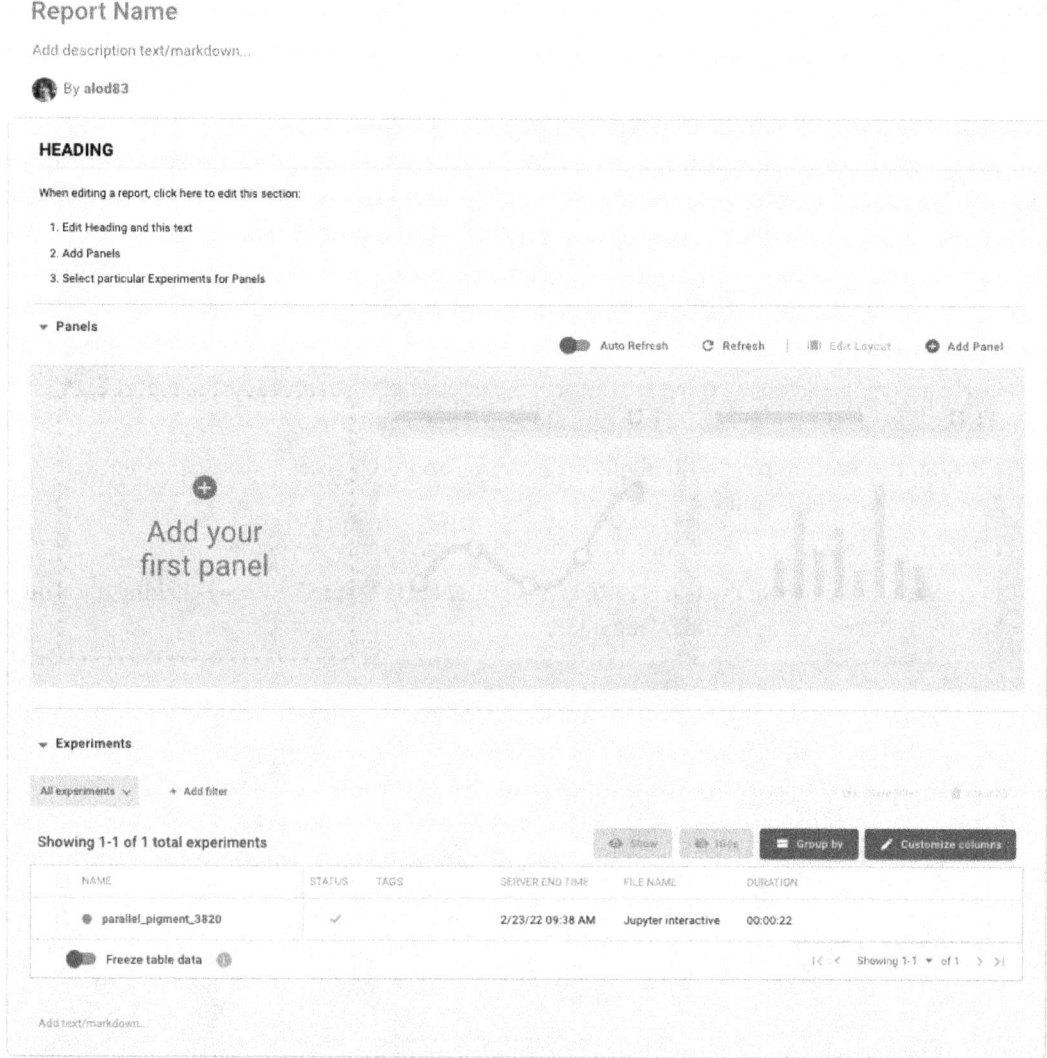

Figure 2.24 – An empty report in Comet

The previous figure shows that the default report is divided into the following parts:

- Title and description
- A section, with the heading, panels, and experiments

We can add other sections by clicking the **Add section here** button, located at the bottom of the default report.

Once our Report is ready, we can save it by clicking the **Save** button. We can also view a preview of the Report by clicking the **Preview** button. We can share the Report by copying its link on the main Report page.

Now, let's look at a practical example to illustrate how we can create a Comet Report. This example uses the features offered by the `pandas-profiling` package to explore the `hotel_bookings` dataset. Let's suppose that we have logged the output of the `pandas-profiling` package relating to the `hotel_bookings` dataset. Thus, we can proceed with report building.

From the project main dashboard, we perform the following steps:

1. Click **New → New Report**.

2. Click on **Report Name** to customize the name – for example, we can name the report EDA for the diabetes dataset. Optionally, you can add also a description.

3. Click on the **Section** box to highlight the heading, and then enter some text, or optionally, simply remove the default text.

4. Click on the area named **Add your first panel**. The panel window opens.

5. In the search box, write the text HTML, and select **HTML Asset Viewer** by clicking the **Add** button, as illustrated in the following figure:

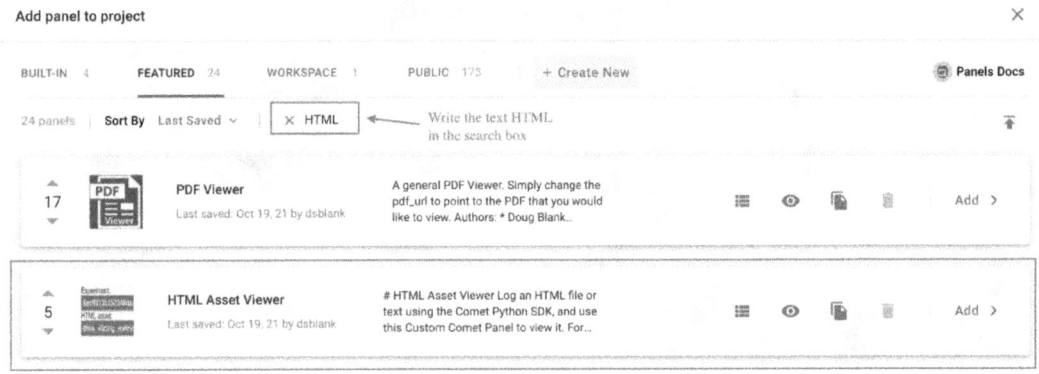

Figure 2.25 – How to select HTML Asset Viewer in the Comet Panel Window

6. The Comet SDK opens with a preview of the HTML Asset Viewer. We click on the **Done** button. The panel is added to the Report, but it is small.

7. To enlarge the panel, click on **Edit Layout** and drag the panel to cover the entire width of the screen. Then, click on the **Done Editing** button.

8. The Report is ready. As a final step, we save it by clicking on the **Save** button, located at the top right of the dashboard.

Now our report is ready with all our statistics. We can access it from the main dashboard under the **Report** tab.

Summary

We just completed the journey of performing EDA in Comet!

Throughout this chapter, we described some general concepts regarding EDA, as well as the main EDA techniques, including visual and non-visual EDA. We also illustrated the importance of EDA in a data science project: EDA permits us to understand our data, correctly formulate the questions we want to solve, and discover hidden patterns.

In the third part of the chapter, we learned which features Comet provides to perform EDA and how we can use them through a practical example. We illustrated the concepts of logs, custom panels, and reports through a practical example.

Throughout this chapter, you learned how easy it is to use Comet to run EDA, as Comet provides very intuitive features that can be combined together to build fantastic reports for EDA.

Now that you have learned how to perform EDA in Comet, we can continue our journey of the discovery of Comet for data science.

In the next chapter, we will review some concepts related to model evaluation and how to perform it in Comet.

Further reading

- Meier, M, Baldwin, D. and Strachnyi, K. (2021). *Mastering Tableau 2021*. Packt Publishing Ltd.

- Mukhiya, S. K., and Ahmed, U. (2020). *Hands-On Exploratory Data Analysis with Python: Perform EDA techniques to understand, summarize, and investigate your data*. Packt Publishing Ltd.

- Swamynathan, M. (2019). *Mastering machine learning with Python in six steps: A practical implementation guide to predictive data analytics using Python*. Apress.

3

Model Evaluation in Comet

Before accepting a data science model, we need to evaluate it, to establish whether it is ready for production or not. Model evaluation is the process of assessing whether a trained model performs as expected. Usually, we perform model evaluation on a different dataset from the one on which the model was trained.

In this chapter, you will review the basic concepts behind model evaluation, such as data splitting, how to choose metrics for evaluation, and basic concepts behind error analysis. In addition, you will see the main model evaluation techniques for the different data science tasks (classification, regression, and clustering).

Finally, you will learn how to perform model evaluation in Comet by deepening some concepts that you already know, such as experiments, panels, and reports, as well as introducing new concepts, including hyperparameter tuning, model registry, and queries.

Throughout the chapter, you will also implement a practical example.

The chapter is organized as follows:

- Introducing model evaluation
- Exploring model evaluation techniques
- Using Comet for model evaluation

Before reviewing the concepts behind model evaluation, let's install all the Python packages needed to run the code and the experiments contained in this chapter.

Technical requirements

We will run all the experiments and code in this chapter using Python 3.8. You can download it from the official website (https://www.python.org/downloads/) and choose the 3.8 version.

The examples described in this chapter use the following Python packages:

- `comet-ml 3.23.0`
- `matplotlib 3.4.3`
- `numpy 1.19.5`
- `pandas 1.3.4`
- `scikit-learn 1.0`

We have already described the first five packages and how to install them in *Chapter 1*, *An Overview of Comet*. So, please refer back to that for further details on installation.

Now that you have installed all the software needed in this chapter, let's move toward how to use Comet for model evaluation, starting from reviewing some basic concepts on model evaluation.

Introducing model evaluation

Model evaluation is the process of assessing the performance of one or more data science models to decide which is the best one to solve a given task. Model evaluation is an iterative task because we run it over and over again, until we reach a satisfactory model.

Model evaluation depends on the task we want to solve. In general, there are two types of tasks:

- **Supervised learning** – You train a model with some labeled data, you test the model on other labeled data, and then you try to predict the target value for unseen and unlabelled data. In this case, model evaluation is simple because, during the testing phase, you can compare the output produced by the model with the labeled testing data.

- **Unsupervised learning** – You do not have any labeled data, but you try to predict the output on the basis of some criteria, such as data similarity. In this case, model evaluation is quite complicated because you do not have any testing data to make comparisons.

In the case of supervised learning, model evaluation involves comparison in terms of committed errors between testing data and values predicted by the model. We can calculate different metrics that depend on a specific task, as we will see in the next sections. In the case of unsupervised learning, model evaluation is not a trivial task because we do not have a reference dataset for comparison. However, we can still also calculate some metrics in this case, as we will see in the following sections.

If we suppose that we already have a set of models to test, model evaluation is composed of the following two steps:

- **Data splitting**
- **Choosing metrics**

Let's investigate each step separately, by starting from the first step – data splitting.

Data splitting

Let's suppose that we want to build an application where users upload pictures and the system recognizes cars in them. To train a classification model, we need a dataset of pictures. We can collect them from the web, split them into training and test sets (for example, 70% of images for the training set and the remaining 30% for the test set), and train a neural network with the training set. Then, we perform model evaluation on the test set, so we calculate the accuracy. Let's suppose that we obtain a good accuracy of 90%. But when we run our system in a real-case scenario, we obtain a bad performance, such as an accuracy of only 50%. What went wrong with our model?

In the previous example, we have split our original dataset into two parts, a training set and a test set, and we have used the training set to train the model and the test set to evaluate the model performance. However, in the era of big data, this practice of dividing data into training and test sets with a ratio of 70–30 is now obsolete because it is sufficient to have enough samples for the test set to evaluate small improvements in performance. Also, the best practice is to use *three* datasets:

- **Training set** – The dataset used to train the model.

- **Dev (development) set** – The dataset used to tune hyperparameters, select features, and perform other decision tasks on our model.

- **Test set** – The dataset used to perform the model evaluation. The results of the evaluation on the test set do not affect the choice of model.

In practice, we use both the dev and test sets to perform model evaluation, but while we exploit the performance calculated on the dev set to improve the model, we use the test set only to assess the final result of the model. In practice, the test set represents the real-case scenario; thus we should not extract it from the original dataset but from the real world, where we will use our system. In the previous example of car recognition, we should extract the test set directly from the application. In the beginning, we will have a few pictures, so our test set will be very small. Then, as users upload pictures, our test set will increase, so we can make more accurate assessments.

All three datasets should have the same distribution. **Data distribution** shows all possible values the data can assume. Although we know many types of data distribution, such as normal, beta, and gamma, in the real world, data does not assume any of them. For example, referring to the previous example of car recognition, if the dev set contains pictures with *racing cars* and the test set contains pictures with *vintage cars*, the two datasets follow different distributions.

Many techniques exist to test whether two variables follow the same distribution. However, they need to test each column of the two datasets independently. In this section, we propose a strategy to check whether the entire datasets follow the same distribution, not just the single columns.

We can use the strategy shown in the following figure.

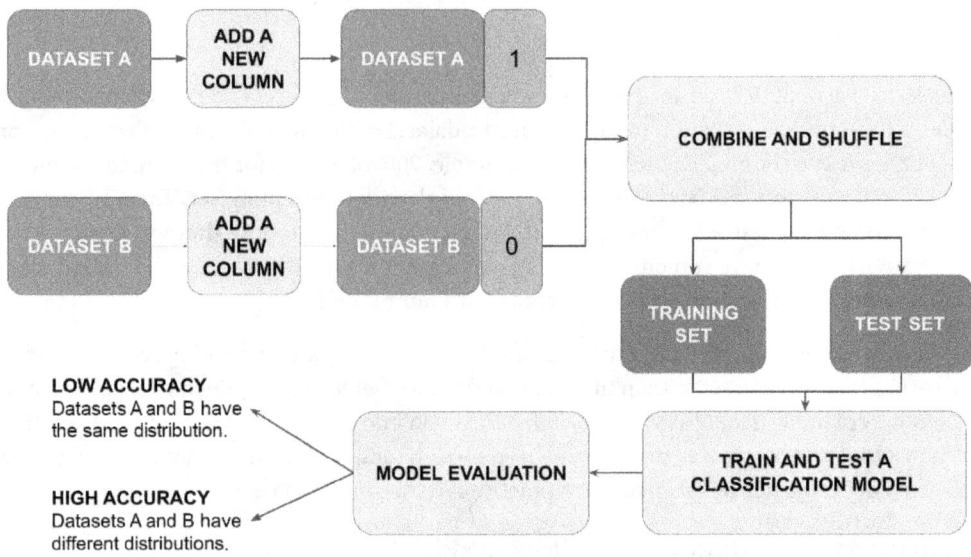

Figure 3.1 – A possible strategy to check whether two datasets have the same distribution

Let's suppose that we have two datasets, namely **Dataset A** and **Dataset B** (for example, the dev and test sets):

1. Firstly, we add a new column to both the datasets, and we name it `target`. For Dataset A, we set the target to 1 for all the records; for Dataset B, we set the target value to 0, for all the records.

2. We combine the two enriched datasets in order to obtain a single dataset. We also shuffle records.

3. We split the obtained dataset into two parts – a training set and a test set.

4. We train a classification model with the training set, and we evaluate it with the test set.

5. We calculate the accuracy of the model. If we obtain a low value for the accuracy, it means that the two original datasets are very similar; thus the model cannot recognize correctly whether a record belongs to the first or the second dataset. If the accuracy is high, it means that the two datasets are quite different.

Note that in the previous strategy, we have transformed the problem of calculating whether two datasets follow the same distribution into a classification problem.

If the dev and test datasets do not follow the same distribution, we have a situation called **covariate shift**. In this case, we can encounter the following problems:

- **Overfitting** – The model performs well on the dev set, but it has poor performance on the test set.

- **Complexity** – The test set could be more complex than the dev set; thus the model is not suitable for managing the complexity of the problem.

- **Diversity** – Simply put, the test set is different from the dev set, so the model solves a different problem from the one we encounter in reality.

Now that you have learned how to split the data into training, dev, and test sets, as well as the general problems of covariate shift, we can move to the next step, choosing metrics.

Choosing metrics

Let's suppose that you have implemented four models that solve the same problem of car recognition in pictures. Now, you want to choose the best model and move it into production. You may decide to calculate different metrics for all the models (for instance, precision and recall) and then choose the model that has the best metrics. However, it may happen that a model performs better than the others for one metric, while another model outperforms the others for another metric, as shown in the following table:

Model	Precision	Recall
Model 1	0.95	0.36
Model 2	0.82	0.94
Model 3	0.83	0.85
Model 4	0.78	0.95

Figure 3.2 – Precision and recall for the four models of the car recognition example

We note that **Model 1** has the best precision value, and **Model 4** has the best recall value. Which one is the best model? According to the calculated metrics, the previous example suggests that there is no absolute best model.

To solve the previous problem, you should combine all the metrics you calculate to define a single metric. Your best model will be that with the best value for your defined metric. You can define your own metric, which depends on your specific task. For example, you can write the following metric:

$$m = w_1 p + w_2 r$$

In the previous formula, m is the combined metric, p and r are precision and recall respectively, and w_1 and w_2 are the weights assigned to p and r respectively. If, for example, precision is more important than recall for your task, you can set $w_1 = 0.7$ and $w_2 = 0.3$.

In the previous example, choosing the best model is simple, as shown in the following table:

Model	Precision	Recall	Combined Metric
Model 1	0.95	0.36	0.773
Model 2	0.82	0.94	0.856
Model 3	0.83	0.85	0.836
Model 4	0.78	0.95	0.831

Figure 3.3 – Precision, Recall, and Combined Metric for the car recognition example

The previous table shows that **Model 2** is the best model.

Once you have defined your combined metric, you should optimize it on the dev set. It may happen that your combined metric is not the best metric to measure the performance of your model. Usually, this occurs when at least one of the following situations occurs:

- **Overfitting** – The performance between the dev set and the test set is totally different.
- **Changing** – The test set has changed.
- **Targeting** – The combined metric measures something other than what the project needs to optimize.

If at least one of the previous cases occurs, it is advisable to change your combined metric.

Now that you have learned how to choose the best metric to evaluate your model, we can move toward the next section, exploring model evaluation techniques.

Exploring model evaluation techniques

Depending on the problem we want to solve, there are different model evaluation techniques. In this section, we will consider three types of problems: regression, classification, and clustering.

The first two problems fall within the scope of supervised learning, while the third method falls within the scope of unsupervised learning.

In this section, you will review the main metrics used for model evaluation in the previously cited problems. We will implement a practical example in Python to illustrate how to calculate each metric. To review the main evaluation metrics, we will use only two datasets: the training and test sets.

Regarding supervised learning, there is also an additional technique to perform model evaluation. This technique is called **cross validation**. The basic idea behind cross validation is to split an original dataset into several subsets. The model trains all the subsets, except one. When the training phase is completed, the model is tested on the remaining subset. This is an iterative procedure, for all the possible subsets of the dataset. We will discuss cross validation and how Comet supports it in detail in *Chapter 8, Comet for Machine Learning*.

This section is organized as follows:

- Loading and preparing the dataset
- Regression
- Classification
- Clustering

Let's start from the first step, loading and preparing the dataset.

Loading and preparing the dataset

We will use the Diamonds dataset, provided by ggplot2 under the MIT licenses (https://ggplot2.tidyverse.org/reference/diamonds.html) and available on Kaggle as a CSV file (https://www.kaggle.com/shivam2503/diamonds):

1. Firstly, we load the dataset as a pandas DataFrame:

```
import pandas as pd
df = pd.read_csv('source/diamonds.csv')
```

The dataset contains 53,940 rows and 11 columns. The following figure shows the first 10 rows of the dataset:

Unnamed: 0	carat	cut	color	clarity	depth	table	price	x	y	z
1	0.23	Ideal	E	SI2	61.5	55	326	3.95	3.98	2.43
2	0.21	Premium	E	SI1	59.8	61	326	3.89	3.84	2.31
3	0.23	Good	E	VS1	56.9	65	327	4.05	4.07	2.31
4	0.29	Premium	I	VS2	62.4	58	334	4.2	4.23	2.63
5	0.31	Good	J	SI2	63.3	58	335	4.34	4.35	2.75
6	0.24	Very Good	J	VVS2	62.8	57	336	3.94	3.96	2.48
7	0.24	Very Good	I	VVS1	62.3	57	336	3.95	3.98	2.47
8	0.26	Very Good	H	SI1	61.9	55	337	4.07	4.11	2.53
9	0.22	Fair	E	VS2	65.1	61	337	3.87	3.78	2.49
10	0.23	Very Good	H	VS1	59.4	61	338	4	4.05	2.39

Figure 3.4 – The first 10 rows of the diamonds dataset

Note that the diamonds dataset contains some categorical columns, including **cut**, **color**, and **clarity**, and some numerical columns (the others).

2. We drop the first column, Unnamed: 0, as follows:

```
df = df.drop(["Unnamed: 0"], axis=1)
```

We use the drop() method provided by pandas with axis=1 to indicate that we want to drop columns.

3. We define two practical functions to transform our data; we will use the first one to convert categorical columns to numerical and the second to scale numerical columns. We define the first function as follows:

```
from sklearn.preprocessing import LabelEncoder
def encode_labels(data):
    categories = (data.dtypes =="object")
    cat_cols = list(categories[categories].index)
    categories = (X.dtypes =="object")
    feature_label_encoder_dict = {}
    for col in cat_cols:
        feature_label_encoder_dict[col] = LabelEncoder()
        X[col] = feature_label_encoder_dict[col].fit_
transform(X[col])
```

The function receives the DataFrame as input. Firstly, we import the LabelEncoder class, which will permit us to convert categorical values into numerical ones. Then, we select all the categorical columns, and we store them in the categories variable. Next, we build a LabelEncoder() object for each category column and store it in a dictionary named feature_label_encoder_dict. Finally, for each categorical column, we fit and transform the built feature_label_encoder_dict object.

4. Now, we define the scale_numerical() function, as follows:

```
from sklearn.preprocessing import StandardScaler
def scale_numerical(data):
    scaler = StandardScaler()
    data[data.columns] = scaler.fit_transform(data[data.
columns])
```

The function receives the DataFrame as input and scales all the numerical columns through a StandardScaler() object.

Now that we have prepared the data, we can move to the next step, evaluation metrics for regression.

Regression

Regression analysis is a type of supervised machine learning that tries to predict a continuous target variable, named Y, on the basis of one or more input variables, named X. To evaluate a regression task, we can calculate many metrics.

As a regression task example, we want to build a model that predicts a diamond's price on the basis of other features. Before calculating these metrics, we build the training and test sets as follows:

1. We split the dataset into input features (X) and a target variable (Y):

    ```
    X = df.drop("price", axis = 1)
    y = df["price"]
    ```

2. We encode labels and scale numerical columns:

    ```
    encode_labels(X)
    scale_numerical(X)
    ```

    Note that we have used the previously defined functions.

3. We split the datasets into training and test sets, through the `train_test_split()` function, provided by `scikit-learn`:

    ```
    from sklearn.model_selection import train_test_split
    X_train, X_test, y_train, y_test = train_test_split(X, y,
    test_size=0.20, random_state=42)
    ```

    We have reserved 20% of samples for the test set and the remaining 80% of samples for the training set. Also, we have set `random_state` to `42` to make the experiment reproducible.

4. We create a new linear regression model, fit it with the training set, and calculate the predicted values on the test set:

    ```
    model = LinearRegression()
    model.fit(X_train, y_train)
    y_pred = model.predict(X_test)
    ```

    In this example, we do not care about model optimization; thus we use the default model.

Now that we have extracted the predicted values, we can calculate the main metrics used to evaluate a regression model. In this section, we calculate three of the most popular metrics: Mean Absolute Error, Root Mean Squared Error, and R Squared:

- **Mean Absolute Error (MAE)** – The average of the difference between the real value and the predicted one. This measures how far the predictions are from the actual output.

The lower the MAE, the better the model. In the previous example, we can calculate MAE as follows:

```
from sklearn.metrics import mean_absolute_error
MAE = mean_absolute_error(y_test,y_pred)
```

- **Root Mean Squared Error (RMSE)** – The square root of **Mean Squared Error** (**MSE**). MSE is similar to MAE. The only difference is that MSE calculates the average of the square of the difference between the real values and the predicted ones. RMSE is the most used metric to evaluate regression models because it gives you an idea of how concentrated the data is around the line predicted by the model.

In the previous example, we can calculate RMSE as follows:

```
from sklearn.metrics import mean_squared_error
import numpy as np
RMSE = np.sqrt(mean_squared_error(y_test, y_pred))
```

- **R Squared** or the **coefficient of determination** – The proportion of variance in Y that can be explained by X. In other words, R Squared describes how well the data fits the regression model. R Squared is a number between 0 and 1. For example, R Squared = 0.80 means that 80% of the data fits the model. In general, the higher the R Squared value, the better the model. However, this is not always true; thus, you should always combine R Squared with other metrics. In the previous example, we can calculate R Squared as follows:

```
from sklearn.metrics import r2_score
R2 = r2_score(y_test, y_pred)
```

Now that you have seen the most popular metrics used to evaluate regression models, we can analyze the most common metrics for classification.

Classification

Classification is a type of supervised learning that tries to predict the target class label, named Y, on the basis of one or more input variables, named X. If the number of class labels is two, we have binary classification; otherwise, if the number of labels is greater than two, we have multiclass classification. In this chapter, we consider only binary classification, but the general concepts described can also be extended to multiclass classification.

As a classification task example, we want to build a model that predicts a diamond's cut on the basis of other features. The diamonds dataset contains five types of diamond cuts (**Ideal**, **Premium**, **Very Good**, **Good**, and **Fair**); thus the problem is multiclass classification. For simplicity, we transform the multiclass classification problem into a binary classification problem.

Before calculating the metrics for binary classification, we prepare the dataset, as follows:

1. We group the cuts into two classes, `Gold` and `Silver`, as shown in the following code:

```
def set_target(x):
    golden_set = ['Ideal', 'Premium', 'Very Good']
    if x in golden_set:
        return 'Gold'
    return 'Silver'
df['target'] = df['cut'].apply(lambda x: set_target(x))
df.drop("cut", axis = 1,inplace=True)
```

We define a function, named `set_target()`, which receives as input a variable, named x, checks whether it belongs to `golder_set` or not, and returns `'Gold'` if true or `'Silver'` otherwise. Then, we create a new column in the original DataFrame, called `target`, which contains the output of the `set_target()` function. We also drop the original `cut` column, which is not needed anymore.

2. Now, we build the training and tests sets, as follows:

```
X = df.drop("target", axis = 1)
y = df["target"]
```

As the input features X, we consider all the columns except the `target` one, which instead is associated with the y output feature.

3. We encode and scale input features:

```
encode_labels(X)
scale_numerical(X)
```

Since we have previously defined the `encode_labels()` and `scale_numerical()` functions, this operation is quite simple.

4. We also encode the `target` labels:

```
label_encoder = LabelEncoder()
y = label_encoder.fit_transform(y)
```

5. We split the dataset into training and test sets, as shown in the following code:

```
X_train, X_test, y_train, y_test = train_test_split(X, y,
test_size=0.20, random_state=42)
```

We use the `train_test_split()` function provided by `scikit-learn`.

6. We build the classification model and train it. In this example, we consider a RandomForestClassifier:

```
model = RandomForestClassifier()
model.fit(X_train, y_train)
y_pred = model.predict(X_test)
```

We fit the model with the training set and predict the values of the test set. Similar to the case of regression, we do not care about model optimization, since in this section, our objective is to show the evaluation metrics.

Now that we have built the model, we can calculate the most popular metrics for classification – confusion matrix, precision, recall, accuracy, the F1-score, and ROC curve:

• **Confusion matrix** – A table with two rows and two columns, as shown in the following figure:

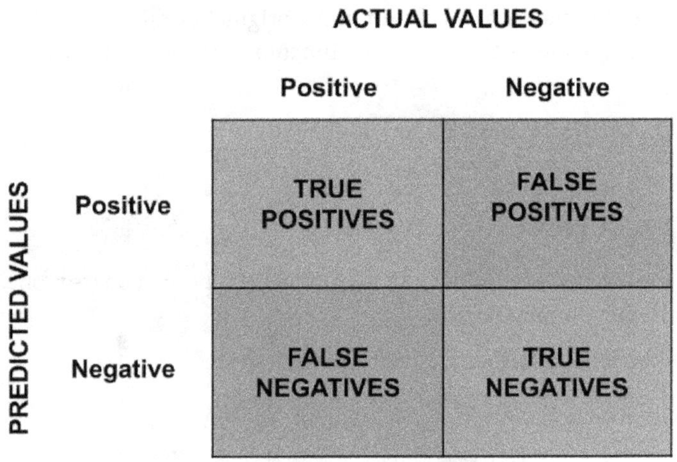

ACTUAL VALUES

	Positive	Negative
Positive	TRUE POSITIVES	FALSE POSITIVES
Negative	FALSE NEGATIVES	TRUE NEGATIVES

PREDICTED VALUES

Figure 3.5 – The confusion matrix

The table shows the predicted values (rows) versus the actual values (columns). Each cell of the table corresponds to the number of correct or wrong classifications:

• **True Positive (TP)** – The actual value is positive, and the predicted value is positive. In this case, the model performed well.

• **True Negative (TN)** – The actual value is negative, and the predicted value is negative. In this case, the model performed well.

• **False Positive (FP)** – The actual value is negative, but the predicted value is positive. In this case, the model committed an error.

• **False Negative (FN)** – The actual value is positive, but the predicted value is negative. In this case, the model committed an error.

7. In `scikit-learn`, we can calculate the confusion matrix as follows:

```
from sklearn.metrics import confusion_matrix
[tp,fp], [fn,tn] = confusion_matrix(y_test, y_pred)
```

8. We use the `confusion_matrix()` function, which receives the test set and the predicted values as input, and returns the TP, FP, FN, and TN. `scikit-learn` also provides a function to directly plot the confusion matrix, as shown in the following piece of code:

```
from sklearn.metrics import plot_confusion_matrix
import matplotlib.pyplot as plt
plot_confusion_matrix(model, X_test, y_test, cmap='GnBu')
plt.show()
```

The function receives the model and the test set as input. As an additional parameter, we can pass the color map (`'GnBu'`, in our case). The following figure shows the output of the `plot_confusion_matrix()` function for our example:

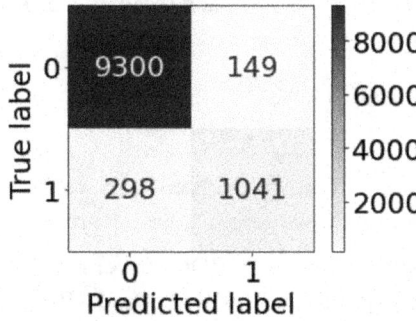

Figure 3.6 – The output of the plot_confusion_matrix() function

The matrix shows the number of records for each cell of the table. For example, the first cell indicates that there are 9,300 true positives. The table also colors the cells according to a gradient of colors.

9. On the basis of the previous measurements, we can define the other metrics:

- **Precision** – Among all the positive predictions (TP and FP), count how many of them are really positive (TP). In `scikit-learn`, we can calculate this as follows:

```
from sklearn.metrics import precision_score
precision = precision_score(y_test, y_pred)
```

We use the `precision_score()` function, which receives the test set and the predicted values as input, and returns a number corresponding to the precision.

- **Recall** – Among all the real positive cases (TP and FN), count how many of them are predicted positive (TP). In `scikit-learn`, we can calculate this as follows:

```
from sklearn.metrics import recall_score
recall = recall_score(y_test, y_pred)
```

We use the `recall_score()` function, which receives the test set and the predicted values as input, and returns a number corresponding to the recall.

- **Accuracy** – Among all the cases (TP + TN + FP + FN), count how many of them have been predicted correctly. In `scikit-learn`, we can calculate this as follows:

```
from sklearn.metrics import accuracy_score
accuracy = accuracy_score(y_test, y_pred)
```

We use the `accuracy_score()` function, which receives the test set and the predicted values as an input, and returns a number corresponding to the accuracy. Many data scientists use accuracy as a single metric to test the validity of their model.

- **F1-score** – The harmonic means of precision and recall. In `scikit-learn`, we can calculate this as follows:

```
from sklearn.metrics import f1_score
f1 = f1_score(y_test, y_pred)
```

We use the `f1_score()` function, which receives the test set and the predicted values as input, and returns a number corresponding to the F1-score.

- **Receiver Operating Characteristics curve (ROC curve)** – A curve showing the **true positive rate** (the recall or **TPR**) against the **false positive rate** (**FRP**). We can think about the FPR as the number of predictions incorrectly classified as positive (FP) among all the negative values (FP and TN). We plot the ROC curve at different threshold values – the threshold = 1, TPR = 1, and FPR = 1. In `scikit-learn`, we can plot the ROC curve as follows:

```
from sklearn.metrics import roc_curve,roc_auc_score
y_pred_proba = model.predict_proba(X_test)[::,1]
fpr, tpr, _ = roc_curve(y_test,  y_pred_proba)
auc = roc_auc_score(y_test, y_pred_proba)
plt.plot(fpr,tpr,label='auc=%.3f' % auc, color='#084081')
axis_ranges = [0,1]
plt.plot(axis_ranges, axis_ranges, linestyle='--',
color='k', scalex=False, scaley=False)
plt.xlabel('False Positive Rate')
plt.ylabel('True Positive Rate')
plt.legend()
```

```
plt.grid()
plt.show()
```

Firstly, we calculate the predicted probabilities for each class through the `predict_proba()` method. Then, we calculate the ROC curve through the `roc_curve()` function, which receives as input the output of the `predict_proba()` method. We also calculate the AUC score, through the `roc_auc_score()` function. The **Area Under the Curve (AUC)** score measures the ability of a classifier to distinguish between the target classes. The higher the AUC, the better the performance of a model in distinguishing between positive and negative targets. The following figure shows the output of the previous code:

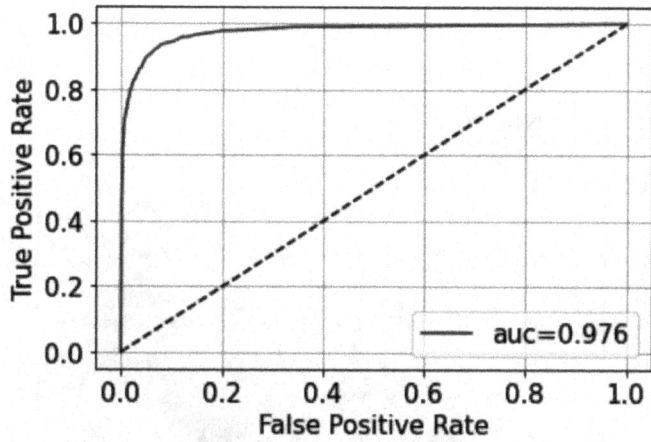

Figure 3.7 – The ROC curve

Note that in our example the model performs quite well because the ROC curve tends to be flattened to the left. The dotted line represents a 0.5 AUC score, which is inherent to the random guessing model.

Now that we have reviewed the most popular metrics used to evaluate classification models, we can analyze the most common metrics for clustering.

Clustering

Clustering analysis is a type of unsupervised machine learning, where there is no training set. Clustering is used to group records according to similarity criteria, such as distance. A clustering model takes a dataset as input and returns a list of labels as output, corresponding to the associated clusters.

Evaluating the performance of a clustering model is not easy because you should verify that each record has been assigned the right cluster. In other words, you should verify that each record is much more similar to the records belonging to its cluster than to the records belonging to the other clusters.

Before calculating the metrics, we will prepare the `diamonds` dataset, as follows:

1. For simplicity, we will consider only two columns of the dataset, `price` and `carat`, as shown in the following code:

```
X = df[['price', 'carat']]
```

2. Then, we plot the dataset, as follows:

```
plt.scatter(X['price'],X['carat'])
plt.xlabel('Price')
plt.ylabel('Carat')
plt.grid()
plt.show()
```

The following figure shows the resulting plot:

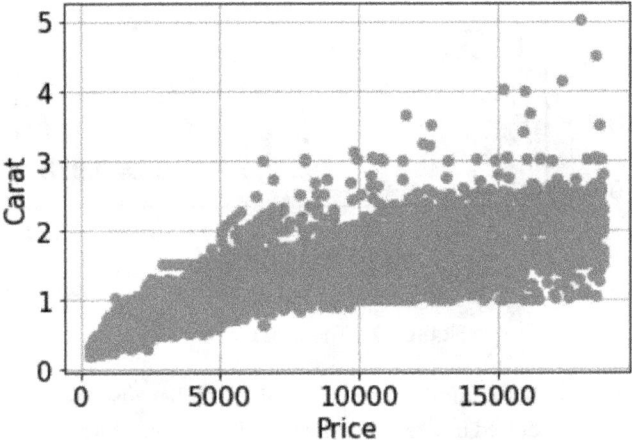

Figure 3.8 – A scatter plot showing Carat against Price

3. We use a K-means model to cluster our data into two clusters, as follows:

```
from sklearn.cluster import KMeans
model = KMeans(n_clusters=2)
labels = model.fit_predict(X)
```

We use the KMeans() class provided by `scikit-learn`, and then we call the `fit_predict()` method to calculate the clusters.

4. We plot the results of clustering, as follows:

```
from matplotlib.colors import ListedColormap
cmap = ListedColormap(['#40B7AD', '#084081'])
```

```
plt.scatter(X['price'],X['carat'], c=labels, cmap=cmap)
plt.xlabel('Price')
plt.ylabel('Carat')
plt.grid()
plt.show()
```

We assign to each point of the scatter plot a color, corresponding to the associated label. The following figure shows the resulting plot:

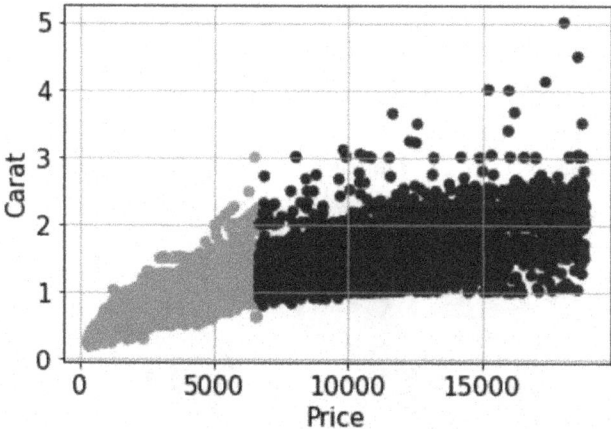

Figure 3.9 – The dataset after clustering

Now that we have clustered our dataset, we can evaluate the model. There are two types of evaluation:

- **Supervised-based evaluation** (or **extrinsic methods**) assumes that there is a ground truth to perform the evaluation. The idea is to compare the ground truth with the result of a clustering algorithm in order to calculate a score. Many metrics fall in to this category, such as the **homogeneity score** and the **Mallows score**. Since calculating a ground truth is very difficult, in this chapter, we will not focus on this type of evaluation.

- **Unsupervised-based evaluation** (or **intrinsic methods**) calculates how well the clusters are separated and how compact the clusters are.

Regarding the intrinsic methods, we can calculate many metrics, including the following ones:

- The **silhouette coefficient** or **silhouette score** measures how similar an object is to its cluster compared to the others. This coefficient can range from –1 to 1. A value equal to 1 means that the object has been correctly classified, while a value of –1 means that the model has inserted the object in the wrong cluster. In `scikit-learn`, we can calculate it as follows:

```
from sklearn.metrics import silhouette_score
score = silhouette_score(X, labels)
```

The function receives the X dataset and the labels as input and returns a number representing the silhouette score. In the previous example, the score is 0.708.

- The **Elbow method** – a visual method to estimate the optimal number of k clusters. We run the algorithm multiple times with an increasing number of clusters, and then we plot the sum of squared distances of samples to their closest cluster center (SSE) as a function of the number of clusters. In Python, we can apply the elbow method as follows:

```
sse = {}
for i in range(2,10):
    model = KMeans(n_clusters=i)
    sse[i] = model.inertia_
```

We define a loop with different values of the number of clusters. After fitting the model, we calculate SSE (`model.inertia_`) for each iteration. We plot the results as follows:

```
plt.plot(sse.keys(), sse.values())
plt.grid()
plt.xlabel('Number of Clusters')
plt.ylabel('SSE')
plt.show()
```

The previous code produces the following figure:

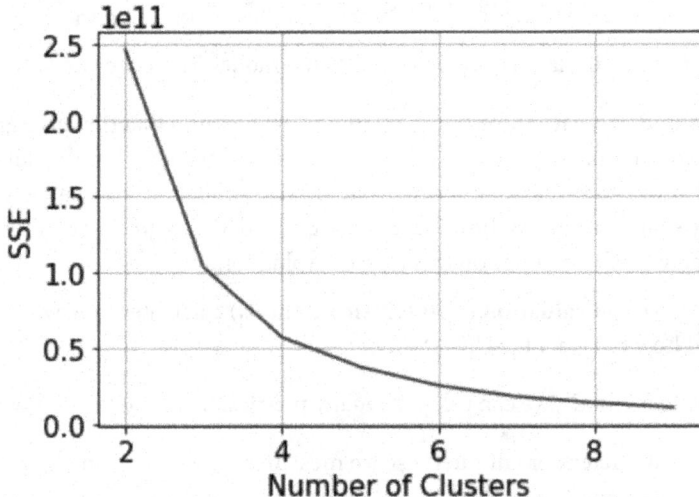

Figure 3.10 – The elbow method

We can identify the correct number of clusters as the point after which the curve begins to decrease rapidly. In our case, the best number of clusters could be two.

Now that we have reviewed all the main techniques for model evaluation, we are ready to move to the next section, using Comet for model evaluation.

Using Comet for model evaluation

Comet provides the following features to deal with model evaluation:

- **Log** – used to store metrics, assets, and objects in Comet
- **Dashboard** – used to compare the results of the experiments
- **Registry** – used to track and store your models
- **Report** – used to show the results

The following figure shows how to combine the features provided by Comet to compare different models and then choose the best one for production:

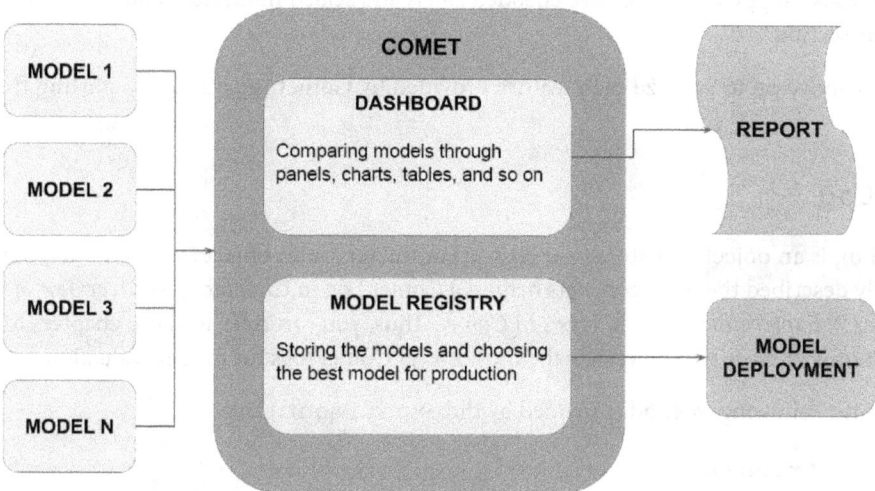

Figure 3.11 – How to use Comet for model evaluation

Let's suppose that you want to compare N models and then choose the best model for deployment. You build your experiments and then you track them in Comet. Through Comet Dashboard, you can compare models by building panels, charts, tables, and other similar objects. You can also store your models in the Comet registry. You can even export a report showing the results of comparison from the Comet platform. Once you have selected the best model, you can export it from the registry and make it available for production.

To show how we can use Comet for model evaluation, you will use the previously defined diamonds dataset, and you will implement four classification models – **Random Forest**, **Decision Tree**, **Gaussian Naive Bayes**, and **K-Nearest Neighbors**. We will use the basic version of these models, without optimizing them, because the objective of this chapter is to show how to perform model evaluation. For more details on how to optimize classification models, you can refer to *Chapter 8*, *Comet for Machine Learning*. You will use the cleaned version of the dataset, described in the previous section, *Classification*, that we built as follows:

```
options = ['Ideal', 'Premium']
df2 = df[df['cut'].isin(options)]
X = df2.drop("cut", axis = 1)
y = df2["cut"]
```

We have supposed that `df` contains the `diamonds` dataset. We have selected only two possible values for `cut`, to deal with binary classification. Then, we have built the input (`X`) and target (`y`) variables. We also suppose that we have encoded labels and scaled numerical values, as described in the previous section.

We can now move on to analyze each feature provided by Comet separately, by starting from the first, Log.

Comet Log

A **Comet Log** is an object that stores a metric, a parameter, or an object in general in Comet. We have already described the basic concepts behind a Comet Log in *Chapter 1*, *An Overview of Comet*, and *Chapter 2*, *Exploratory Data Analysis in Comet*. Thus, you can refer to those chapters for basic concepts. In this section, we will review the most useful Comet logs for model evaluation.

We will use the following methods provided by the `experiment` class:

- `log_metrics()`
- `log_curve()`
- `log_confusion_matrix()`

Let's suppose that we have already split our dataset into training and test sets, as described in the previous section:

1. Firstly, we define an auxiliary function, named `compute_metrics()`, which calculates all the evaluation metrics for our experiment:

   ```
   from sklearn.metrics import precision_score, recall_
   score, f1_score, accuracy_score
   def compute_metrics(y_pred, y_true):
   ```

```
metrics = {}
    metrics['precision'] = precision_score(y_true, y_pred)
    metrics['recall'] = recall_score(y_true, y_pred)
    metrics['f1-score'] = f1_score(y_true, y_pred)
    metrics['accuracy'] =  accuracy_score(y_true, y_pred)
    return metrics
```

The function returns a dictionary, named `metrics`, which stores all the calculated metrics. We calculate precision, recall, the F1-score, and accuracy. Following the discussion in the *Choosing metrics* section, we will use just one metric to perform the comparison among the different models: accuracy. However, for completeness, we also calculate the other metrics.

2. Now, we define another auxiliary function, which runs a single experiment. The function receives the model class and the model's name as input, then it trains the model with the training set, and finally, it calculates the metrics, as well as the confusion matrix and the ROC curve. The following code implements the described function:

```
from sklearn.metrics import roc_curve
def run_experiment(ModelClass, name):
    experiment = Experiment()
    experiment.set_name(name)
    experiment.add_tag(name)

    model = ModelClass()
    with experiment.train():
        model.fit(X_train, y_train)
        y_pred = model.predict(X_train)
        metrics = compute_metrics(y_pred, y_train)
        experiment.log_metrics(metrics)
        experiment.log_confusion_matrix(y_train, y_pred)

    with experiment.validate():
        y_pred = model.predict(X_test)
        metrics = compute_metrics(y_pred, y_test)
        experiment.log_metrics(metrics)
        experiment.log_confusion_matrix(y_test, y_pred)
        fpr, tpr, _ = roc_curve(y_test, y_pred)
        experiment.log_curve(name, fpr, tpr)
```

Firstly, we create a new `experiment` object to permit communication with Comet. To configure the `experiment` parameters, including the workspace and the API key, you can refer to *Chapter 1, An Overview of Comet*. We also set the experiment name to a name passed as input, to make the experiment recognizable from the Comet Dashboard. We use the `set_name()` method of the `Experiment` class to perform this operation. In addition, we add a new tag to the experiment through the `add_tag()` method. Then, we create our model through the `model = ModelClass()` statement. Note that `ModelClass()` is a variable, which you will set when calling the function.

Once we have created the model, we can train it. We use the `with experiment.train()` statement to let Comet know that all the logged objects belong to the training phase. We calculate metrics through the `compute_metrics()` function, and then we log them in Comet through the `log_metrics()` method provided by the `Experiment` class. In addition, we log the confusion matrix through the `log_confusion_matrix()` method. Once training is complete, we can test the model. Similar to the training phase, we use the `with experiment.validate()` statement to let Comet know that all the logged objects belong to the training phase. We call the `predict()` method provided by the model on the test set, and we calculate all the metrics, as already done during the training phase. In addition, in this phase, we calculate the ROC curve through the `roc_curve()` function, and we log it in Comet through the `log_curve()` method, provided by the `Experiment` class.

3. Finally, we can run the experiments. We test four classifiers, as follows:

```
from sklearn.ensemble import RandomForestClassifier
from sklearn.tree import DecisionTreeClassifier
from sklearn.naive_bayes import GaussianNB
from sklearn.neighbors import KNeighborsClassifier


run_experiment(RandomForestClassifier, 'RandomForest')
run_experiment(DecisionTreeClassifier,
'DecisionTreeClassifier')
run_experiment(GaussianNB, 'GaussianNB')
run_experiment(KNeighborsClassifier,
'KNeighborsClassifier')
```

Simply, we call the `run_experiment()` function for each of the classifiers we want to test. In the example, we test Random Forest, Decision Tree, Gaussian Naive Bayes, and K-Nearest Neighbors.

The `log_metrics()` method permits us to also log epochs and steps. An **epoch** is a hyperparameter available only for certain types of algorithms, such as those based on gradient descent. An epoch corresponds to the number of times the algorithm will work with the training dataset.

Setting the epoch to 1 means that each sample of the training set has just one opportunity to update the algorithm during the training phase. A **step** (or batch) defines the number of samples to use before updating the model parameters. We can set the step equal to the training set size or a smaller number.

4. To deal with epochs, we can modify the run_experiment () function, as follows:

```
def run_experiment(ModelClass, name, n_epochs):
    ...
    with experiment.train():
        for i in range(n_epochs):
            model = ModelClass(max_iter=n_epochs)
            model.fit(X_train, y_train)
            y_pred = model.predict(X_train)
            metrics = compute_metrics(y_pred, y_train)
            experiment.log_metrics(metrics, epoch = i)
            experiment.log_confusion_matrix(y_train, y_
pred, epoch=i)
    ...
```

We add another argument to the function, which is the number of epochs (n_epochs). Then, during the training phase, we build a loop over the number of epochs, and we build a model for each epoch. The previous function works only with models supporting the number of epochs, such as the SGD classifier:

```
from sklearn.linear_model import SGDClassifier
run_experiment_with_epoch(SGDClassifier, 'SGD',1000)
```

We build a classifier and set the number of epochs up to 1,000.

Similar to the number of epochs, we can set the number of steps. For example, we can decide to train our model with different sizes of the dataset. Thus, we can modify the run_experiment () function, as follows:

```
import numpy as np
def run_experiment(ModelClass, name):
    step_size = len(X_train)
    min_steps = 20

    ...
    with experiment.train():
        for i in np.arange(min_steps, step_size+1, step =
5000):
            model = ModelClass()
```

```
          X_t = X_train[0:i]
          y_t = y_train[0:i]
          model.fit(X_t, y_t)
          y_pred = model.predict(X_t)
          metrics = compute_metrics(y_pred, y_t)
          experiment.log_metrics(metrics, step = i)
          experiment.log_confusion_matrix(y_t, y_pred,
   step=i)
     . . .
```

We set the step size to the length of the training set and the minimum number of samples in a step equal to 20. Then, we loop over the different steps and train different models, each with an increasing number of samples.

Now that you have logged all the needed metrics, we can analyze them in the Comet Dashboard.

Comet Dashboard

Comet Dashboard is the Comet online website, which stores all your experiments. Let's suppose that you have run all four experiments, described before. Under the **Panels** or **Experiments** tab, you will see them, as shown in the following figure:

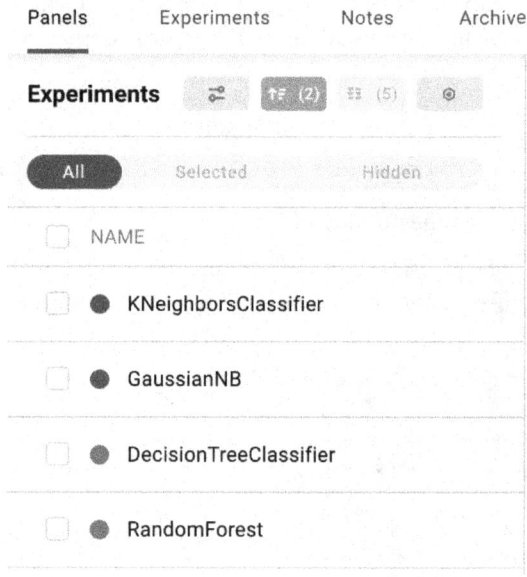

Figure 3.12 – The four experiments shown in Comet

We can easily identify each experiment because we have set the experiment name for each model.

We can perform the following types of comparison among the experiments:

- Ordering
- Raw comparison through a table
- Filtering
- Grouping

Let's look at each type of comparison separately, starting with the first one, ordering.

We can order experiments by a specific parameter or metric:

1. We select the ordering button, as shown in the following figure:

Figure 3.13 – The ordering button in Comet Dashboard

2. A new pop-up window opens, showing the default criteria for ordering the experiments. There are two criteria as default. We can add as many criteria as we want. In this example, we set just one criterion. We can remove one of the default criteria by clicking the **X** button.

3. Then, we can change the remaining criterion by opening the drop-down menu and selecting **validate_accuracy**, as shown in the following figure:

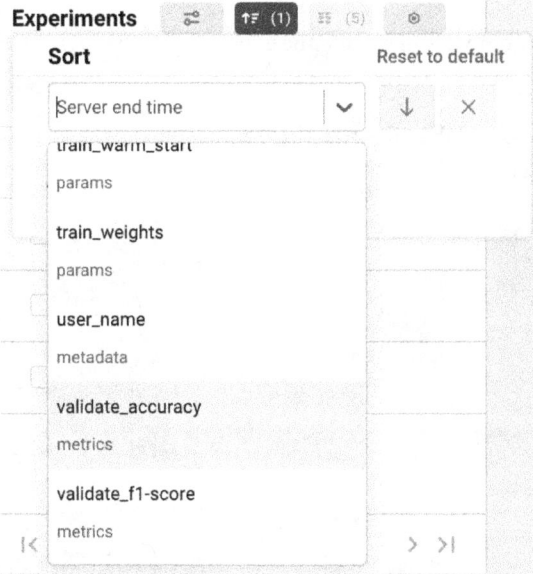

Figure 3.14 – How to select the ordering criterion in the drop-down menu

We can also choose whether to order the experiment by ascending (the top arrow) or descending (the down arrow). In our case, we select the descending order. Now, the four models are ordered, as shown in the following figure:

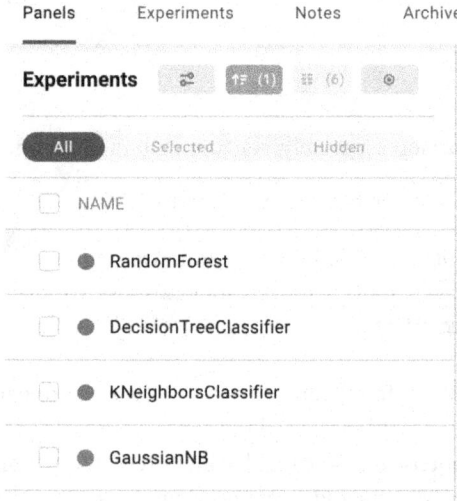

Figure 3.15 – The four experiments after sorting by validate_accuracy

Thanks to this simple operation, we know which is the best model, according to our evaluation metric.

Now, we can perform a raw comparison among the experiments through a table. We can see detailed information about each evaluation metric.

1. We click the **Experiments** tab and see all the information about all the experiments, as shown in the following figure:

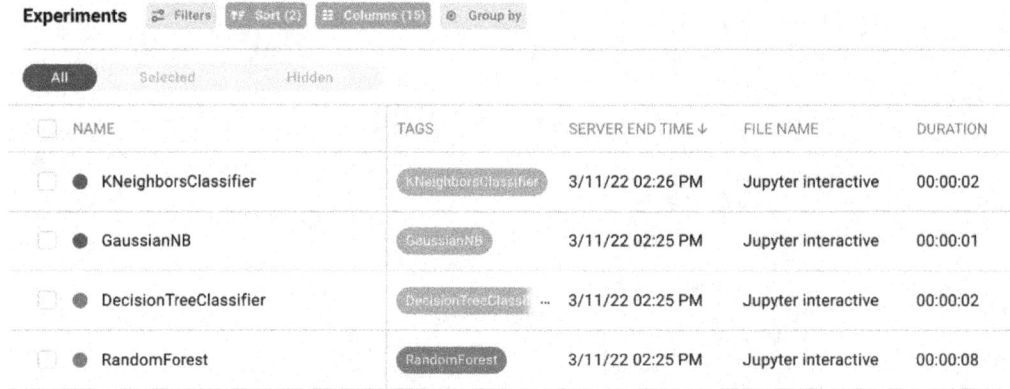

Figure 3.16 – A portion of the Experiments tab

The figure shows only some details about each experiment, such as **TAGS**, **SERVER END TIME**, **FILE NAME**, and **DURATION**. However, Comet provides the user with all the logged parameters and metrics, 15 columns in our case, as shown in the **columns** button at the top part of the previous figure.

2. If we click the **columns** button, we can choose which columns we want to view, as shown in the following figure:

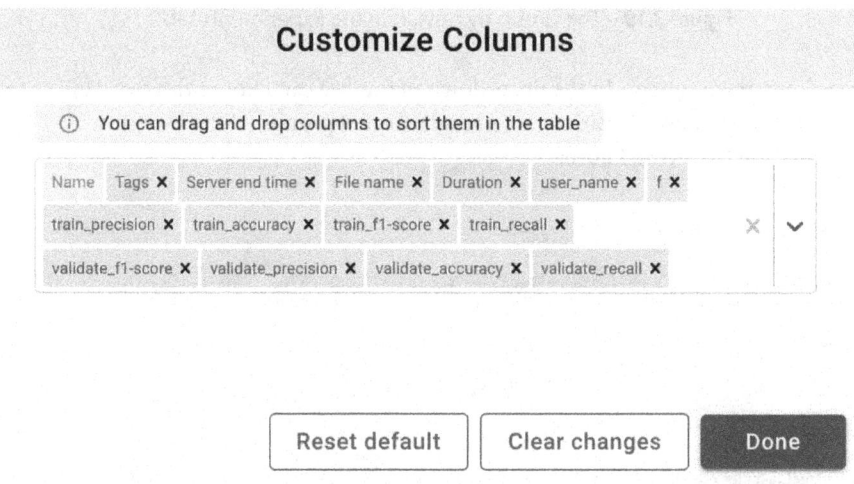

Figure 3.17 – The pop-up window to customize columns

3. We choose the following columns – **validate_accuracy**, **validate_precision**, **validate_recall**, and **validate_f1-score**. As a result, the **Experiments** tab shows a table, as shown in the following figure:

Experiments   ⇄ Filters   ↑↓ Sort (1)   ☰ Columns (6)   ⊘ Group by

NAME	VALIDATE_ACCUR... ↓	VALIDATE_PRECISION	VALIDATE_RECALL	VALIDATE_F1-SCORE	DURATION
● RandomForest	0.957	0.872	0.772	0.819	00:00:10
● DecisionTreeClassifier	0.942	0.762	0.777	0.769	00:00:04
● KNeighborsClassifier	0.929	0.794	0.587	0.675	00:00:04
● GaussianNB	0.904	0.66	0.481	0.557	00:00:01

Figure 3.18 – The Experiments tab after selecting some columns

We note that the Random Forest model reaches an accuracy of 0.957, followed by the Decision Tree model with an accuracy of 0.942, and then the other two models. Through the Comet Dashboard, it is very simple to compare experiments.

Now, we can group experiments by some criteria:

1. We can click the **Group by** button, as shown in the following figure:

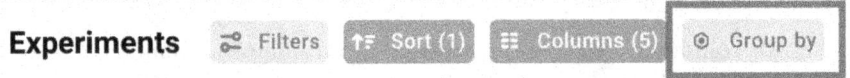

Figure 3.19 – The Group by button in the Experiments tab

2. A drop-down menu opens. In the drop-down menu, we can choose the grouping criteria – for example, we can choose **Duration**. As a result, Comet groups all the experiments by duration, as shown in the following figure:

NAME	VALIDATE_ACCUR... ↓	VALIDATE_PRECISION	VALIDATE_RECALL	VALIDATE_F1-SCORE
▶ Duration: 00:00:01				
▼ Duration: 00:00:04				
● DecisionTreeClassifier	0.942	0.762	0.777	0.769
● KNeighborsClassifier	0.929	0.794	0.587	0.675
▶ Duration: 00:00:10				

Figure 3.20 – The result of grouping by duration

There are three groups – 1, 4, and 10 seconds. In the 4-seconds group, there are two models, while in the other groups, there is just one model.

Finally, we can filter experiments by some criteria:

3. We can click the **Filters** button, as shown in the following figure:

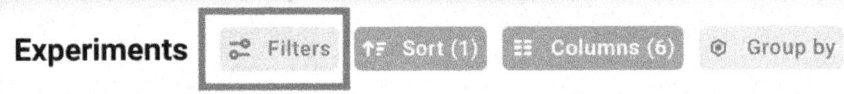

Figure 3.21 – The Filters button in the Experiments tab

4. A pop-up window opens. We can click on the **Add Filter** button | **validate_accuracy** | **greater than** | **0.92** | **Done** | **Close**. As a result, Comet Dashboard shows only the experiments satisfying the filter, as shown in the following figure:

NAME	VALIDATE_ACCUR.. ↓	VALIDATE_PRECISION	VALIDATE_RECALL	VALIDATE_F1-SCORE	DURATION
● RandomForest	0.957	0.872	0.772	0.819	00:00:10
● DecisionTreeClassifier	0.942	0.762	0.777	0.769	00:00:04
● KNeighborsClassifier	0.929	0.794	0.587	0.675	00:00:04

Figure 3.22 – The Experiments tab after applying the validate_accuracy greater than 0.92 filter

Note that Random Forest, Decision Tree, and K-Nearest Neighbors satisfy the applied filter.

For all the described operations, you can apply as many criteria as you want from the pop-up window.

Now that you have learned how to compare experiments in Comet, we can move onto the next step, model registry.

Registry

A **Comet Registry** or **Model Registry** is a place available in the Comet platform that stores all the registered models. There are at least two advantages of registering models in Comet. Firstly, you can keep track of all the stages of your project, and secondly, you can use the registry as secure storage.

To make a model available in the Model Registry, firstly, you need to register it during an experiment by using one of the following methods provided by the Experiment class:

- log_model(name, file_name) – logging the model as an artifact and then, manually, adding it to the registry.

- register_model(MODEL_NAME) – registering the full experiment and adding it to the registry

To log the model, we can modify the previously defined run_experiment() function, as follows:

1. First, we define an auxiliary function that saves a model both in the local file system and in Comet:

```
import pickle
def save_file_to_comet(obj, obj_name, file_name,
experiment):
    with open(file_name, 'wb') as file:
        pickle.dump(obj, file)
    file.close()
    experiment.log_model(obj_name, file_name)
```

The function receives the object (obj) as input to save its name, the output filename, and the Comet experiment object. We can save the model as a pickle file, through the dump() function provided by the pickle library. Then, we log the saved model through the log_model() method of the Experiment class.

2. Then, we can use this function in the run_experiment() function, as follows:

```
def run_experiment(ModelClass, name):

   ...

   with experiment.train():

       ...

       file_name = 'model.pkl'
       save_file_to_comet(model, name, file_name,
experiment)

           ...
```

3. You also should keep in mind to save the scalers and the encoders used to build the training and test set. You can do it simply by using the save_file_to_comet() function previously defined. For example, you can pass the scaler and encoder objects as additional parameters to the run_experiment() function, and then call the save_file_to_comet() function to save them in Comet, as shown in the following piece of code:

```
def run_experiment(ModelClass, name, feature_label_
encoder_dict, scaler,label_encoder):

       ...

       if feature_label_encoder_dict:
       for k,v in feature_label_encoder_dict.items():
           obj_name = f"{k}FeatureLabelEncoder"
           file_name = f"{obj_name}.pkl"
           save_file_to_comet(feature_label_encoder_
dict[k], obj_name, file_name, experiment)

    if label_encoder:
        obj_name = "labelEncoder"
        file_name = f"{obj_name}.pkl"
        save_file_to_comet(label_encoder, obj_name, file_
name, experiment)
```

```
if scaler:
    obj_name = "scaler"
    file_name = f"{obj_name}.pkl"
    save_file_to_comet(scaler, obj_name, file_name,
experiment)
```

We save each object separately. For each object, we define its name (obj_name) and the output filename (file_name) in the local filesystem.

Once we have run all the experiments, we can access the saved models, as follows:

1. Under the **Experiments** tab of each experiment (for example, Random Forest), we select the **Assets** and **Artifacts** menu items, as shown in the following figure:

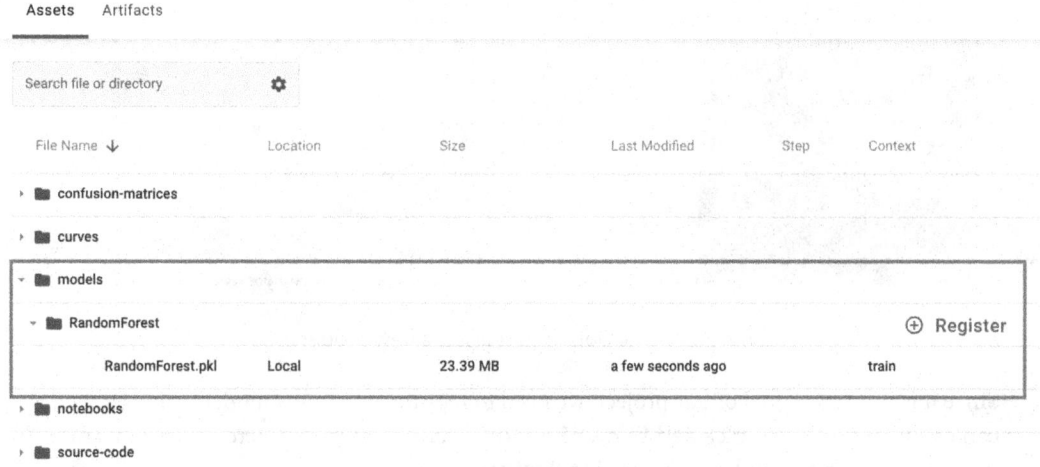

Figure 3.23 – The models directory under the Assets and Artifacts menu items

The figure shows all the logged objects, with a focus on the models directory. We can download the model if we want.

2. On the right part of the screen, there is a button called **Register**. We click on it to add the model to the Registry. The following window opens:

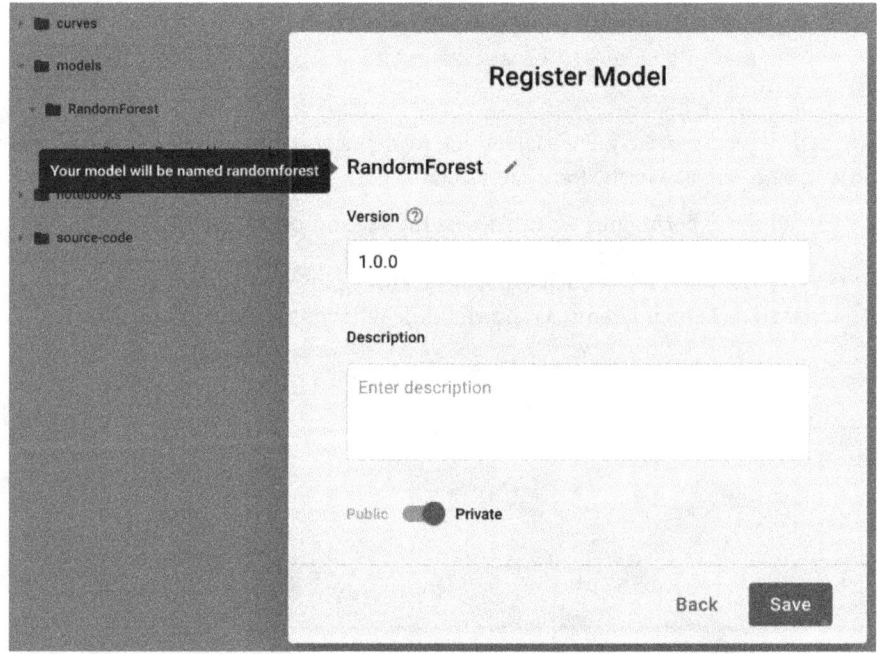

Figure 3.24 – The popup to register a new model

Since it is the first model of our project, we need to register it as a new model – for example, we can call it `Diamonds classification`. If we have already registered a previous model, we can add the new model to an existing Registry.

3. We can repeat the same procedure for the other experiments, but when we need to add the `experiment` model to the Registry, we save it to the existing model. To add it, we need to increase the model version – for example, 1.0.1. In this example, we save only experiments with an accuracy greater than 0.92.

4. We can now access the Registry from the main Comet Dashboard. We need to exit the current project, as shown in the following figure:

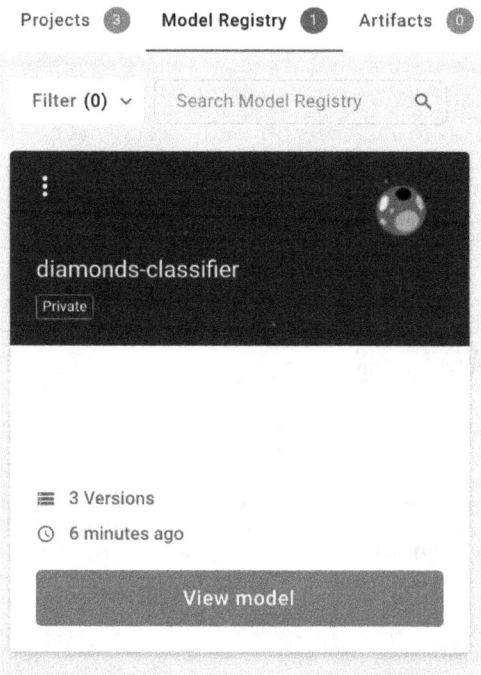

Figure 3.25 – A view of the Model Registry

5. Click the **View model** button. We have three models, corresponding to the registered models. We can choose the best model for production (1.0.0 in our case, which corresponds to Random Forest), as shown in the following figure:

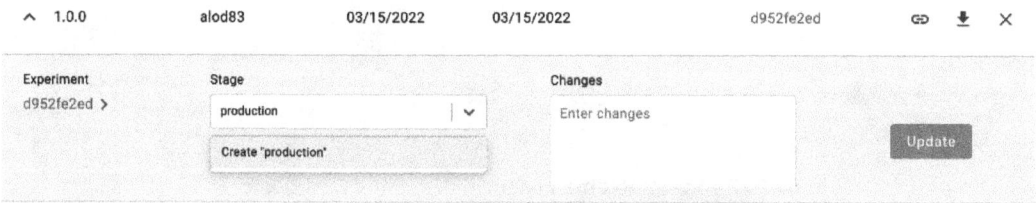

Figure 3.26 – How to set the stage as an experiment

6. We click the arrow on the left, near the version, and then we set the stage as **production**. Finally, we click the **Update** button. If you do not remember which is the best experiment, you can click on the experiment key and retrieve the original experiment.

Now that you have learned how to save your models in Comet, we can move towards the next step, a report.

Reports

A Comet Report is an interactive document that contains experiments, panels, and text. You already learned the basic concepts behind the Comet Report in *Chapter 2, Exploratory Data Analysis in Comet*. In this section, you will learn how to build a report for model evaluation. The report will contain the following panels:

- Precision, recall, the F1-score, and accuracy graphs
- The ROC curve

Let's start from the first panel, precision, recall, the F1-score, and accuracy graphs. Let's suppose that we have logged the involved metrics using the `step` parameter, as described in the previous sections:

1. Firstly, we create a new report, as described in *Chapter 2, Exploratory Data Analysis in Comet*. We set its name as `Diamonds Report`. The reporting dashboard appears with empty content.

2. Then, we return to the main dashboard by clicking the project name at the top left of the screen.

3. We access the **Panels** section, where we can see some of the logged metrics, as shown in the following figure:

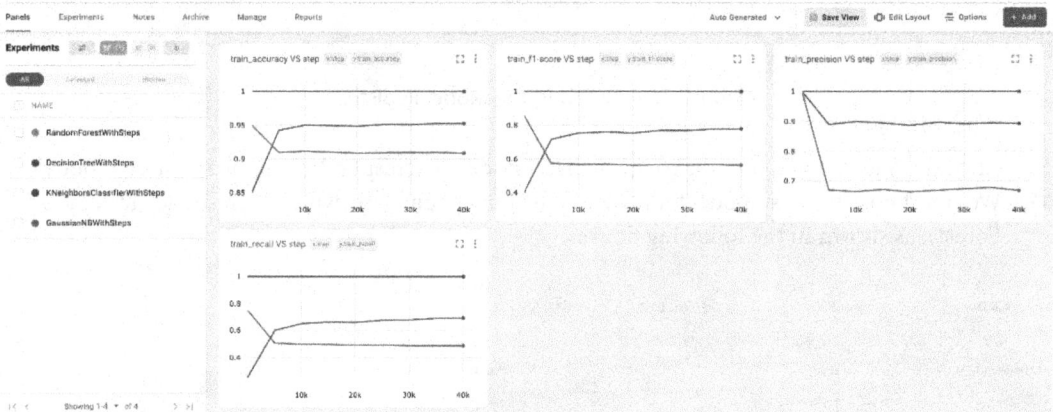

Figure 3.27 – The Panels menu item in the Comet Dashboard

The window contains four line charts, respectively of training accuracy, a training F1-score, training precision, and training recall, all against the step.

4. We can add as many panels as we want simply by clicking **Add | New Panel**. In our case, we add a new bar chart, with validation accuracy. Once **New Panel** is selected, we choose **Bar Chart | Add | Y Axis | validate_accuracy | Done**.

5. Now, we save the current view by clicking the **Save View** button | **Create new view**. We set the view name as `metrics`.

6. Finally, we can add the panels to the report by clicking **Add | Add to Report | diamonds-report**. The Comet Dashboard now shows the report. We can click on the **Save** button to save it.

Now that we have added the basic panels to your report, we can move onto the next step, the ROC curve. Since Comet does not provide a default panel for the ROC curve, the idea is to build a custom panel that shows it:

1. Firstly, we open the Comet online SDK, and we select the Python language. You have already learned how to build a custom panel in *Chapter 2, Exploratory Data Analysis in Comet*; thus you can refer to it for further details on how to access the Comet online SDK.

2. We import all the needed libraries, including `matplotlib`, which we will use to plot a graph:

```
from comet_ml import API, ui
import matplotlib.pyplot as plt
```

3. We retrieve all the `experiment` keys, as follows:

```
api = API()
experiment_keys = api.get_panel_experiment_keys()
```

We build an `API()` object, and we get all the experiment keys through the `get_panel_experiment_keys()` method.

4. For each experiment, we extract the logged curves, and we plot them:

```
colors = iter(['#40B7AD', '#A1CDB3','#508DED','#454372'])
for experiment_key in experiment_keys:
    for curve in api.get_experiment_curves(experiment_
key):
        curve_json = api.get_experiment_asset(experiment_
key,curve['assetId'], return_type='json')
        plt.plot(curve_json['x'],curve_
json['y'],color=next(colors), label=curve_json['name'])
```

We use the `get_experiment_curve()` method to access all the logged curves, and the `get_experiment_asset()` method to retrieve each curve as JSON, which contains the TPR (x) and FPR (y). We plot the curve through the `plot()` function provided by `matplotlib`.

5. We set the plot layout, as follows:

```
plt.plot([0, 1], [0, 1], color="navy", linestyle="--")
plt.xlim([0.0, 1.0])
plt.ylim([0.0, 1.05])
plt.xlabel("False Positive Rate")
```

```
plt.ylabel("True Positive Rate")
plt.legend()
```

Firstly, we plot the line `y = x`, and then we set the axis ranges through the `xlim()` and `ylim()` functions. Finally, we define the axis titles through `xlabel()` and `ylabel()`, as well as the legend.

6. We use the `display()` function to show the graph in the panel:

```
ui.display(plt)
```

7. Now, we save the panel as `ROC curve`. The following figure shows the produced graph:

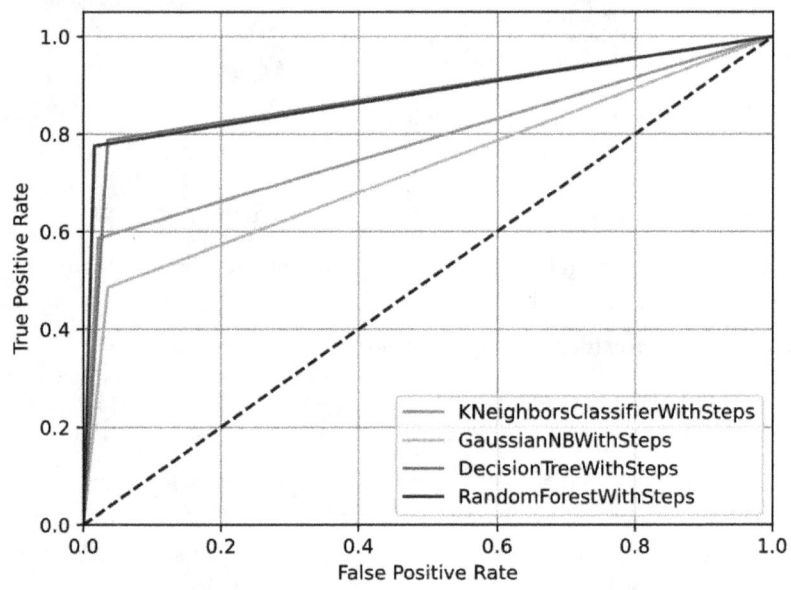

Figure 3.28 – The custom panel showing the ROC curve for all the experiments

8. Finally, we are ready to add the custom panel to our report. We open the report, and then we click on **Add Panel** | **Workspace** | **ROC curve** | **Add** | **Done**.

The following figure shows the final report:

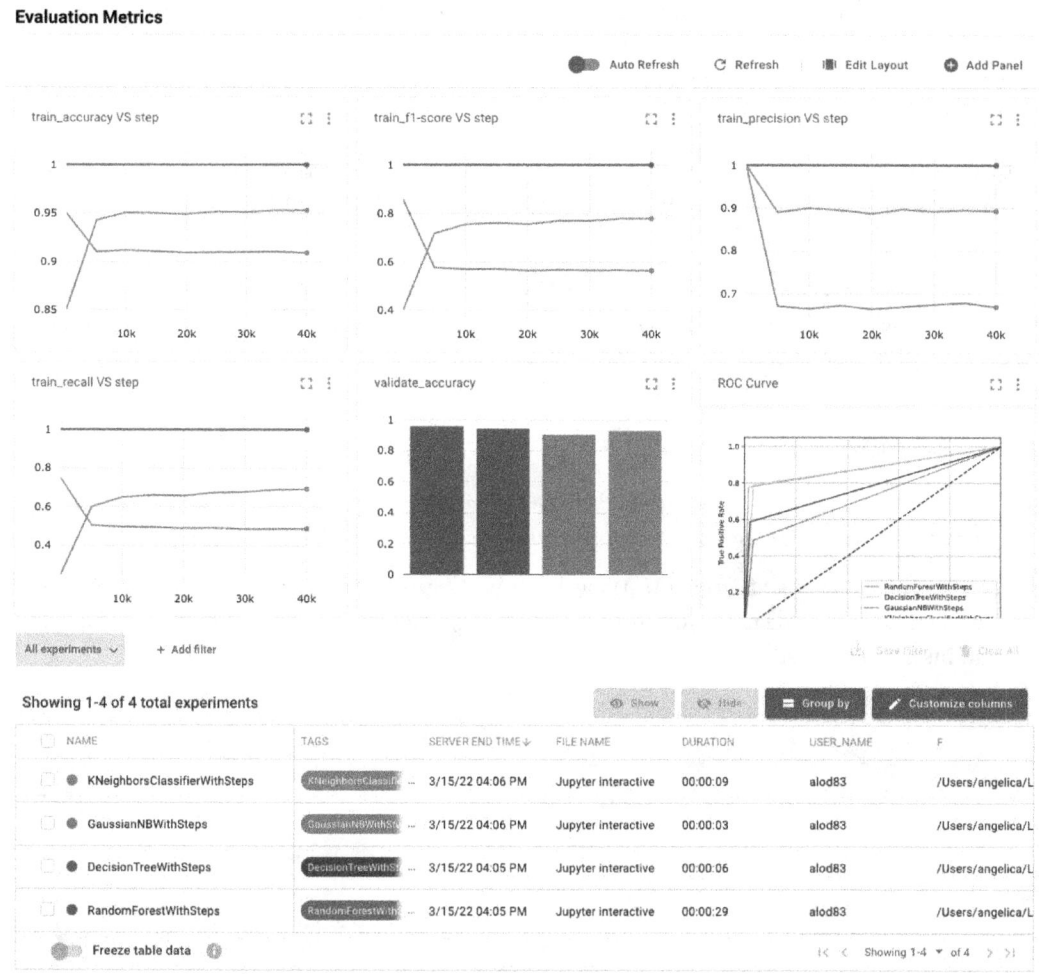

Figure 3.29 – The final report

The report is interactive; thus, we can modify it dynamically by selecting the involved experiments in the **Experiment** tab.

Summary

We have just completed the journey to perform model evaluation in Comet!

Throughout this chapter, we described some general concepts regarding model evaluation, as well as the main techniques to evaluate regression, classification, and clustering. We also illustrated the importance of model evaluation in a data science project; model evaluation permits us to define some metrics to choose the best model for production.

In the third part of the chapter, you learned which features Comet provides to perform model evaluation and how you can use them through a practical example. We deepened the concepts of logs and reports, which you already knew about, and illustrated two new concepts, the Comet Dashboard and the Model Registry.

Throughout this chapter, you learned how easy it is to use Comet to run model evaluation, as Comet provides very intuitive features that can be combined to build fantastic reports, as well as how to keep track of the best model for production.

Now that you have learned how to perform model evaluation in Comet, we can continue our journey toward the discovery of Comet for data science. In the next chapter, we will learn about some advanced concepts regarding workspaces, projects, experiments, and models in Comet.

Further reading

- Bonaccorso, G. (2018). *Mastering Machine Learning Algorithms: Expert techniques to implement popular machine learning algorithms and fine-tune your models.* Packt Publishing Ltd.

- Ng, A. (2017). *Machine Learning Yearning:* https://www.deeplearning.ai/machine-learning-yearning/. To download this book, you need to go to the bottom of the page and insert your email.

Section 2 – A Deep Dive into Comet

In this section, you will learn some advanced concepts behind Comet, related to experiments, models, and panels, as well as some basic strategies to use Comet in Java and R (*Chapter 4, Workspaces, Projects, Experiments, and Models*).

You will also learn how to use Comet to implement the last two steps of a data science project life cycle: how to transform your data into a story (*Chapter 5, Building a Narrative in Comet*), and how to deploy your data science project into a deployed product (*Chapter 6, Integrating Comet into DevOps*, and *Chapter 7, Extending the GitLab DevOps Platform with Comet*).

To get more familiar with the described concepts, you will be guided to implement some advanced use cases, through step-by-step and commented examples in Python, as well as some basic examples in Java and R.

The main focus of this section is to enable you to work with all the advanced features provided by Comet.

This section includes the following chapters:

- *Chapter 4, Workspaces, Projects, Experiments, and Models*
- *Chapter 5, Building a Narrative in Comet*
- *Chapter 6, Integrating Comet into DevOps*
- *Chapter 7, Extending the GitLab DevOps Platform with Comet*

Workspaces, Projects, Experiments, and Models

Comet is an experimentation platform that permits you to track, monitor, and compare experiments within a data science project. So far, you have learned some basic concepts, including how to create and deal with workspaces, projects, experiments, panels, and reports. You have also learned how to compare experiments, customize panels, and store models in the Comet Registry.

In this chapter, you will deepen your understanding of some concepts regarding Comet, including how to add collaborators to your workspaces or projects, how to publish your projects, advanced techniques to manage experiments, and how to perform parameter optimization in Comet. In addition, you will learn how to implement a Comet experiment using R or Java as the main programming language. Finally, you will extend the basic examples implemented in *Chapter 1, An Overview of Comet*, with the advanced concepts learned in this chapter.

In detail, the chapter is organized as follows:

- Exploring the Comet user interface (UI)

- Using experiments and models

- Exploring other languages supported by Comet

- First use case – offline and existing experiments

- Second use case – model optimization

Before describing the advanced concepts behind Comet, let's install all the packages needed to run the code and the experiments contained in this chapter.

Technical requirements

We will run the experiments and code in this chapter using Python, R, and Java. We'll describe the required packages for each programming language separately, starting with Python.

Python

For Python, we will use Python 3.8. You can download it from the official website at `https://www.python.org/downloads/` and choose the 3.8 version.

The examples described in this chapter use the following Python packages:

- `comet-ml 3.23.0`
- `matplotlib 3.4.3`
- `numpy 1.19.5`
- `pandas 1.3.4`
- `scikit-learn 1.0`

We have already described the first five packages and how to install them in *Chapter 1*, *An Overview of Comet*, so please refer back to that for further details on installation.

R

R is a very popular language for statistical computing. You can use it as a valid alternative to Python since it also provides many libraries for **machine learning** (**ML**) and data science in general. You can download it from the R official website, available at this link: `https://cran.rstudio.com/index.html`. You should choose the version available for your operating system, and install it.

We will use the following R packages:

- `caret`
- `cometr`
- `Metrics`

`caret` (short for **Classification And REgression Training**) is an R package for ML. You can install it by running the following command from the R terminal:

```
install.packages('caret')
```

cometr is the official package provided by Comet to use Comet in R. You can install it by running the following command from the R terminal:

```
install.packages("cometr")
```

For more details on how to install cometr, you can refer to the Comet official documentation, available at this link: https://github.com/comet-ml/cometr.

Metrics is an R package that implements some evaluation metrics. You can install it by running the following command from the R terminal:

```
install.packages("Metrics")
```

For more details on how to install Metrics, you can refer to the Metrics official documentation, available at this link: https://cran.r-project.org/web/packages/Metrics/index.html.

Java

Java is a programming language for building applications. Although it does not support data science natively, it can also be used to implement data science projects. Many implementations of Java exist. In this chapter, we will use the Java **Standard Edition** (**SE**) **Software Development Kit 17.0.2** (**SDK 17**), provided by Oracle. It can be used under the Oracle **No-Fee Terms and Conditions** (**NFTC**) license. You can download the Java SDK from this link: https://www.oracle.com/java/technologies/downloads/#java17. To make it work, you should follow the installation instructions available at this link: https://docs.oracle.com/en/java/javase/17/install/overview-jdk-installation.html#GUID-8677A77F-231A-40F7-98B9-1FD0B48C346A.

To facilitate the installation of the various packages used in this chapter, we also install **Apache Maven**, which is a tool to build, manage, and run Java applications. You can download the last release of Apache Maven from this link: https://maven.apache.org/download.cgi. To make Apache Maven work properly, please make sure that your JAVA_HOME environment variable is properly set.

We will use the following Java packages:

- comet-java-sdk-1.1.10
- weka 3.8.6

comet-java-sdk-1.1.10 is a Java package provided by Comet to interact with the Comet platform. You can download it from the Comet official repository, available at the following link: https://github.com/comet-ml/comet-java-sdk/releases. You should download the source file.

Once downloaded, you can do the following:

1. Place the file wherever you like in your filesystem.

2. Unzip the file.

3. Enter the unzipped directory.

4. Run the following command:

   ```
   mvn clean install
   ```

   Installation may fail, due to some dependencies on some external files in the example directory. If this is the case for you, you can edit the pom.xml file by commenting on the following line:

   ```
   <module>comet-examples</module>
   ```

5. Finally, you save the file and run the previous command again.

weka is a Java package for ML. To install it, you can follow the weka official documentation available at this link: https://waikato.github.io/weka-wiki/maven/. Then, proceed as follows:

1. You can download the weka **Java ARchive (JAR)** file from Maven Central, available at this link: https://search.maven.org/search?q=a:weka-stable.

2. Then, from the directory where you placed the JAR file, you can run the following command in a terminal:

   ```
   mvn install:install-file \
       -Dfile=weka-stable-3.8.6.jar \
       -DgroupId= nz.ac.waikato.cms.weka:weka-stable:3.8.6 \
       -DartifactId=weka-stable \
       -Dversion=3.8.6 \
       -Dpackaging=jar \
       -DgeneratePom=true
   ```

   This command will add the weka libraries to the Maven repository.

Now that you have set up the environment by installing all the needed packages, we can move to the next step: exploring the Comet UI—including workspaces and projects.

Exploring the Comet UI

The Comet UI provides a very useful dashboard that you can use to organize and track experiments. As already described in *Chapter 1, An Overview of Comet*, the Comet UI is organized into workspaces and projects. Conceptually, a workspace is a container for similar projects, while a project is a container for experiments involving the same task.

In this section, you will learn some advanced concepts regarding workspaces and projects. So, let's start with the first one: workspaces.

Workspaces

A Comet workspace is a collection of projects. In *Chapter 1, An Overview of Comet*, you have already learned how to create a new workspace and how to add a new project to an existing workspace. In this section, you will learn how to add collaborators to a workspace.

To add a collaborator to a workspace, you need to sign up for at least a **Teams** plan or, alternatively, an **Academic** plan, which is free, as explained in *Chapter 1, An Overview of Comet*. Depending on your plan, you can add a number of of collaborators to your workspace.

To add a new collaborator, proceed as follows:

1. In the Comet dashboard, you can click your username avatar in the top-right part of the screen. You should make sure that your workspace is the current workspace. If not, you can change it, by following the procedure described in *Chapter 1, An Overview of Comet*.

2. Click **Settings** from the drop-down menu. Then, click **Collaborators | +Collaborators**. A popup should open, like the one shown in the following screenshot:

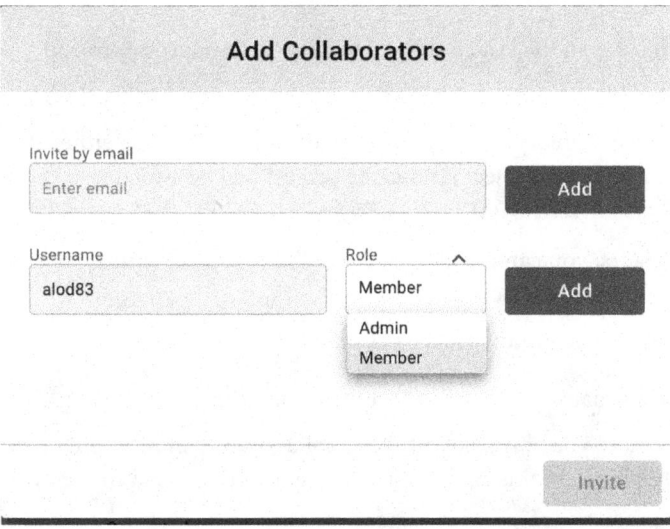

Figure 4.1 – Popup to add new collaborators

You have two options to add a new collaborator: by inviting them by email or by searching for them by username. In this last case, you can also specify their role: **Admin** or **Member**.

3. Click the **Invite** button. The collaborator will receive a notification of the invite. When they accept it, they are ready to work with you.

Now that you have learned how to add new collaborators to a workspace, let's move on to review some advanced concepts on projects.

Projects

A Comet project is a collection of experiments. In *Chapter 1*, *An Overview of Comet*, you have already learned how to create a new project. In this section, we will describe how to do the following:

- Set the project visibility.
- Share a project.

To set the project visibility, proceed as follows:

1. Access the workspace containing the project.
2. Select the gear shape button corresponding to your project, as shown in the following screenshot:

Figure 4.2 – Gear shape button corresponding to the pandas-profiles project

The button is located in the last column of the row defining your project.

3. Select the **Edit** menu item.
4. Choose the project visibility and click on **Update**. You can choose either **Private** or **Public**. If you choose to make your project public, the project will be available at this link: `https://www.comet.ml/<WORKSPACE NAME>/<PROJECT NAME>/view`.

As with workspaces, you can decide to share a single project with a collaborator. Again, you need at least a **Teams** or an **Academic** plan.

To share a project with a collaborator, proceed as follows:

1. From the project's main dashboard, select the **Manage** tab.
2. If you want to share your project with your collaborators in read-only mode, you can click **Create Sharable Link**. In this case, they will be able only to read the project; they will not be able to edit it.
3. Alternatively, you can click the **+Collaborators** button to share the project with one or more collaborators, who will also be able to edit your project. The procedure is similar to that described for workspaces.

Now that you have learned some advanced concepts on projects and workspaces, we can move on to investigate some other aspects of experiments and models.

Using experiments and models

Experiments and models are the core of a Comet project because they permit you to track and monitor all your data science projects. You have already learned the basic concepts behind experiments and models in *Chapter 1*, *An Overview of Comet*, and *Chapter 3*, *Model Evaluation in Comet*. In this section, you will learn some advanced topics involving offline and existing experiments, as well as model optimization.

This section is organized as follows:

- Experiments
- Models

Let's start from the first point: experiments.

Experiments

A Comet experiment is a process that permits you to track your variables while the underlying conditions change. You have already learned the basic concepts behind Comet experiments in *Chapter 1*, *An Overview of Comet*. You have already seen the `Experiment` class, which permits you to connect directly with Comet through an available internet connection. However, it may happen that at a certain time your internet connection is not available for some reason, or you need to stop your experiments and then continue them after a given period of time. To deal with these situations, Comet provides three additional experiment classes, as outlined here:

- **Offline experiment**—If you do not have an internet connection, you can create an `OfflineExperiment()` object, which permits you to store your experiment locally on your filesystem. Then, you can upload the experiment to Comet in a second instance. You can create an offline experiment like this:

  ```
  from comet_ml import OfflineExperiment
  experiment = OfflineExperiment(offline_directory="PATH/
  TO/THE/OUTPUT/DIRECTORY")
  ```

  The `offline_directory` parameter specifies the output directory where the experiment will be saved. Once you have created an experiment, you can use it as you usually do with the standard `Experiment` class. When the experiment ends, the output directory will contain a zipped file. You can upload it by running it through the command-line utility, like so:

  ```
  comet upload PATH/TO/THE/ZIP/FILE
  ```

- **Existing experiment**—You can continue an existing experiment by creating an ExistingExperiment() object, which receives as input the key of the experiment to continue. This is particularly useful when you want to separate the training and test phases of an experiment, or if you want to improve a previous experiment. You can create an existing experiment like this:

```
from comet_ml import ExistingExperiment
experiment = ExistingExperiment(previous_experiment=
"EXPERIMENT_KEY ")
```

To use the ExistingExperiment() object, you need to know the experiment key in advance, which you can specify through the previous_experiment input parameter.

- **Offline existing experiment**—You can continue an existing experiment offline. This is a combination of the previous two cases. In this case, you should create an ExistingOfflineExperiment() object, as specified in the following piece of code:

```
from comet_ml import ExistingOfflineExperiment
experiment = ExistingOfflineExperiment(offline_directory=
"PATH/TO/THE/OUTPUT/DIRECTORY", previous_experiment=
"EXPERIMENT_KEY ")
```

You must specify both the offline_directory parameter and the experiment key. If you want to upload an experiment in Comet, you should run the following command from a terminal:

```
comet upload PATH/TO/THE/ZIP/FILE
```

You will practice with the different types of experiments at the end of this chapter, when we will extend the first use case described in *Chapter 1, An Overview of Comet*.

Now that you have learned some advanced concepts regarding experiments, we can move to the next point: models.

Models

A Comet model is an algorithm that learns a pattern from known data and uses it to make predictions on unknown data. You can build your model for different purposes, including data classification, regression, **natural language processing (NLP)**, time-series forecasting, and so on.

You have already learned how to track and organize models in Comet in *Chapter 3, Model Evaluation in Comet*. Comet also provides an additional feature that permits you to optimize your models. This feature is called an **Optimizer**. In this section, we will describe how to build an Optimizer in Comet, while in *Chapter 8, Comet for Machine Learning, Chapter 9, Comet for Natural Language Processing, Chapter 10, Comet for Deep Learning*, and *Chapter 11, Comet for Time Series Analysis*, you will review the main techniques for model optimization.

You can use a Comet Optimizer for tuning the hyperparameters of your model, by choosing one of the supported optimization algorithms: **Grid**, **Random**, and **Bayes**. If you want to learn more details about the optimization algorithms, you can refer to the Comet official documentation, available at the following link: https://www.comet.ml/docs/python-sdk/introduction-optimizer/#optimizer-algorithms.

Follow these next steps:

1. You can create a Comet Optimizer like this:

    ```
    from comet_ml import Optimizer
    optimizer = Optimizer(config)
    ```

    We create an Optimizer object, which receives as input some configuration parameters that include the metric to maximize/minimize, the number of trials, the optimization algorithm, and the parameters to test.

2. You can define the configuration parameters like so:

    ```
    config = {"algorithm": <MY_OPTIMIZATION_ALGORITHM>,
            "spec": {
            "objective": <MINIMIZE/MAXIMIZE>,
            "metric": <METRIC>,
                },
        "trials":   <NUMBER OF TRIALS>,
        "parameters": [LIST OF PARAMETERS],
        "name": <OPTMIZER NAME>
    }
    ```

    The config object is a dictionary that includes all configuration parameters. In addition to the parameters listed in the previous piece of code, you can define other parameters, as specified in the Comet official documentation.

3. The list of parameters includes all specific parameters to test, and for each of them, you must specify a range of possible values (integer, categorical, and so on). Depending on the type of parameter, the syntax changes. For an integer type, you should specify the minimum and maximum values, as well as the scaling type, as shown in the following piece of code:

    ```
    {<PARAMETER-NAME>:
      {"type": "integer",
        "scalingType": "linear" | "uniform" | "normal" |
    "loguniform" | "lognormal",
        "min": <MIN-VALUE>,
        "max": <MAX-VALUE>,
      }
    ```

The PARAMETER-NAME value depends on the specific algorithm. For example, for a **K-Nearest Neighbors (KNN)** classifier, you may need to hypertune the number of neighbors.

4. If you want to hypertune a categorical parameter, you can use the following syntax:

```
{<PARAMETER-NAME>:
  {"type": "categorical",
   "values": ["LIST", "OF", "CATEGORIES"]
  }
}
```

The "values" key contains a list of all possible categories to test.

When you create an Optimizer object, the system will create an experiment for each combination of parameters included in the configuration. You can access a list of experiments by executing the following code:

```
optimizer.get_experiments():
```

By iterating over the list of experiments, you can access every single parameter, and use it to fit a different model, as follows:

```
for experiment in opt.get_experiments():
    param1 = experiment.get_parameter("param1")
    # create, fit and test the model with param 1
```

The previous code shows how to retrieve each parameter, which you can use as you want to create, fit, and test your preferred model. The commented line indicates that the code you should write after retrieving a parameter depends on the model you want to implement.

Once you have run all the experiments, you will see the results in Comet. You will implement a practical use case at the end of this chapter, in the *Second use case – model optimization* section.

Now that you have learned some advanced concepts regarding experiments and models, we can move to the next step: other languages supported by Comet.

Exploring other languages supported by Comet

So far, you have learned how to use Comet in Python. However, Comet also supports other programming languages. The concepts learned so far on experiments, panels, and so on are also valid with the other languages supported by Comet.

In this section, you will apply the concepts already acquired in the previous chapters to the R and Java languages. So, if you are not interested in programming in these languages, you can skip this section and go directly to the next one, *First use case – offline and existing experiments*.

In this section, you will learn how to build and run an experiment in the following languages:

- R

- Java

Let's start with the first language: R.

R

Let's suppose that you have already created a new project in Comet and obtained an **application programming interface (API)** key, as described in *Chapter 1, An Overview of Comet*. As described in that chapter, you should define a configuration file named .comet.yml that should contain all configuration parameters, as shown in the following piece of code:

```
COMET_WORKSPACE: MY_WORKSPACE
COMET_PROJECT_NAME: MY_PROJECT_NAME
COMET_API_KEY: MY_API_KEY
```

The configuration file should contain the workspace name, the project name, and the API key. You should save the file either in the working directory or in your home directory.

In this section, you will use the caret package to perform an ML task in R. The section is divided into the following parts:

- Exploring the cometr package

- Reviewing the caret package

- Running a practical example

Let's start with the first part: exploring the cometr package.

Exploring the cometr package

The cometr package provides functions to interact with the Comet platform. For a list of available functions, you can refer to the Comet official documentation, available at this link: https://www.comet.ml/docs/r-sdk/getting-started/. Here is a list of the most common functions:

- create_experiment()/create_project()—To create a new experiment/project

- get_experiments()/get_projects()/get_workspaces()—To get a list of experiments/projects/workspaces

You can use the `create_experiment()` function to create an `experiment()` object. Once you have created an `experiment()` object, you can log parameters, metrics, and objects, as you usually do in Python. For a list of available methods, you can refer to the Comet official documentation, available at this link: `https://www.comet.ml/docs/r-sdk/Experiment/`. Here is a list of the most common methods available for the `experiment()` object:

- `log_metric()`/`log_parameter()`/`log_html()`/`log_code()`/`log_graph()`/`log_other()`—To log a metric, a parameter, a **HyperText Markup Language** (HTML) page, code, a graph, or other objects

- `upload_asset()`—To upload an asset to the Comet platform

Now that you have learned the basic functions and methods provided by the `cometr` package, we can briefly review the `caret` package.

Reviewing the caret package

The `caret` package provides the following main features to perform ML in R:

- **Preprocessing**, which permits you to clean, normalize, and center your dataset, as well as performing other preprocessing operations. To perform preprocessing, you can use the `preProcess()` function, as follows:

```
X_preprocessed <- preProcess(X, method = c("center",
"scale"))
```

The function takes a dataset as input, as well as a list of operations to perform. In the example, we performed centering and scaling.

- **Data splitting**, which permits you to split your data into training and test sets. You can use the `createPartition()` function to split a dataset into training and test sets, as illustrated in the following piece of code:

```
index <- createDataPartition(dataset, p = .8,
          list = FALSE,
          times = 1)
```

The `p` parameter specifies the training set size in terms of probability. Setting the `list = FALSE` parameter avoids returning the data as a list, and the `times` parameter sets the number of splits to return. The function returns a list of indices belonging to the first partition. So, you need to create two variables—one for the training set and the other for the test set containing the selected indices, as follows:

```
training <- df[index,]
test <- df[-index,]
```

`df` is the original dataset loaded as a DataFrame.

- **Model training** permits you to train a specific model. The `caret` package provides the `train()` method to train a model, as shown in the following piece of code:

```
model <- train(class ~ ., method='knn', data = training,
metric='Accuracy')
```

The first argument is the predictor, which permits you to select the output. You can also specify the type of model through the `method` argument, the input data, and the metric to calculate. You can even set a training control for hyperparameter tuning. For more details on this aspect, you can refer to the `caret` official documentation, available at the following link: `https://topepo.github.io/caret/model-training-and-tuning.html#model-training-and-parameter-tuning`.

- **Model prediction**, which permits you to predict the output for new unseen data. You can use the `predict()` function like so:

```
y_pred <- predict(model, X_test)
```

- **Model evaluation**, which permits you to evaluate the performance of the model. Depending on the specific task, you can calculate different metrics. For more details, you can refer to the `caret` official documentation, available at the following link: `https://topepo.github.io/caret/measuring-performance.html`. Since the `caret` package does not provide a direct method to calculate the accuracy, in the example described in the next section, we will use the `Metrics` package to calculate it.

Now that you have learned the basic concepts to build a model in `caret`, you can implement a practical use case.

Running a practical example

We will use the *Mushrooms Classification* dataset, available on Kaggle (`https://www.kaggle.com/uciml/mushroom-classification`) under the **Creative Commons Zero (CC0)** public license. Our objective is to build a classification model that predicts whether a mushroom is edible or poisonous. (You can find the full code of this example in the GitHub repository of the book, available at the following link: `https://github.com/PacktPublishing/Comet-for-Data-Science/tree/main/04/r-example`.)

Here are the steps we'll take:

1. Firstly, we import all the needed libraries, as follows:

```
library(cometr)
library(caret)
library(Metrics)
```

2. Then, we load the dataset as a DataFrame, like this:

```
df <- read.csv('mushrooms.csv')
```

The dataset contains 8,124 rows and 23 columns. The following screenshot shows the first 10 rows of the dataset:

class	cap-shape	cap-surface	cap-color	bruises	odor	gill-attachment	gill-spacing	gill-size	gill-color	stalk-shape	stalk-root	stalk-surface-above-ring	stalk-surface-below-ring	stalk-color-above-ring	stalk-color-below-ring	veil-type	veil-color	ring-number	ring-type	spore-print-color	population	habitat
p	x	s	n	t	p	f	c	n	k	e	e	s	s	w	w	p	w	o	p	k	s	u
e	x	s	y	t	a	f	c	b	k	e	c	s	s	w	w	p	w	o	p	n	n	g
e	b	s	w	t	l	f	c	b	n	e	c	s	s	w	w	p	w	o	p	n	n	m
p	x	y	w	t	p	f	c	n	n	e	e	s	s	w	w	p	w	o	p	k	s	u
e	x	s	g	f	n	f	w	b	k	t	e	s	s	w	w	p	w	o	e	n	a	g
e	x	y	y	t	a	f	c	b	n	e	c	s	s	w	w	p	w	o	p	k	n	g
e	b	s	w	t	a	f	c	b	g	e	c	s	s	w	w	p	w	o	p	k	n	m
e	b	y	w	t	l	f	c	b	n	e	c	s	s	w	w	p	w	o	p	n	s	m
p	x	y	w	t	p	f	c	n	p	e	e	s	s	w	w	p	w	o	p	k	v	g
e	b	s	y	t	a	f	c	b	g	e	c	s	s	w	w	p	w	o	p	k	s	m

Figure 4.3 – An extract of the mushrooms dataset

3. The first column of the dataset contains the target class. We can also note that all columns of the dataset contain categorical variables. Thus, we convert them into numerical values, as shown in the following piece of code:

```
for(i in 2:ncol(df)) {
    df[ , i] <-  as.numeric(factor(df[ , i], levels =
unique(df[ , i]), exclude = NULL))
    }
```

In the for loop, we start from the second column, because the first one is the target class.

4. Now, we encode the target class, as follows:

```
df$class <- as.factor(df$class)
```

With respect to the previous code, we do not convert the class to numeric values.

5.   We create a Comet experiment by executing the following code:

```
experiment <- create_experiment()
```

6.   Now, we are ready to create a model. The idea is to split the dataset into batches, and track the performance of the model for each step. We use a KNN classifier to perform the classification task. The following code snippet shows how we train and test the model, as well as how we log all metrics in Comet:

```
set.seed(10)
n <- dim(df)[1]
burst <- 1000
for (i in seq(200, n+burst, by=burst)) {
  if(i > n)
    i = n
  dft <- df[c(1:i),]
  index <- createDataPartition(y = dft$class, times = 1,
p = 0.7, list = FALSE)
  training <- dft[index,]
  test <- dft[-index,]

  model <- train(class ~ ., method='knn', data =
training, metric='Accuracy')
  test$pred <- predict(model, test)

  acc <- accuracy(test$class, test$pred)
  test$factor_pred <- as.factor(test$pred)
  test$factor_truth <- as.factor(test$class)

  precision <- posPredValue(test$factor_truth,
test$factor_pred)
  recall <- sensitivity(test$factor_truth, test$factor_
pred)

  F1 <- (2 * precision * recall) / (precision + recall)
  experiment$log_metric("accuracy", acc, step=i)
  experiment$log_metric("precision", precision, step=i)
  experiment$log_metric("recall", recall, step=i)
  experiment$log_metric("F1", F1, step=i)
}
```

Firstly, we set the seed to 10 to make the experiment reproducible. For each batch, we split the dataset into training and test through the `createDataPartition()` function, provided by the `caret` library. Then, we train the model, through the `train()` function. We also specify that we want to optimize the accuracy. Now, we test the model performance through the `predict()` function, and we calculate the accuracy, the precision, the recall, and the F1-score. Finally, we log all the calculated metrics in Comet, through the `log_metric()` method of the `experiment` class.

7. Finally, we terminate the experiment, as follows:

```
experiment$stop()
```

8. We run the code from the R console, as follows:

```
setwd('/PATH/TO/THE/DIRECTORY/CONTAINING/YOUR/CODE')
source('script.R')
```

Firstly, we need to change directory through the `setwd()` function, then we run the code, through the `source()` function.

9. Now, the experiment is available in Comet, as shown in the following screenshot:

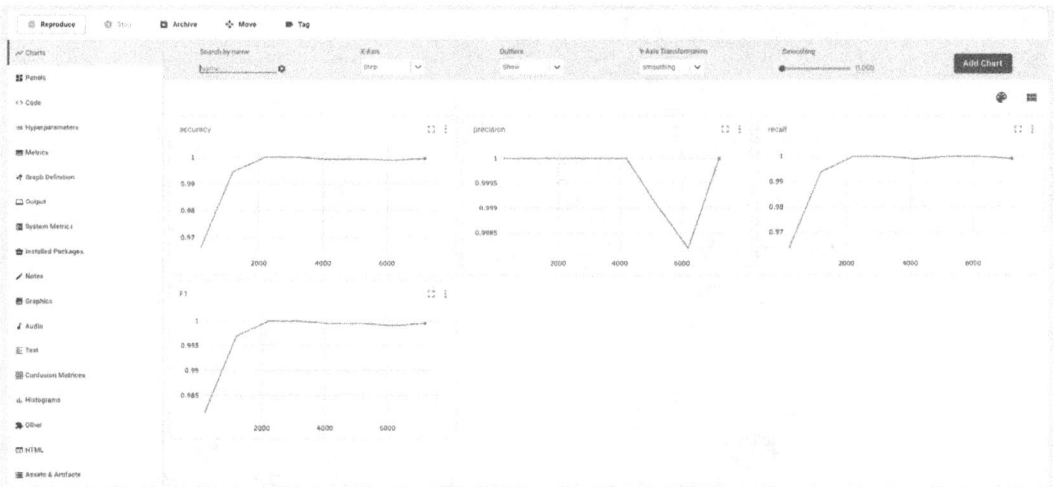

Figure 4.4 – Output of the experiment in Comet

Figure 4.4 shows the output of the experiment in Comet, with a focus on the accuracy, precision, recall, and F1-score curves.

Now that you have learned how to build a Comet experiment in R, we can move on to the next step: building a Comet experiment in Java.

Java

Let's suppose that you have already created a new project in Comet and you already have an API key, as described in *Chapter 1, An Overview of Comet*. Let's further suppose that you have already created a Maven project with a pom.xml file Take the following steps:.

1. You should add the Comet library to the dependency section of your pom.xml file, as follows:

```xml
<dependency>
        <groupId>ml.comet</groupId>
        <artifactId>comet-java-client</artifactId>
        <version>1.1.10</version>
    </dependency>
```

In the previous code snippet, we have added comet-java-client version 1.1.10 to our project.

2. You should also configure a file named application.conf, as follows:

```
comet {
        baseUrl = "https://www.comet.ml"
        apiKey = "YOUR API KEY"
        project = "YOUR PROJECT NAME"
        workspace = "YOUR EXPERIMENT"
}
```

The file contains configuration parameters to access the Comet platform.

3. You should add the application.conf file to the project classpath in the pom.xml file, as follows:

```xml
<resources>
        <resource>
            <directory>PATH/TO/THE/CONF/FILE</directory>
        </resource>
</resources>
```

In this section, you will use the weka package to perform an ML task in Java. The section is divided into the following parts:

*   Exploring the ml.comet package
*   Reviewing the weka package
*   Running a practical example

Let's start from the first part: exploring the ml.comet package.

Exploring the ml.comet package

The ml.comet package provides functions to interact with the Comet platform. For a list of available classes and functions, you can refer to the Comet official documentation, available at this link: https://www.comet.ml/docs/java-sdk/getting-started/. Comet provides the OnlineExperimentBuilder class to create experiments, and the OnlineExperiment interface to manage experiments. You can create a new experiment by running the following code:

```
import ml.comet.experiment.ExperimentBuilder;
import ml.comet.experiment.OnlineExperiment;
OnlineExperiment experiment = ExperimentBuilder.
OnlineExperiment().build();
```

The previous code creates a new experiment() object you can use to log and track your experiments.

Similar to the experiment class defined for the other programming languages, the OnlineExperiment interface also provides the following methods:

- logMetric()/logModel()/logArtifact()/logParameter()—To log a metric, a model, an artifact or a parameter
- setEpoch()/setStep()—To set the current epoch/step of the experiment
- nextEpoch()/nextStep()—To increment the current epoch/step of the experiment

The previous list of methods contains the most important methods provided by the OnlineExperiment interface. You can refer to the Comet official documentation for a complete list of methods.

Now that you have learned the basic functions and methods provided by the ml.comet package, we can briefly review the weka package.

Reviewing the weka package

The weka package contains many subpackages to perform ML tasks. Among them, we will use the following ones:

- core, which contains the main classes and functions to prepare data for further analysis. In particular, you can load a **comma-separated values (CSV)** file, as follows:

```
CSVLoader loader = new CSVLoader();
loader.setSource(new File("/path/to/csv/file.csv"));
Instances dataset = loader.getDataSet();
```

- `classifiers`, which contains all algorithms used to perform **supervised learning** (**SL**). Algorithms related to regression are available under the `classifiers.functions` subpackage. To build a classifier and train it, you can run the following code:

```
MyModel model = new MyModel();
model.buildClassifier(training);
```

The `MyModel()` class is one of the classes provided by the `classifiers` subpackage, such as `IBk()` to implement a KNN classifier. The `classifiers` package also contains a class named `Evaluation` that you can use to perform model evaluation. You will see a practical example of how to use it in the next section.

Now that you have reviewed the basic packages provided by the `weka` package, we can move to the next step: implementing a practical use case.

Running a practical example

We will use the *Mushrooms Classification* dataset, available on Kaggle (`https://www.kaggle.com/uciml/mushroom-classification`) under the CC0 public license. We have already used it in the previous section. We will implement the same use case as the previous section—that is, build a classification model that predicts whether a mushroom is edible or poisonous. We will use the `weka` package to perform ML tasks.

(You can find the full code of this example in the GitHub repository of the book, available at the following link: `https://github.com/PacktPublishing/Comet-for-Data-Science/tree/main/04/java-example`.)

Here are the steps we'll take:

1. Firstly, we create a new Java script named `KNN.java` and we set the package name, as follows:

```
package packt.comet;
```

2. Then, we import all the needed classes, like so:

```
import ml.comet.experiment.ExperimentBuilder;
import ml.comet.experiment.OnlineExperiment;
import weka.core.Instances;
import weka.core.converters.CSVLoader;
import weka.classifiers.Classifier;
import weka.classifiers.lazy.IBk;
import weka.classifiers.Evaluation;
import java.util.Random;
import java.io.File;
```

From `ml.comet`, we have imported the classes needed to interact with Comet; from `weka`, we have imported the classes needed to load the CSV file and to perform classification, and from `java`, we have imported some utility functions.

3. Now, we define a class named `KNN` that will contain the code, as follows:

```
public class KNN {
    public static void main( String[] args )
    {...}
}
```

We have also defined a `main()` method, which will contain all the following code.

4. Within the `main()` method, we create a new experiment, as follows:

```
OnlineExperiment experiment = ExperimentBuilder.
OnlineExperiment().build();
experiment.setExperimentName("KNN");
```

We also set the experiment name through the `setExperimentName()` method.

5. We load the file as an `Instances` object, as follows:

```
try {
    CSVLoader loader = new CSVLoader();
    loader.setSource(new
        File("src/main/resources/mushrooms.csv"));
    Instances data = loader.getDataSet();
    data.setClassIndex(0);
} catch (Exception ex) {
    System.err.println("Exception occurred! " + ex);
}
```

We also set the target class to the first column, through the `setClassIndex()` method provided by the `Instances` class.

6. Now, we are ready to create a model. Similar to the previous section, we split the dataset into batches, and for each batch, we calculate the performance of the model. We use a KNN classifier to perform the classification task. The following code shows how we train and test the model, as well as how we log all metrics in Comet:

```
int n = data.numInstances();
for (int i = 200; i < n+1000; i+=1000) {
```

```
    if(i > n) i = n;
    Instances current_data = new Instances(data, 0, i);
    // train test splitting
    int seed = 10;
    current_data.randomize(new Random(seed));
    int trainSize = (int) Math.round(current_data.
numInstances() * 0.7);
    int testSize = current_data.numInstances() - trainSize;
    Instances train = new Instances(current_data, 0,
trainSize);
    Instances test = new Instances(current_data, trainSize,
testSize);
    // train the model
    IBk model = new IBk();
    model.buildClassifier(train);
    // evaluate the model
    Evaluation eval = new Evaluation(test);
    eval.evaluateModel(model,test);

    double accuracy = eval.pctCorrect()/100;
    experiment.logMetric("accuracy", accuracy);
    experiment.setStep(i);
}
```

We use the IBk class provided by weka to create a KNN classifier, and the Evaluation class to evaluate the model.

7. Finally, we terminate the experiment, as follows:

```
experiment.end();
```

8. From the project root directory, we run the code, as follows:

```
mvn clean compile exec:java -Dexec.mainClass="packt.
comet.KNN"
```

In the previous code, firstly we compile the project, and then we run it.

9.   We can access the results of the experiment directly in Comet, as shown in the following screenshot:

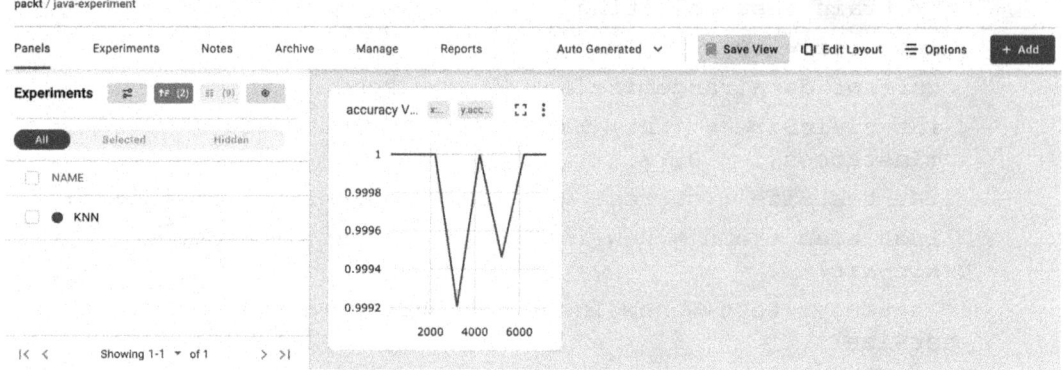

Figure 4.5 – The Comet dashboard after running the Java experiment

Figure 4.5 shows the Comet main dashboard with an accuracy graph related to the KNN experiment.

Now that you have learned some advanced concepts on Comet, you can practice with them through two practical use cases. Let's start with the first use case: offline and existing experiments.

First use case – offline and existing experiments

In *Chapter 1, An Overview of Comet*, you built a simple use case that permitted you to track images in Comet. The example used 52 time series indicators related to **gross domestic product (GDP)** in Italy, built 52 images, and uploaded them to Comet.

During the experiment, you will surely have noticed that the loading of the images in Comet was quite slow, depending on the bandwidth available in your internet connection. In this example, you will see how to use the concepts of offline and existing experiments to make the loading process smoother. You will also see how an existing experiment can be improved at a later time.

In this example, we suppose that the code implemented in *Chapter 1, An Overview of Comet*, for the first use case is running. Thus, please refer to it for further details.

The full code of this example is available in the GitHub repository, at the following link: `https://github.com/PacktPublishing/Comet-for-Data-Science/tree/main/04/first-use-case-advanced`.

In detail, the example is organized as follows:

- Running an offline experiment

- Continuing an existing experiment
- Improving an existing experiment offline

Let's start from the first phase: running an offline experiment.

Running an offline experiment

The idea here is to transform the online experiment described in *Chapter 1, An Overview of Comet,* into an offline experiment. This will permit you to upload the images in Comet in the background, once the experiment is completed.

To perform this operation, you can just change the following line of code:

```
from comet_ml import Experiment
experiment = Experiment()
```

Your code should now look like this:

```
from comet_ml import OfflineExperiment
experiment = OfflineExperiment(offline_directory="output")
```

Obviously, you need to create a directory named output in your current working directory. After running the experiment, your output directory will contain a file similar to the one shown in the following screenshot:

Figure 4.6 – Directory containing the offline experiment

Now, you can upload the offline experiment to Comet, as explained in the *Using experiments and models* section of this chapter. You should also remember to configure the .comet.config file to make the experiment work, as explained in *Chapter 1, An Overview of Comet.*

Now that you have built an offline experiment, we can move to the next step: continuing an existing experiment.

Continuing an existing experiment

Dividing the experiment into two parts could be an alternative to the previous strategy to save internet bandwidth. Thus, we could build an experiment that uploads to Comet the first *N* images, and then continue it and upload to Comet the remaining images.

Here's how we'll go about this:

1. Firstly, we wrap the existing code to build graphs into a single function, like so:

```
def run_experiment(df, experiment, indicators =
df.columns):
    for indicator in indicators:
        ts = df[indicator]
        ts.dropna(inplace=True)
        ts.index = ts.index.astype(int)
        fig = plot_indicator(ts,indicator)
        experiment.log_image(fig,name=indicator, image_
format='png')
```

The run_experiment() function receives the df DataFrame, the experiment, and a list of indicators as input, and for each indicator, it builds and logs the corresponding plot.

2. Now, you build an experiment as you usually do, as illustrated here:

```
from comet_ml import Experiment, ExistingExperiment
experiment = Experiment()
experiment.set_name('Track Indicators - first part')
```

I also set the name of the experiment through the set_name() function.

3. Then, you run the experiment by tracking the first 10 experiments, as follows:

```
N = 10
run_experiment(df,experiment, indicators=df.columns[0:N])
```

4. You retrieve the experiment key to continue the experiment later, as illustrated in the following code snippet:

```
experiment_key = experiment.get_key()
```

If you access the experiment in Comet, you can see an experiment called Track Indicators – first part, and, under the **Graphics** menu item, you will see the first 10 graphs. You will note that the uploading process was quite fast.

5. Now, you can continue the previous experiment by defining an ExistingExperiment() object, as follows:

```
from comet_ml import ExistingExperiment
experiment = ExistingExperiment(previous_
experiment=experiment_key)
experiment.set_name('Track Indicators - final')
```

You have also changed the name of the experiment, to track changes in the Comet dashboard.

6. Now, you can run the experiment with the remaining indicators, as follows:

```
run_experiment(df,experiment, columns=df.columns[N:])
```

If you now access the Comet dashboard, you can see the experiment with a different name, Track Indicators - final, and under the **Graphics** section, you will see all the graphs.

Now that you have learned how to continue an existing experiment, we can move to the next step: improving an existing experiment offline.

Improving an existing experiment offline

Let's suppose that for each indicator, you want to also plot a trendline that shows whetherthe indicator has an increasing or decreasing trend. In practice, for each indicator, you should calculate a linear regression model and then plot the resulting line. Since this operation could be time-consuming, you could decide to perform it offline and then upload the results in Comet in the background. Let's also suppose that you know the experiment key associated with your experiment.

Here are the steps we'll take:

1. Firstly, we create an ExistingOfflineExperiment() object, as follows:

```
from comet_ml import ExistingOfflineExperiment
experiment = ExistingOfflineExperiment(offline_
directory="output",previous_experiment=experiment_key)
```

We have specified the offline directory and the experiment key.

2. Then, we modify the plot_indicator() function defined in *Chapter 1, An Overview of Comet,* to also plot the trendline, as follows:

```
def plot_indicator(ts, indicator,trendline):
    fig_name = 'images/' + indicator.replace('/', "") +
'.png'
    xmin = np.min(ts.index)
    xmax = np.max(ts.index)
    plt.figure(figsize=(15,6))
    plt.plot(ts)
    plt.plot(ts.index,trendline)
    plt.title(indicator)
    plt.grid()
    plt.savefig(fig_name)
    return fig_name
```

3. Afterward, for each indicator, we implement a linear regression model, and we plot the time series and the output of prediction through the `plot_indicator()` function previously defined. The code is illustrated in the following snippet:

```
from sklearn.linear_model import LinearRegression
for indicator in df.columns:
    ts = df[indicator]
    ts.dropna(inplace=True)
     ts.index = ts.index.astype(int)

    X = ts.index.factorize()[0].reshape(-1,1)
    y = ts.values
    model = LinearRegression()
    model.fit(X,y)
    y_pred = model.predict(X)

    fig = plot_indicator(ts,indicator,y_pred)
    experiment.log_image(fig,name=indicator, image_
format='png', overwrite = True)
```

Note that we have use the `LinearRegression()` class provided by `scikit-learn` to implement the linear regression model. For each indicator, we build a graph similar to the one shown in the following screenshot:

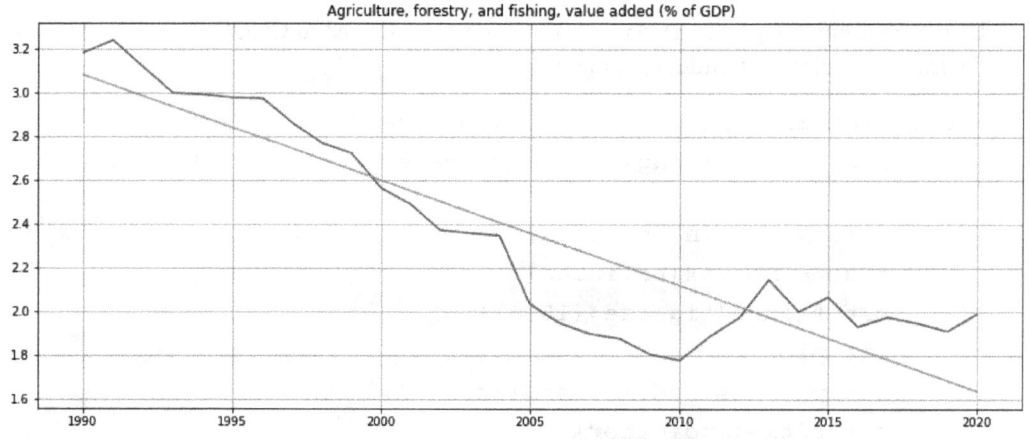

Figure 4.7 – Graph produced for the "Agriculture, forestry, and fishing, value added" indicator

Figure 4.7 shows the trendline in orange and the indicator line in blue.

4. Finally, once the experiment is completed, we upload it to Comet, like so:

```
comet upload PATH/TO/THE/ZIP/FILE
```

If you access the Comet dashboard, corresponding to your experiment, under the **Graphics** menu item, you will see all updated graphs, as shown in the following screenshot:

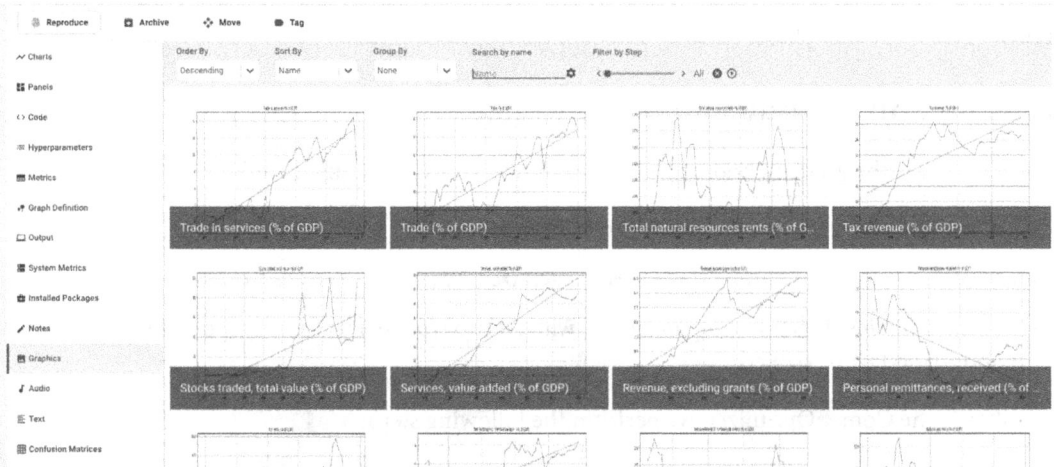

Figure 4.8 – Updated graphs under the Graphics menu item

Now that you have completed the first use case, we can move on to the second use case.

Second use case – model optimization

In *Chapter 1, An Overview of Comet*, you built a simple use case that permitted you to define a simple regression model and show the results in Comet. The example used the diabetes dataset provided by the scikit-learn library and calculated the **mean squared error** (**MSE**) for different values of seeds.

During the experiment, you will surely have noticed that the average MSE was about 3,000. In this example, we show how to use the concept of Optimizer to reduce the MSE value. Since the linear regression model does not provide any parameters to optimize, in this example, we will build a gradient boosting regressor model, and we will tune some of the parameters it provides.

In this example, we suppose that the code implemented in *Chapter 1, An Overview of Comet,* for the second use case is running. Thus, please refer to it for further details.

The full code of this example is available in the GitHub repository, at the following link: https://github.com/PacktPublishing/Comet-for-Data-Science/tree/main/04/second-use-case-advanced.

In detail, the example is organized like this:

- Creating and configuring an Optimizer
- Optimizing the model
- Showing the results in Comet

Let's start from the first phase: creating and configuring an Optimizer.

Creating and configuring an Optimizer

We will optimize the gradient boosting regressor model by hypertuning the following parameters:

- n_estimators—The number of trees in the forest. We will test values from 100 to 110.
- max_depth—The number of leaves in each tree. We will test values from 4 to 6.
- loss—The loss function to optimize. We will test two types of loss functions: squared_error and absolute_error.

To configure the Comet Optimizer, we perform the following steps:

1. We define a list of parameters, as follows:

```
params = {
    'n_estimators':{
        "type"        : "integer",
        "scalingType" : "linear",
        "min"         : 100,
        "max"         : 110
    },
    'max_depth':{
        "type"        : "integer",
        "scalingType" : "linear",
        "min"         : 4,
        "max"         : 6
    },
    'loss': {
        "type"        : "categorical",
        "values"      : ['squared_error', 'absolute_
error']
    }
}
```

For each parameter, we specify the type and other properties that depend on the type of parameter (categorical or integer).

2. We define a configuration variable that will be given as input to the Comet `Optimizer` class, as follows:

```
config = {
    "algorithm": "grid",
    "spec": {
        "maxCombo": 0,
        "objective": "minimize",
        "metric": "loss",
        "minSampleSize": 100,
        "retryLimit": 20,
        "retryAssignLimit": 0,
    },
    "trials": 1,
    "parameters": params,
    "name": "GB Optiimizer"
}
```

We choose `grid` as the optimization algorithm, and we specify that we want to minimize the loss function. We also set the `parameters` key to the `params` variable previously defined.

3. We build the Comet Optimizer, as follows:

```
from comet_ml import Optimizer
opt = Optimizer(config)
```

We pass as the input parameter to the `Optimizer` class the `config` variable previously defined.

Now that we have set up the Comet Optimizer, we are ready to optimize the model.

Optimizing the model

Let's suppose that you already have loaded the diabetes dataset, as described in *Chapter 1*, *An Overview of Comet*. Now, we'll build an experiment for each combination of parameters returned by the Optimizer. For each experiment, we will calculate the MSE for different values of seed, as already performed in *Chapter 1*, *An Overview of Comet*.

Here are the steps we'll take:

1. For each experiment contained in the list of experiments returned by the Comet Optimizer through the get_experiments() method, we build a GradientBoostingRegressor() object, initialized with the parameters defined for the current experiment.

2. Then, for each seed instance in the list of seeds, we split the dataset into training and test sets, and we fit the current model.

3. Finally, we calculate the MSE and log it in Comet through the log_metric() method.

The following code implements the previously described steps:

```
for experiment in opt.get_experiments():
    model = GradientBoostingRegressor(
            n_estimators=experiment.get_parameter("n_
estimators"),
            max_depth=experiment.get_parameter("max_depth"),
            loss=experiment.get_parameter("loss"),
    )


    for seed in seed_list:
        X_train, X_test, y_train, y_test = train_test_split(X,
y, test_size=0.20, random_state=seed)

        model.fit(X_train, y_train)
        y_pred = model.predict(X_test)
        mse = mean_squared_error(y_test,y_pred)
        experiment.log_metric("MSE", mse, step=seed)
```

After running the previous code, you are ready to see the results in Comet. So, let's move to the final step: showing the results in Comet.

Showing the results in Comet

In total, we performed 40 experiments. We can order the experiments by increasing MSE, as described in *Chapter 3, Model Evaluation in Comet*. The result is shown in the following screenshot:

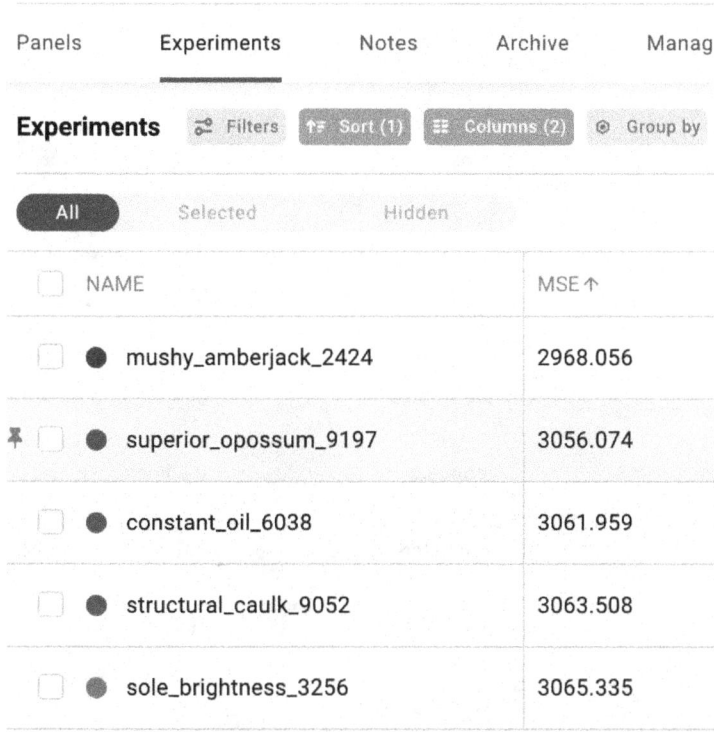

Figure 4.9 – The Comet dashboard after ordering the experiments by increasing MSE

`mushy_amberjack_2424` is the experiment with the lowest MSE. If we click on this experiment, we can view its parameters under the **Hyperparameters** section, as shown in the following screenshot:

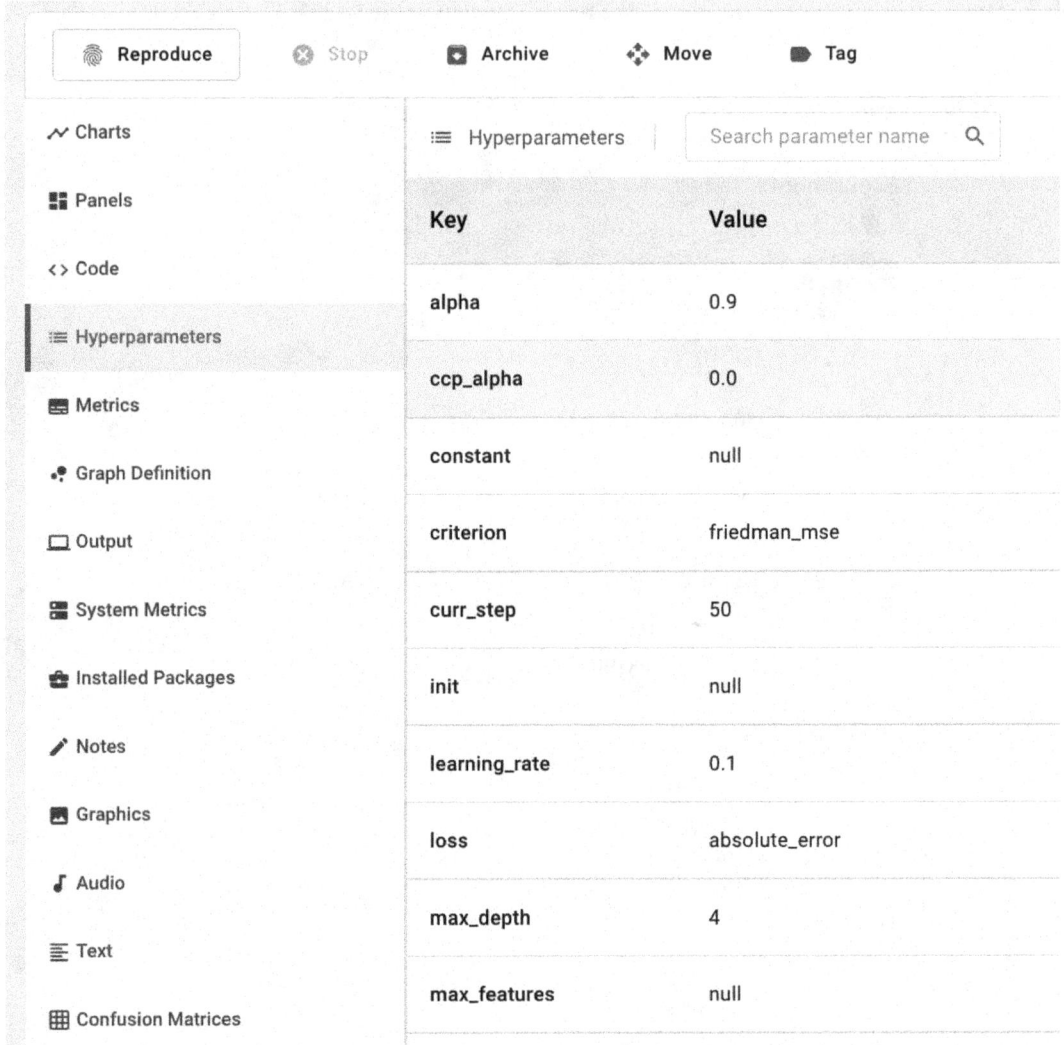

Figure 4.10 – Hyperparameters menu item in the Comet dashboard

Figure 4.10 shows only a subset of the parameters.

Optionally we could log all the models, save them in the Registry, and then choose the best model to send to production, as described in *Chapter 3, Model Evaluation in Comet*.

This example also plots the MSE metric versus the seed, as shown in the following screenshot:

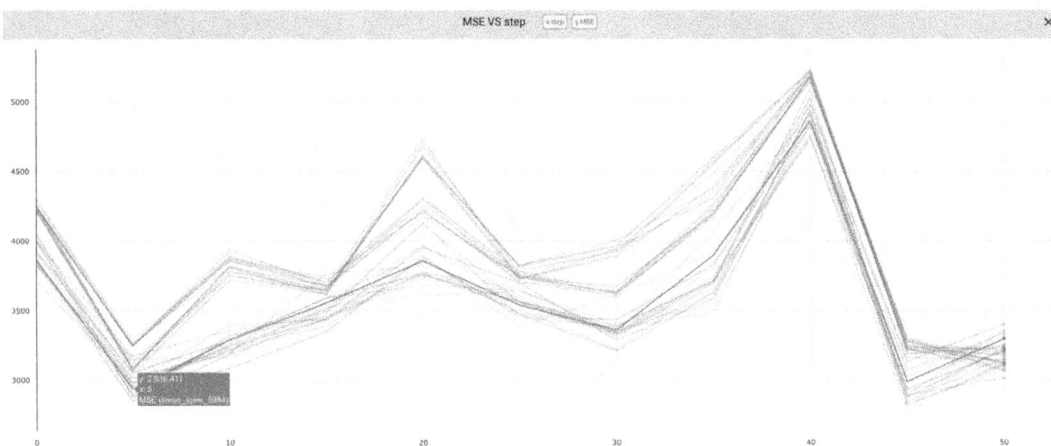

Figure 4.11 – MSE metric for the different experiments

Comet produced the previous output automatically during the running process.

Summary

We just completed the journey through advanced concepts in Comet!

Throughout this chapter, you learned how to share projects and workspaces with your collaborators, as well as how to make them public or private. In addition, you learned some advanced concepts on experiments and models.

Regarding experiments, you learned about the concept of offline experiments, which permitted you to run an experiment if the internet connection is not available. In addition, you learned about the concept of existing experiments, which permitted you to continue an experiment—for example, by enriching or extending it.

Regarding models, you learned how to optimize model parameters, through the concept of the Comet Optimizer. Using the Comet Optimizer makes it simple to choose the best parameters for a given model.

Finally, you extended the basic examples defined in *Chapter 1, An Overview of Comet,* with the concepts described in this chapter.

In the next chapter, you will review some concepts about data narrative and how to perform it in Comet.

Further reading

- Bhatia, A. & Kaluza, B. (2018). *Machine Learning in Java: Helpful techniques to design, build, and deploy powerful machine learning applications in Java.* Packt Publishing Ltd.
- Lantz, B. (2019). *Machine Learning With R: Expert Techniques For Predictive Modeling.* Packt Publishing Ltd.

5

Building a Narrative
in Comet

Data narrative, also known as **data storytelling**, is the art of telling stories starting from data. It is not simply a matter of summarizing the data but of building compelling stories, which can attract not only the attention of the audience they are aimed at but also arouse emotions that push the audience to action.

Data narrative is one of the final processes of the data science project life cycle and can be implemented either in parallel with the model deployment phase or immediately after.

In this chapter, you will review the basic concepts and techniques to build a narrative from data, including an overview of the Data, Information, Knowledge and Wisdom (DIKW) pyramid, and learn how to turn your data into a story. Then, you will learn how to build a narrative in Comet, using the concepts you are already familiar with, such as panels and reports. You will also implement two practical examples.

In detail, the chapter is organized as follows:

- Discovering the DIKW pyramid
- Moving from data to wisdom
- Choosing the correct chart type
- Using Comet to build a narrative

Before moving to the first step, let's install the technical requirements needed to run the code described in this chapter.

Technical requirements

We will run all the experiments and code in this chapter using Python 3.8. You can download it from the official website, `https://www.python.org/downloads/`, choosing the 3.8 version.

The examples described in this chapter use the following Python packages:

- `comet-ml 3.23.0`
- `matplotlib 3.4.3`
- `pandas 1.3.4`

We already described the first five packages and how to install them in *Chapter 1, An Overview of Comet*, so please refer to that for further details on installation.

In this chapter, you will also implement some code in JavaScript, by using some online libraries, which do not require any offline installation.

Now that you have installed all the libraries needed in this chapter, we can learn the concept of DIKW pyramid.

Discovering the DIKW pyramid

When you want to build a story from data, you first need to explore your data to understand which questions it can answer, as well as which data is relevant for your project. You already learned how to perform EDA in *Chapter 2, Exploratory Data Analysis in Comet*, so in this chapter, we suppose that you already have relevant data and, in general, have an idea of which questions your data can answer.

To build a story from data, you first need to think about the **audience** that will read your story. When you write a story, your preliminary purpose should be one of the following:

- Entertaining the audience
- Informing the audience
- Teaching something to the audience

The effect of your story should be calling the audience to action. To achieve your goal, you need to transform your data by interpreting it, enriching it with contextual information, and finally, linking it to an ethical model that calls the audience to action.

The **Data Information Knowledge Wisdom (DIKW)** pyramid helps you to understand how to move from raw data to the final message, which encourages the audience to action. More formally, the DIKW pyramid is a hierarchical representation of the relationships between data, information, knowledge, and wisdom, as shown in the following figure:

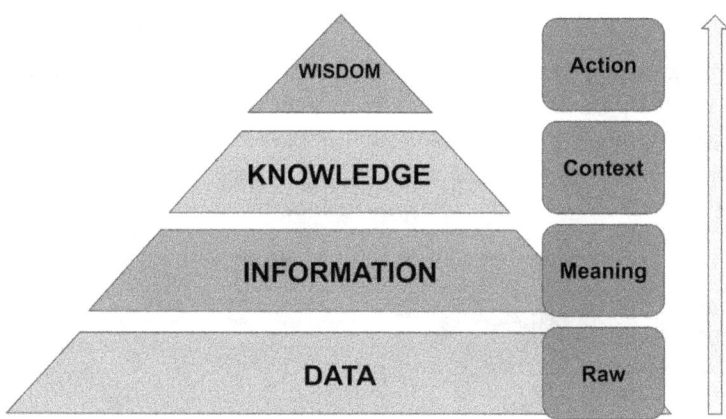

Figure 5.1 – The DIKW pyramid

The DIKW pyramid involves the following four steps:

- **Data**

- **Information**

- **Knowledge**

- **Wisdom**

Let's investigate each step separately, starting with the first – data.

Data

Data is at the bottom of the pyramid. It is the basis of everything – without data, you cannot build a story. To proceed with the other steps of the pyramid, you need to prepare your data by cleaning it, and enriching it, if needed. In a final report, you should not present your data as it is because usually, it is raw data. The following table shows an example of data:

Gender	Percentage
Male	78%
Female	21%
Prefer to not say	1%

Figure 5.2 – An example of data

The table specifies the output of a survey, where users should indicate their gender. Data is raw and still needs to be elaborated to transmit something to your audience.

Information

Information involves extracting *meaning* from data; it is about interpreting your data. In this step, the data is transformed into information that can be used by the common user in the form of readable content, including graphics, videos, images, and plain text. To achieve your goal, you need to perform EDA. However, this is not sufficient because you also need to generate readable content for the final user. The following figure shows a possible interpretation of the table shown in *Figure 5.2*:

Figure 5.3 – Extracted information from the table in Figure 5.2

We have removed the people who preferred to not say their gender, since they were only 1%. Then, we have extracted the following information: *out of five people, four are men and one is a woman*. Note that we have rounded the values.

Knowledge

Knowledge permits you to add context to your data. Data context is the set of circumstances that surrounds data and influences the data trending and behavior. The context should explain why a certain phenomenon happens. Data context can include the following:

- **Events** – something that happens
- **Environment** – an external or internal constraint
- **Time** – a chronological order in the data

Through a context, you can connect data to other data, discover causes and effects among it, and explain why some data behaves in a certain way. The following figure adds a possible context to the male/female example:

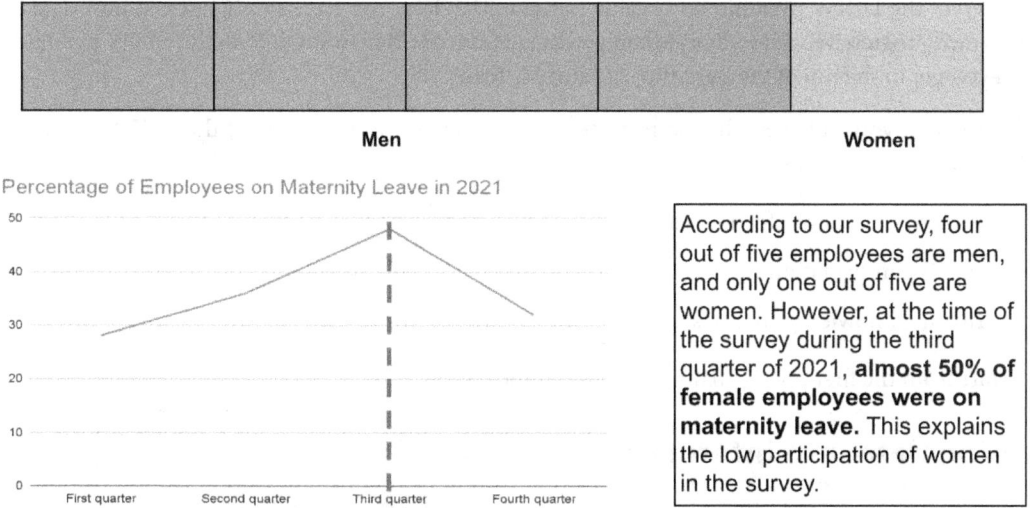

Figure 5.4 – Adding context to the extracted information of Figure 5.3

The context explains why data behaves in a certain way. In the previous figure, we can see that the survey refers to the third quarter of 2021, when there was a peak in maternity leave. This could explain why the percentage of women participating in the survey is so low.

Wisdom

Wisdom involves a call to action. In this phase, you should decide what is the best strategy to follow for the future and why you should choose it. All the actions involved in this phase should follow a specific ethical evaluation framework, including but not limited to the following ones:

- **Virtues** – the best choice follows a set of predefined values
- **Fairness** – the best choice optimizes equity
- **Common good** – the best choice optimizes societal well-being
- **Utilitarian** – the best choice optimizes global happiness

Referring to the previous example, a possible call to action could be incentivizing women to answer the survey, although they are on maternity leave. How could you achieve this objective? For example, if you followed a utilitarian approach, you could give a reward to women participating in the survey.

Now that you have learned the main steps of the DIKW pyramid, we can move to the next step, to build the final narrative.

Moving from data to wisdom

Each step of the DIKW pyramid adds value to the initial data. You have surely noticed how data is transformed progressively into a story when you move from one step to another on the DIKW pyramid. When you get to the top of the pyramid, the story is ready.

In this section, you will learn the main strategies for moving from one step of the DIKW pyramid to the other:

- Turning data into information
- Turning information into knowledge
- Turning knowledge into wisdom

Let's start from the first point, turning data into information.

Turning data into information

Often, the datasets we are dealing with are organized in a tabular form, so they already have a structure. Our task is therefore to select the relevant data that answers our questions. The principle is that the more data we have, the less meaning we can extract from a single piece of data. This is because, the more data we have, the less our brain is able to process it in order to extract meaning from it.

Turning data into information involves trying to give meaning to data. Data is a fact, something that is present and available, while information is the data enriched with meaning.

You might think that turning data into information can just be done by transforming it into a graphic form, but in reality, this is not exactly true, as shown in the following figure:

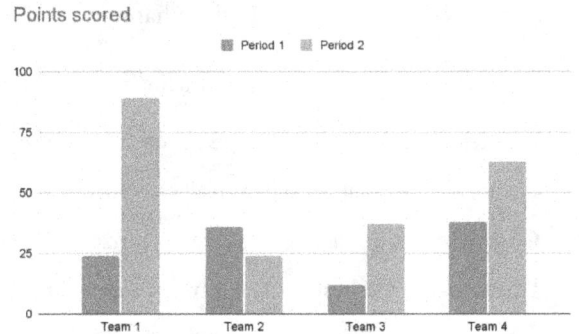

	Period 1	Period 2
Team 1	24	89
Team 2	36	24
Team 3	12	37
Team 4	38	63

Figure 5.5 – Different representations of the same data

The previous figure shows some data in tabular form on the left and the form of a graph on the right, as a bar chart. Looking at the two representations of the same data, you can see how the graph adds nothing to what is already expressed by the table.

Therefore, that graph does not carry any information. Conversely, the table is clearer than the graph because it makes the raw data immediately accessible.

To turn data into information, you should apply the following strategies:

- **Focusing on a single message**: if your message brings everything, it brings nothing. Your audience gets confused if you try to communicate more than a single message. But it is very common to try to say everything with your data, thus saying little at all. Although your data may bring more than a single message, you should represent just one piece at a time.

- **Simplifying**: You should not give all the details relating to the data – for example, the thousandth degree of precision – unless it is explicitly requested. The idea is to abstract data as much as possible in order to enrich it with meaning. For example, it is easier for an audience to understand that one in five people play sport than 22.38% of the population plays sport. We have made a simplification, but surely the reader better grasps the meaning of that data. The simplification also involves the use of colors in graphs. It is advisable to use a maximum of three colors in the graphs.

The following figure shows a possible way to turn the data contained in *Figure 5.5* into information:

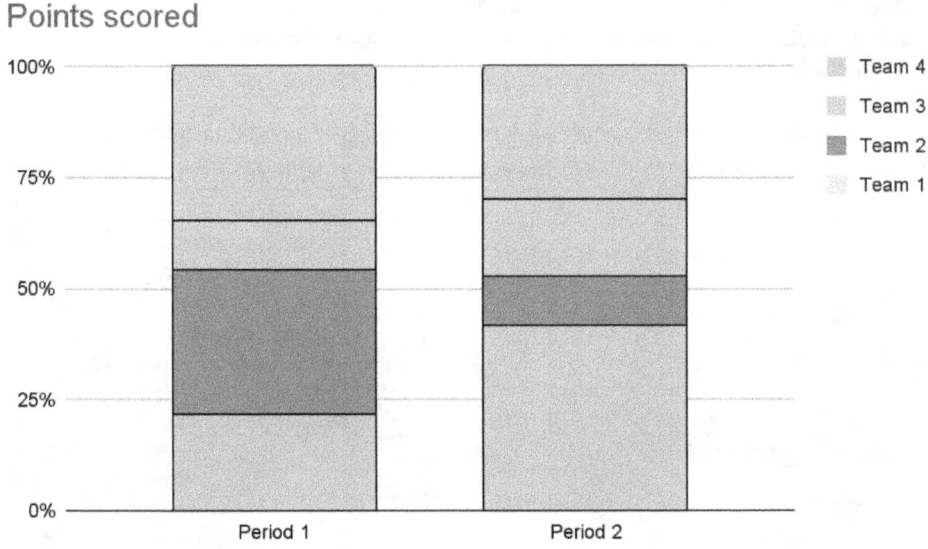

Figure 5.6 – Possible information extracted from data contained in Figure 5.5

We have adopted the following strategy:

- Focusing on a single message – *the points earned by Team 2 drastically decreased in the second period*

- Blacking out all the other teams, by coloring them in gray and focussing only on Team 2
- Simplifying the graph, showing it as a stacked bar

Once you have extracted information from data, you are ready to move on to the next step, turning information into knowledge.

Turning information into knowledge

Turning information into knowledge means adding context to your data that's already enriched with meaning. Adding context to your data permits your audience to understand your message. Obviously, different contexts produce different knowledge; thus, you should pay attention to the type of context you would like to add to your data.

To turn information into knowledge, you should apply the following strategies:

- *Defining communication goals* by defining clearly what you want to communicate to your audience.
- *Choosing only information that permits you to achieve your communication goals* and removing all the other information.
- *Adding annotations* in terms of a story, the description of an environment, a statistic, a metric, and so on. Within an annotation, you can use terms that address a position, such as first, second, and third, which are easily understood by the human brain.

Let's consider again the previous example, shown in *Figure 5.6*. Depending on the context you add, the message totally changes. The following figure adds a possible context to the previous graph:

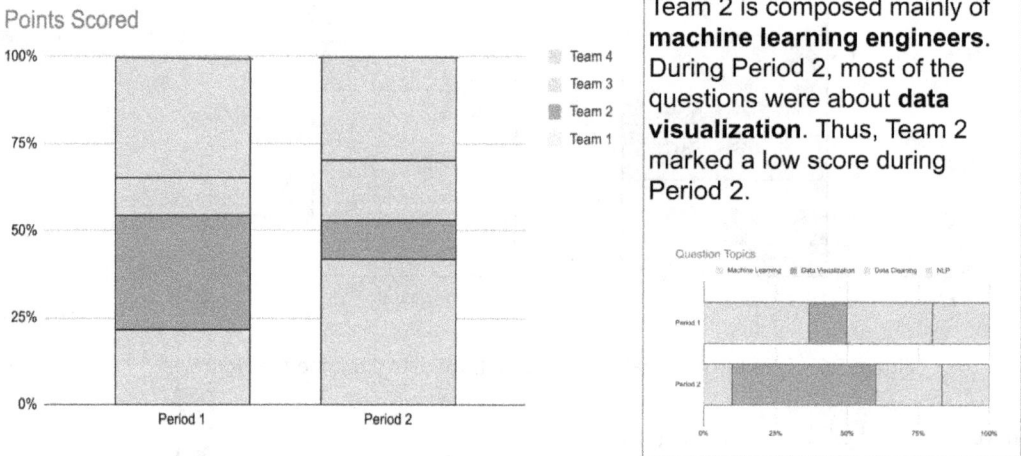

Figure 5.7 – A possible context for the information described in Figure 5.6

The annotation on the right explains why Team 2 marked a low score in Period 2. The explanation is that Team 2 is composed of machine learning engineers, who have little knowledge about data visualization. During Period 1, most of the questions applied machine learning and similar topics; thus, they achieved a high score. During Period 2, instead, most of the questions involved data visualization; thus, Team 2 achieved a low score.

A totally different context can produce a different interpretation of the same information, as shown in the following figure:

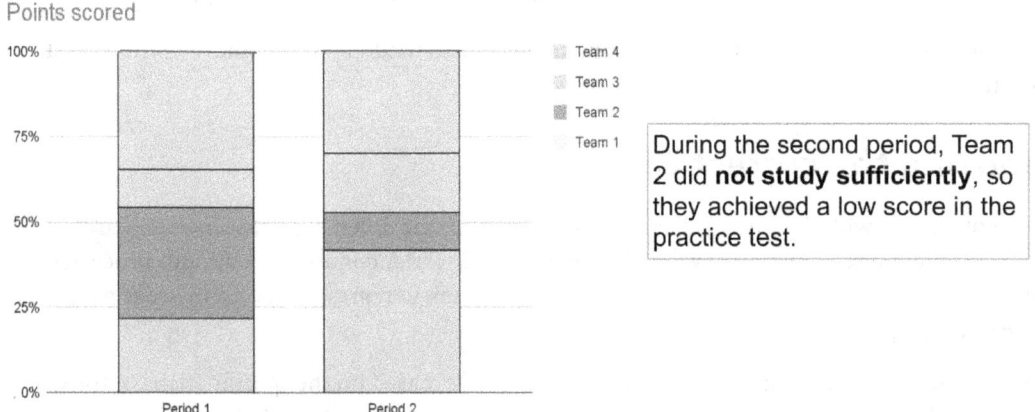

Figure 5.8 – Another possible context for the information described in Figure 5.6

In this case, the annotation simply states that in Period 2, Team 2 marked a low score because they did not study sufficiently.

You should pay attention to the type of context you add to your information because it can be misunderstood by the audience, thus producing misinterpretations of the message and bad decision-making. For example, during intercultural communication, the context could be incorrectly interpreted, due to different cultural bias.

Once you have added context to your information, you are ready to move to the next step, turning knowledge into wisdom.

Turning knowledge into wisdom

This step consists of involving the audience to make decisions, and to act. It is the final step, which projects the data that typically concerns the past into the future. You turn knowledge into wisdom when you apply your knowledge to make the right decisions.

If you have developed the previous steps well, the call to action is automatic and is typically expressed with one of the following questions:

- What can be done to improve the results?

- What opportunities do you have?

- What scenarios can be outlined?

In this phase, it is not enough to just invite the audience to action; you should also listen to their proposals and their answers to your questions. This is the discussion phase.

Referring to the example shown in *Figure 5.7*, a possible call to action could be to incentivize machine learning engineers to learn the data visualization principles. Instead, referring to the example shown in *Figure 5.8*, a possible call to action could be the organization of recovery courses.

Now that you have learned how to move from knowledge to wisdom, we can move to the next step, choosing the correct chart type.

Choosing the correct chart type

Representing data with the correct chart type is what makes the difference between a standard graph and an excellent one. You may have the best data in the world, context-specific and processed to convey an important message, but if you use the wrong graph to represent it, your message will likely not be fully grasped.

In this section, we briefly discuss which chart types to use, based on the specific shape of the data. These are guidelines that you will have to adapt from time to time to your needs.

The section describes the most common graphs and when you should use them. We will review the following chart types:

- A line chart

- A bar chart

- An area chart

- A pie chart

Let's start with the first chart, a line chart.

A line chart

A line chart compares data values that are sequentially connected. Usually, you can use a line chart to represent time series, as shown in the following figure:

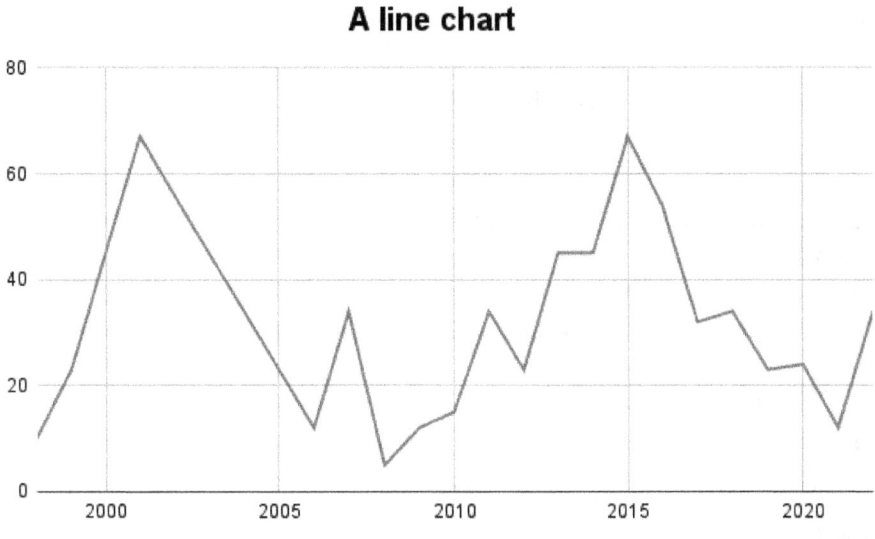

Figure 5.9 – A line chart

The previous graph shows the trend line of a generic quantity over time. The graph is very clear. Usually, you use a line chart to compare one or more series, as shown in the following figure:

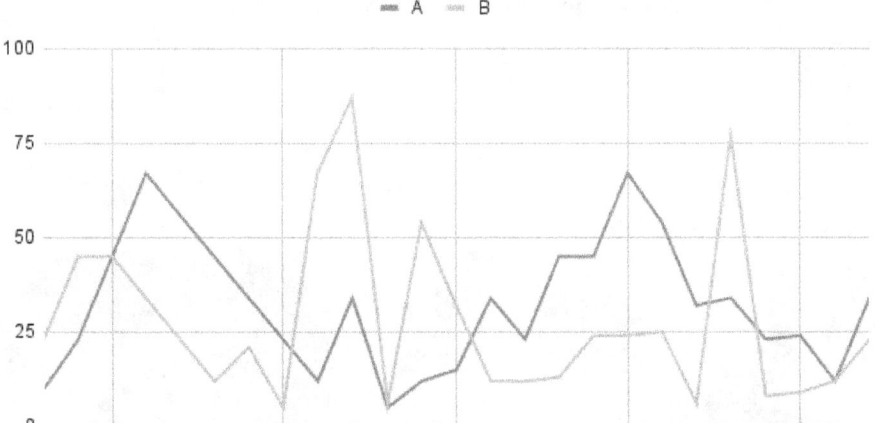

Figure 5.10 – A line chart for multiple series

The previous graph does not focus on any series in particular. However, it is a good practice to focus on a single series by highlighting it to make the graph more readable, as described in the previous sections.

Now that you have learned when you should use a line chart, we can move on to the next chart, a bar chart.

A bar chart

A bar chart compares data values for different groups of data, or categories. They are very similar to line charts. Similar to line charts, you can have different series of data.

> **Tip**
>
> In general, you can use a bar chart whenever you want to compare large changes or differences in data among categories.

There are different types of bar charts, including the following ones:

- A vertical bar chart
- A horizontal bar chart
- A stacked bar chart
- A 100% stacked bar chart
- A diverging bar chart

A vertical bar chart

The following figure shows an example of a vertical bar chart:

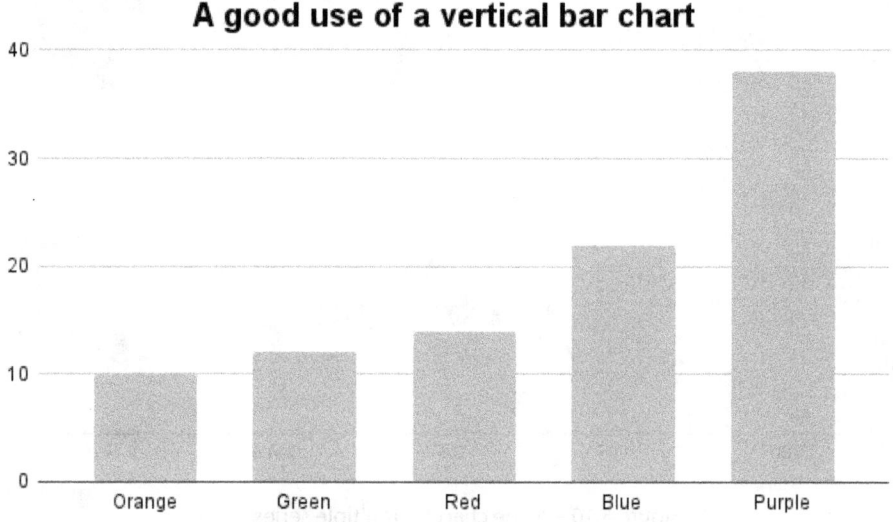

Figure 5.11 – A vertical bar chart

The graph is clear and easy to understand. When the difference between the values in categories is small, the bar chart is not appropriate, as shown in the following figure:

A bad use of a vertical bar chart

Figure 5.12 – Bad use of a vertical bar chart

In the previous graph, all the values range from 10 to 11; thus, it is not easy to understand the gaps between categories. In addition, the graph is not ordered.

> **Tip**
> A good practice is to order a bar chart.

The following figure shows an example of a vertical bar chart for multiple series:

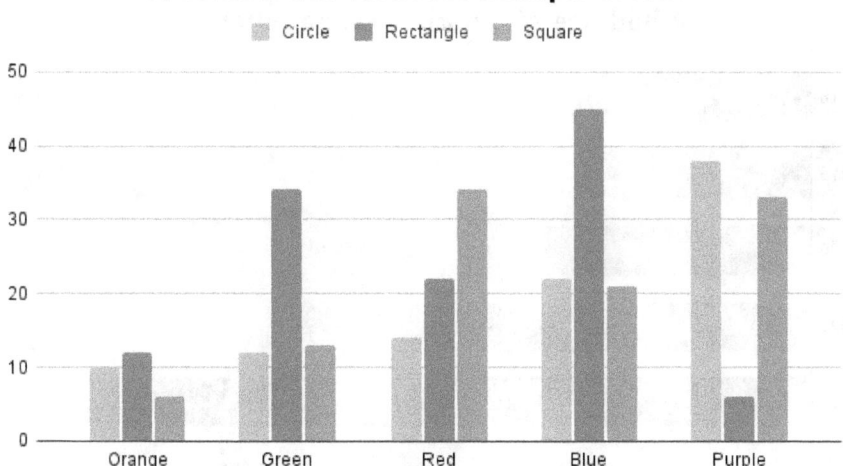

A vertical bar chart for multiple series

Figure 5.13 – A vertical bar chart for multiple series

There are three series in the graph: circle, rectangle, and square. Note that the previous graph is difficult to read because it uses too many colors. If you want to use this type of graph, you should focus on a single series and highlight only that.

A horizontal bar chart

A horizontal bar chart is an alternative to the vertical bar chart, as shown in the following figure:

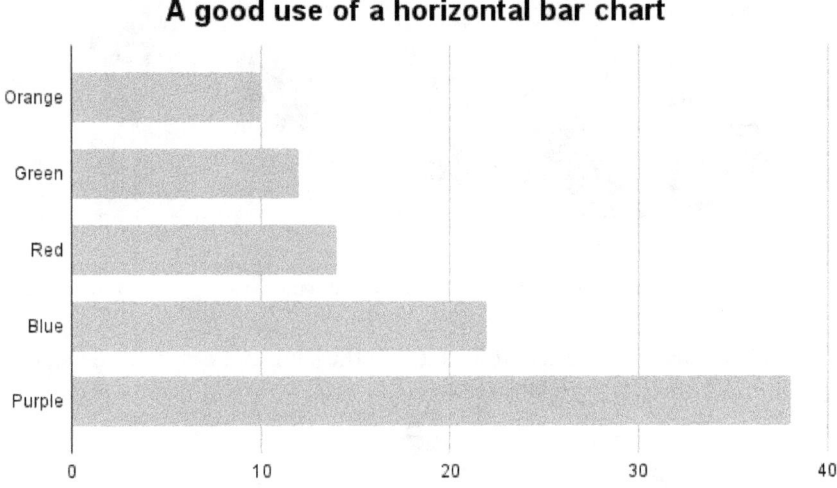

Figure 5.14 – A horizontal bar chart

You can use the vertical bar chart and horizontal bar chart interchangeably. However, you should always use a vertical bar chart for categories that represent time spans. The following figure shows an example of the misuse of the horizontal bar chart:

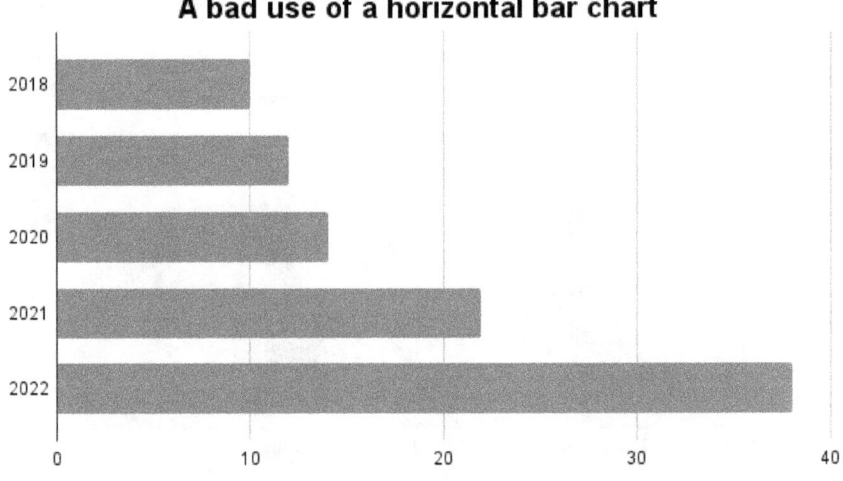

Figure 5.15 – Bad use of a horizontal bar chart

Dates are represented as categories on the ordinate axis, and it is difficult to understand the temporal progression.

A stacked bar chart

The objective of a stacked bar chart is to show how members of a category contribute to the total. You can use a stacked bar chart to represent multiple series. The following figure shows an example of a stacked bar chart:

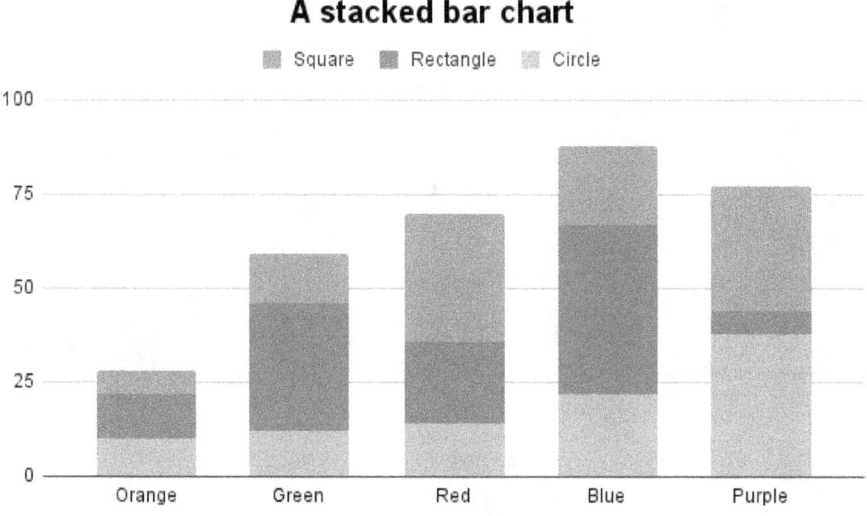

Figure 5.16 – A stacked bar chart

The graph shows the same data shown in *Figure 5.13* in a much clearer way. In fact, for each category (orange, green, red, blue, and purple), you can read the contribution of each series.

You can build a stacked bar chart either vertically, as shown in the previous figure, or horizontally.

A 100% stacked bar chart

A 100% stacked bar chart is an alternative to the stacked bar chart. In the 100% stacked bar chart, the contribution of each series is scaled to 100%, as shown in the following figure:

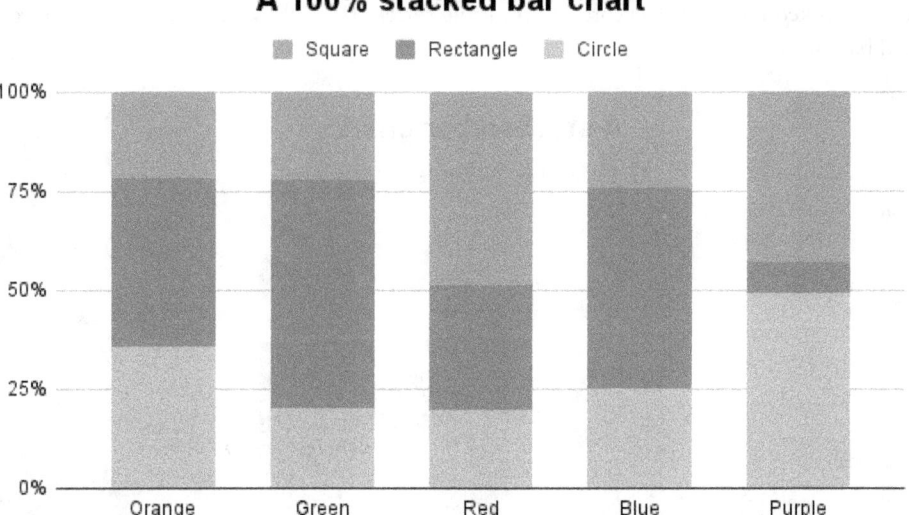

Figure 5.17 – A 100% stacked bar chart

You can use the 100% stacked bar chart and the stacked bar chart interchangeably. The only difference between the two graphs is that the stacked bar chart represents absolute values, while the 100% stacked bar chart represents relative values.

> **Tip**
> You can use a 100% stacked bar chart to represent small changes in temporal data, such as changes between different quarters of a year, as shown in *Figure 5.6*.

A diverging bar chart

A diverging bar chart compares two (opposite) series along a common scale, as shown in the following figure:

A diverging bar chart

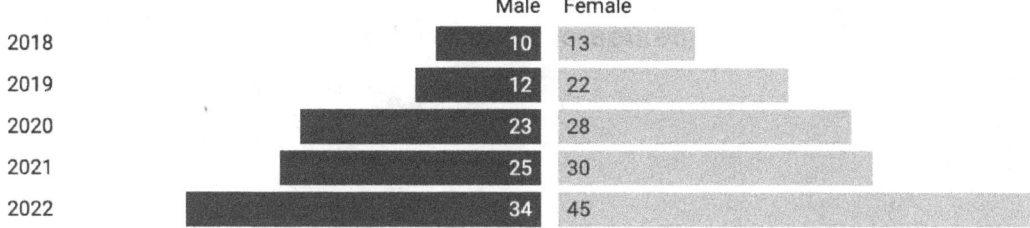

Figure 5.18 – A diverging bar chart

The previous graph compares Male and Female categories. When using the diverging bar chart, you should pay attention that the two series represent two divergent concepts, such as positive and negative, male and female, and so on.

Now that we have learned when we should use a bar chart, we can move on to the next chart, an area chart.

An area chart

An area chart is a line chart that fills the area under a line, as shown in the following figure:

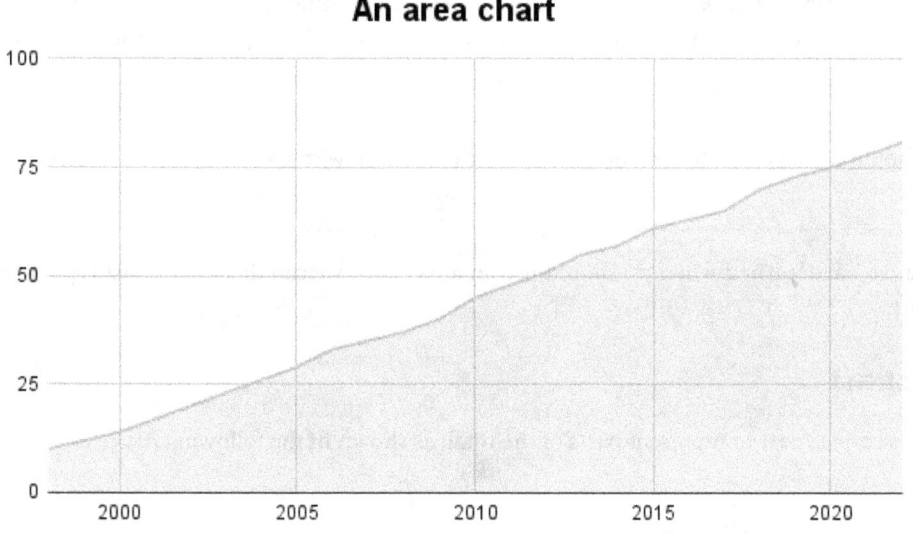

Figure 5.19 – An area chart

You can use an area chart to show the cumulative trend over time of data values. You can use an area chart to plot more than one series, as shown in the following figure:

A bad use of an area chart

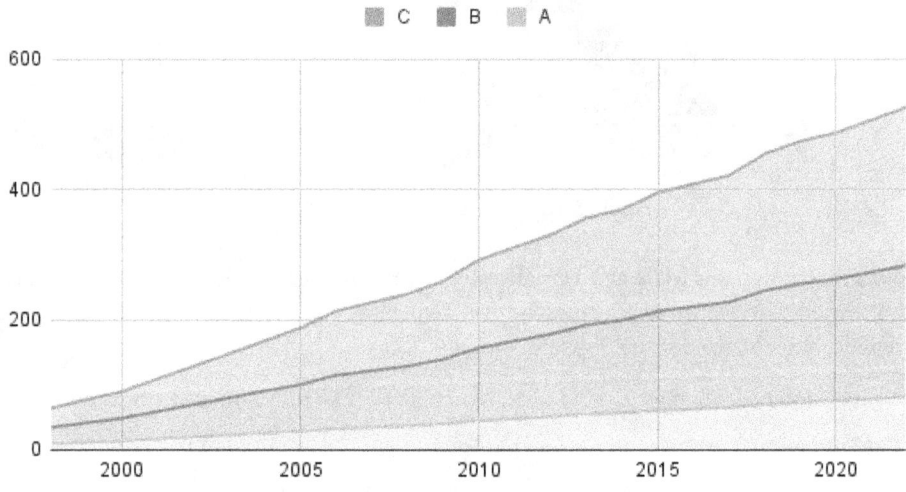

Figure 5.20 – Bad use of an area chart

You may find the areas of series other than the lowermost series in this graph difficult to understand because the contribution of each series is given by the sum of the other series beneath it. Thus, you may not immediately recognize the cumulative value associated with the uppermost series in the graph.

> **Tip**
> You should use an area chart to compare two series at most, with an emphasis to the lowermost series.

Now that we have learned when we should use an area chart, we can discuss the last type of graph, a pie chart.

A pie chart

You can use a pie chart to represent parts of the total, as shown in the following figure:

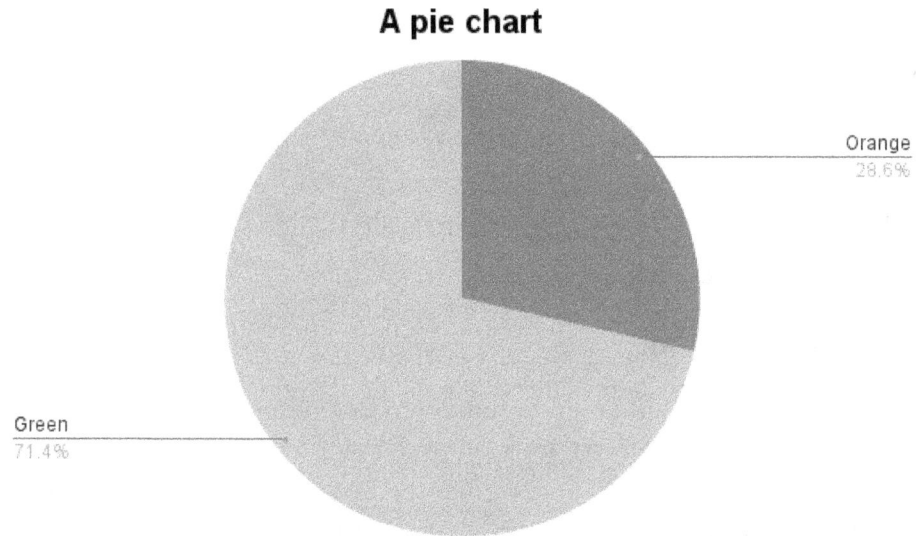

Figure 5.21 – A pie chart

A best practice is to use pie charts to compare only two or three series of data. The idea is to think of a pie chart as a **Pac-Man chart**, where there are two different values, one much larger than the other, making the chart resemble the videogame character Pac-Man.

> **Tip**
>
> If you have multiple series of data, you should avoid using a pie chart because you can always replace it with a bar chart.

Now that you have learned how to choose the correct graph to represent your data, we can move to the next step, using Comet to build a narrative.

Using Comet to build a narrative

Comet provides two main features to build a narrative – panels and reports. You have already learned some basic concepts related to panels in *Chapter 1*, *An Overview of Comet*, and other advanced concepts relating to panels and reports in *Chapter 2*, *Exploratory Data Analysis in Comet*.

Regarding panels, you have already learned how to implement them in Python using the SDK provided by Comet. Comet also allows you to implement panels in JavaScript. In this section, you will learn the main classes and functions provided by Comet to implement a panel in JavaScript. In addition, you will learn some advanced concepts about reports. We will create two examples, which will allow you to practically learn how Comet can be used to transform data into a story.

The section is organized as follows:

- JavaScript panels
- Advanced reports
- Example one
- Example two

Let's start with the first point, JavaScript panels.

Using JavaScript panels

A JavaScript panel is a Comet panel written in JavaScript. Comet defines the `Comet.Panel` class to build a panel in JavaScript. This class provides many methods, including, but not limited to, the following ones:

- `setup()` – to configure the environment, including default options for the panel.
- `draw(experimentKeys, projectId)` – to build a panel. This receives the list of experiment keys and the project ID as input. You should modify this method to build your panel.
- `drawOne(experimentKey)` – to build a panel for a single experiment, passed as an input parameter.

For further details on the methods provided by the `Comet.Panel` class, you can refer to the Comet official documentation, available at the following link: `https://www.comet.ml/docs/javascript-sdk/getting-started/`.

To build your own panel, you should extend the `Comet.Panel` class by defining at least the `draw()` method, as follows:

```
class MyPanel extends Comet.Panel {
  draw(experimentKeys, projectId) {
    // code to draw the panel
  }
}
```

You should name your panel `MyPanel`.

The `Comet.Panel` class also provides an interface to the Comet experiments and projects through the `this.api` object. The `this.api` object is an instance of the `API` class, which implements all the methods to access logged metrics, parameters, objects, and so on.

Here is the list of the most important methods provided by the `API` class:

- `experimentMetricsForChart()` – to get the logged metrics for all the experiments passed as an input parameter
- `experimentMetric()` – to get a specific metric for a specific experiment passed as an input parameter
- `experimentParameters()` – to get the logged parameters for a given experiment passed as an input parameter
- `experimentAssets()` – to get the logged assets for a given experiment passed as input parameter

You can find the list of all the methods provided by the API class in the Comet official documentation, available at this link: `https://www.comet.ml/docs/javascript-sdk/api/`.

Now that you have learned how to build a panel in JavaScript, we can move on to the next step, advanced reports.

Building advanced reports

The Comet report is the main tool provided by Comet to build a story from your data because it wraps panels and text together. You already learned how to build a report in *Chapter 2, Exploratory Data Analysis in Comet*. In this section, we will explore the report options and how to integrate external media into a report.

Report options include the following features:

- **Downloading a report as a PDF**: You can click the download button, located in the top-right corner of the screen, once your report is open, as shown in the following figure:

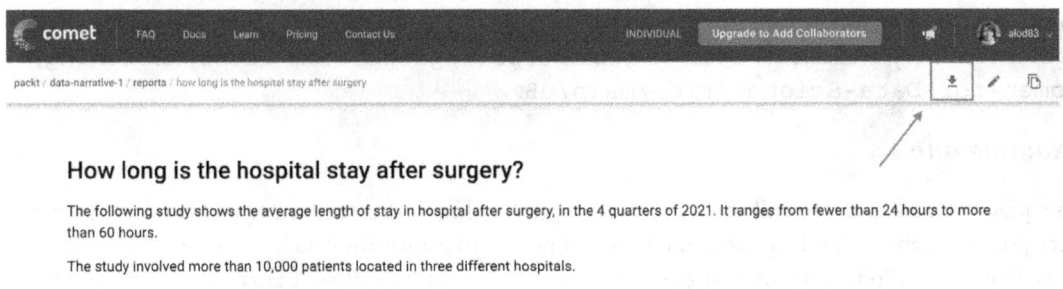

How long is the hospital stay after surgery?

The following study shows the average length of stay in hospital after surgery, in the 4 quarters of 2021. It ranges from fewer than 24 hours to more than 60 hours.

The study involved more than 10,000 patients located in three different hospitals.

Figure 5.22 – The position of the download button for a report

Saving the report as a PDF creates a static file, which you can print for further discussions.

- **Sharing a report**: You can click one of the sharing options, as shown in the following figure:

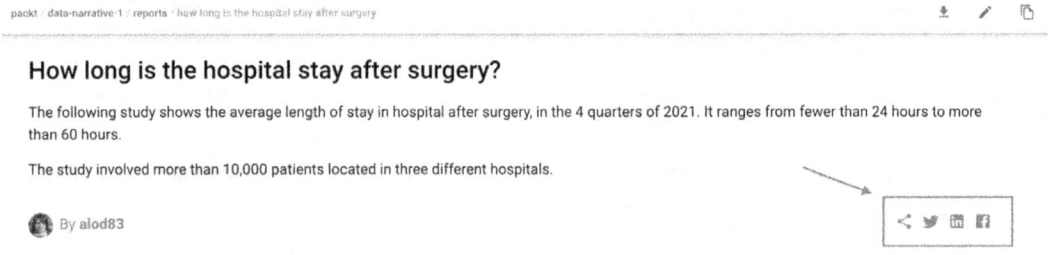

How long is the hospital stay after surgery?

The following study shows the average length of stay in hospital after surgery, in the 4 quarters of 2021. It ranges from fewer than 24 hours to more than 60 hours.

The study involved more than 10,000 patients located in three different hospitals.

By alod83

Figure 5.23 – The position of the sharing buttons in a report

Sharing a report permits you to maintain the report interactivity but requires that the people you share the report with have a Comet account.

In a report, you can include external media, such as short videos, images, and audio. For example, you could shoot a video showing your model and then include it in your report. To include media in a report, you should perform the following steps:

1. Firstly, you should upload your media to cloud storage, which is publicly available on the web. If the media is already on the web, you can simply copy its URL.

2. Then, you can add the media in a textual section of your report as you usually do in Markdown. For example, if you want to add an image, you can use the following syntax:

```
![Alternate text](/path/to/the/image)
```

Since each textual section of the Comet report is a Markdown cell, you can use it to add every type of media.

Now that you have learned some advanced concepts on reports, you can apply the learned concepts to two practical examples. You can download the code of the two examples from the book official GitHub repository, available at the following link: https://github.com/PacktPublishing/Comet-for-Data-Science/tree/main/05.

Example one

As a practical example, you will solve the exercise provided by storytellingwithdata.com, a very popular website that helps communities and people to transform data into stories. The exercise is available at the following link: https://community.storytellingwithdata.com/exercises/how-can-we-improve-this-graph.

The dataset contains hospital stay lengths after surgery, as shown in the following table:

	<=24	24 and 36	36 and 48	48 and 59	>=60	Unknown
2019/Q1	12.2%	53.9%	6.3%	18.9%	6.9%	1.8%
2019/Q2	14.3%	58.3%	4.1%	16.7%	5.9%	0.7%
2019/Q3	19.5%	52.2%	2.8%	17.4%	7.0%	1.1%
2019/Q4	25.4%	50.3%	2.7%	14.4%	5.6%	1.7%

Figure 5.24 – The hospital stay dataset

The objective of this example is to transform the previous dataset into a narrative shown in a Comet report. You will also build a custom panel using the D3.js library. In this example, we assume that you are familiar with the D3.js library. If you are not, you can refer to the D3.js official documentation to get started, which is available at this link: https://d3js.org/.

To achieve our objective, we will perform the following steps:

1. Firstly, we export the previous table as a CSV file – for example, named data.csv.

2. Then, we create a Comet experiment, which logs the CSV file as an asset. To create a new experiment, you can refer to *Chapter 1, Overview of Comet*. The following code shows how to log the CSV file in Comet:

```
import pandas as pd
from comet_ml import Experiment
df = pd.read_csv('data.csv')
experiment = Experiment()
experiment.log_table('data.csv', tabular_data=df)
```

We read the CSV file as a pandas DataFrame, and then we log it in Comet through the log_table() method provided by the Experiment class. The log_table() method receives as the first parameter the name of the uploaded file in Comet (in our case, it is the same as that of the original file in our local filesystem) and the DataFrame as the second parameter.

3. You can access the CSV file in Comet under the **Assets** and **Artifacts** menu item, as shown in the following figure:

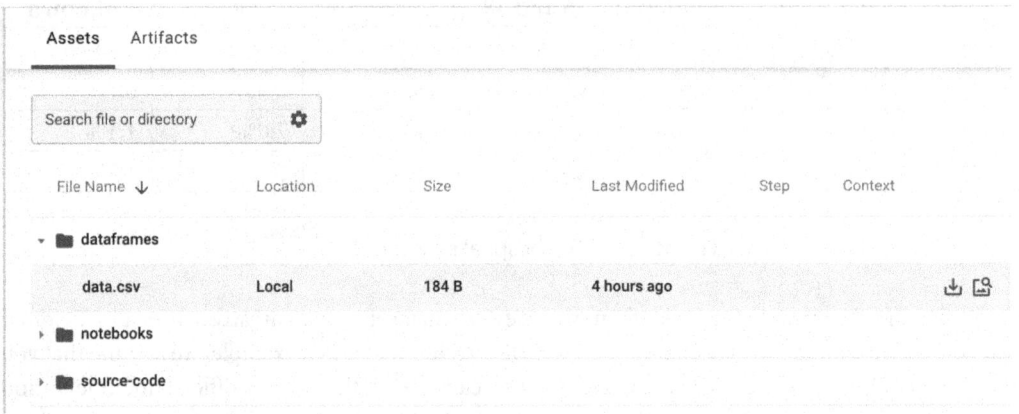

Figure 5.25 – The CSV file available in Comet under the Assets and Artifacts menu item

4. Now, you can create a custom panel, which loads the CSV file and builds a chart. The idea is to use the Comet panel to transform data into information. The panel will contain a stacked bar chart, which shows the trend of each period of stay over the different quarters, as shown in the following figure:

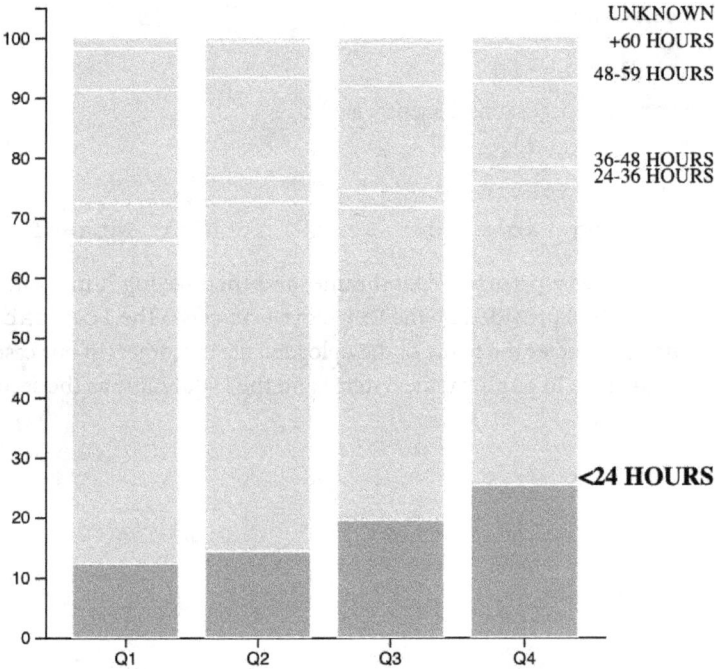

Figure 5.26 – A stacked bar showing the data described in Figure 5.9

The graph also focuses on a single message: since **Q1**, there is a progressive increase in the short stay (less than 24 hours) after surgery.

To build the panel, you can access the online SDK, as described in *Chapter 2, Exploratory Data Analysis in Comet*, and then select JavaScript as the main programming language.

5. Under the **Resources** tab, you should add the link to the D3.js library. In this specific example, we will use the following version of D3.js, https://d3js.org/d3.v4.js, so you should add it.

6. Under the **HTML** tab, you should create a new div container, which will contain the graph, as follows:

```
<div id="stacked_bar"></div>
```

7. Under the **Code** tab, the editor already shows some code. You should modify the setup() method to include the default options, as follows:

```
setup() {
    this.options = {
        highlight : "<24 HOURS",
        width : 860,
        height : 350,
    margin : {top: 10, right: 80, bottom: 20, left: 50},
        };
    }
```

The options include the width, the height, and the margins of the graph, as well as the column name to highlight in the graph.

8. Now, you can modify the draw() method, as follows:

```
async draw(experimentKeys, projectId) {
        experimentKeys.forEach(async experimentKey => {
            const data = await
            this.api.experimentAssetByName(
                        experimentKey,
                        'data.csv');
            this.drawGraph(data);
        });
    }
```

We loop over all the experiments, and for each experiment key, we retrieve the asset named data.csv. We use the experimentAssetByName() method provided by the API to access a single asset. Then, we call the drawGraph() method to draw the graph.

9. The `drawGraph()` method contains the code to build the graph in D3.js:

```
async drawGraph(data_string){}
```

10. The method receives as input the CSV file parsed as a string by `experimentAssetByName()`. Within the `drawGraph()` method, first, you can define the graph size:

```
let highlight = this.options.highlight;
var margin = this.options.margin,
width = this.options.width - margin.left - margin.right,
height = this.options.height - margin.top - margin.
bottom;
```

We retrieve the parameters from the `options` variable.

11. Then, you can append an SVG object at the end of the `div` defined in the HTML section:

```
var svg = d3.select("#stacked_bar")
    .append("svg")
    .attr("width", width + margin.left + margin.right)
    .attr("height", height + margin.top + margin.bottom)
    .append("g")
    .attr("transform",
        "translate(" + margin.left + "," + margin.top +
")");
```

12. Now, you need to convert the data string passed as input to an object, which can be parsed by D3.js:

```
var data = await d3.csvParse(data_string, function(d) {
  return {
    Period : d.Period,
    '<24 HOURS' : +d['<=24'],
    '24-36 HOURS' : +d['24 and 36'],
    '36-48 HOURS' : +d['36 and 48'],
    '48-59 HOURS' : +d['48 and 59'],
    '+60 HOURS' : +d['>=60'],
    'UNKNOWN' : +d['Unknown'],
  };
});
data.columns = ['Period', '<24 HOURS','24-36 HOURS','36-
48 HOURS','48-59 HOURS','+60 HOURS','UNKNOWN'];
```

We use the `csvParse()` method provided by `D3.js` to perform such a conversion. In addition, we rename all the columns for better visualization, and we convert strings to numbers through the + symbol before each column.

13. You prepare data for graphical representation as follows:

```
var subgroups = data.columns.slice(1);
var groups = d3.map(data, function(d){return(d.
Period);}).keys();
var stackedData = d3.stack().keys(subgroups)(data);
```

Firstly, we extract the list of subgroups from the columns by removing the first column, which contains `Period`. This list contains the length of stay (`<24 HOURS`, `24-36 HOURS`, and so on). Then, we extract the list of groups, which includes the periods (Q1, Q2, Q3, and Q4). Finally, we build a `D3.js` stack generator, which we will use to build the graph.

14. You can add the *x* axis as `scaleBand()`:

```
var x = d3.scaleBand()
    .domain(groups)
    .range([0, width])
    .padding([0.2]);
svg.append("g")
    .attr("transform", "translate(0," + height + ")")
    .call(d3.axisBottom(x).tickSizeOuter(0));
```

15. Then, you can add the y axis as a linear scale:

```
var y = d3.scaleLinear()
    .domain([0, 105])
    .range([ height, 0 ]);
svg.append("g")
    .call(d3.axisLeft(y));
```

16. Now, you can draw the stacked bar, as follows:

```
svg.append("g")
    .selectAll("g")
    .data(stackedData)
    .enter().append("g")
    .attr("fill", function(d) { if(d.key == highlight)
return '#40b7ad';return '#D9D9D9'; })
    .selectAll("rect")
```

```
    .data(function(d) { return d; })
    .enter().append("rect")
    .attr("x", function(d) { return x(d.data.Period); })
    .attr("y", function(d) { return y(d[1]); })
    .attr("height", function(d) { return y(d[0]) -
y(d[1]); })
    .attr("width",x.bandwidth())
    .attr("stroke", "#FFFFFF")
    .attr("strokewidth", 3);
```

We build a rectangle for each group and subgroup, and then we set the position, the color, and the size.

17. In the last part of the drawGraph() method, we should add the annotations. We prepare data as follows:

```
var q4 = data[3];
var sum = 0;
var ann_data = [];
for(var i = 0; i < subgroups.length; i++){
var index = subgroups[i];
sum += q4[index];
ann_data.push({'key' : index, value : sum });
}
```

The ann_data array stores for each subgroup the position in the graph. The current position is given by the sum of the previous positions plus the current one.

18. Finally, we append text to the SVG object for each group, as follows:

```
svg.append("g")
    .selectAll('text')
    .data(ann_data)
    .enter()
    .append("text")
    .text(function(d){return d.key;})
    .attr('x', width + 65)
    .attr("y", function(d) {if (d.key == 'UNKNOWN') return
y(d.value)-10; return y(d.value); })
    .text(function(d){return d.key;})
```

```
    .attr('font-size', function(d) {if (d.key ==
highlight) return 15; return 12;})
    .attr('text-anchor', 'end')
    .attr('font-weight', function(d) {if (d.key ==
highlight) return 'bold'; return 'normal';});
```

We also set some properties, including font size, font weight, and position. With respect to the *y* position, we shift it slightly if the key is equal to UNKNOWN, to avoid text overlapping.

19. Now, your graph is ready. You can click the **Run** button, and you will see the graph shown in *Figure 5.26* in the right part of your JavaScript SDK.

Once the panel is ready, you can build a report, showing the results:

1. Firstly, you can create a new report, as described in *Chapter 2, Exploratory Data Analysis in Comet*.

2. Then, you can add the title, which should already highlight the message you want to communicate, such as the following one: **How long is the hospital stay after surgery?**.

3. Next, you can add an image to get the audience's attention, as well as a short introduction to the problem, as shown in the following figure:

How long is the hospital stay after surgery?

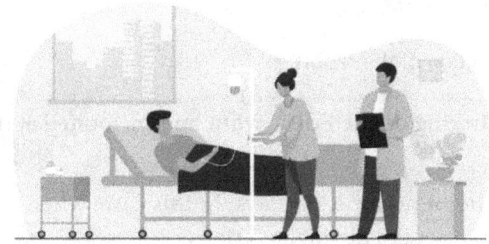

Hospital patient vector created by pch.vector - www.freepik.com

The following study shows the average length of stay in hospital after surgery, in the 4 quarters of 2021. It ranges from fewer than 24 hours to more than 60 hours. The study involved more than 10,000 patients located in three different hospitals.

Figure 5.27 – The report header

We include an image about patients, created by **pch.vector - www.freepik.com**, and available under the Freepik license, which permits use and redistribution, provided that the source is properly cited. To include the image shown in the previous figure, you can write the following Markdown code:

```
![](https://img.freepik.com/free-vector/patient-lying-
bed-during-intensive-therapy_74855-7774.jpg)
```

4.   Now, you can add the panel showing your graph, as described in *Chapter 2, Exploratory Data Analysis in Comet*. You can add some words, which add some meaning to it, such as the following ones: **Thanks to the new hires in the day hospital wards in the previous period, there is a progressive reduction in hospitalization times after surgery**. The following figure shows the produced section of the report, which includes the panel:

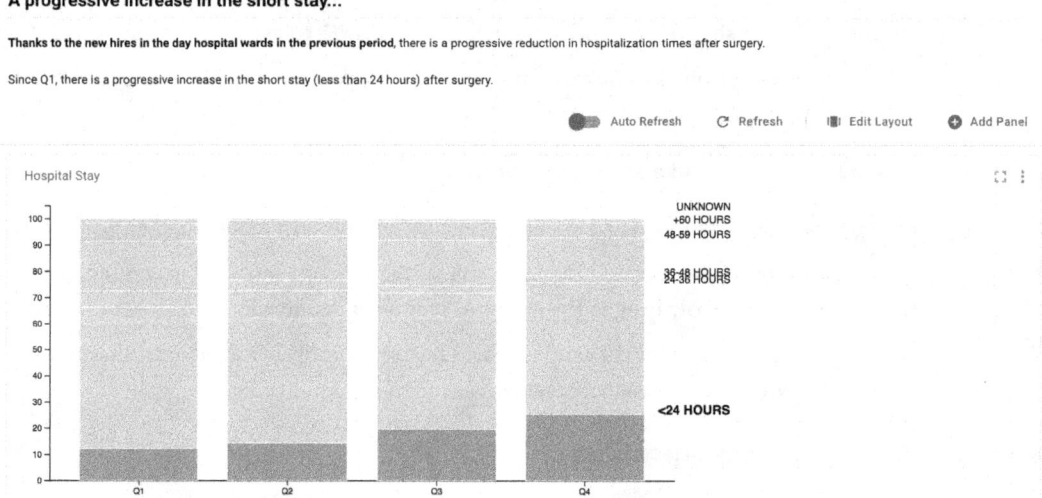

A progressive increase in the short stay...

Thanks to the new hires in the day hospital wards in the previous period, there is a progressive reduction in hospitalization times after surgery.

Since Q1, there is a progressive increase in the short stay (less than 24 hours) after surgery.

Figure 5.28 – The first section of the report

You can personalize the size of the panel by clicking on the **Edit Layout** button, located at the top-right part of the section.

5.   You can add a second panel, which shows the trend for long stays. In this case, when you configure the panel, on the **options** tab, you can set the highlight option as follows:

```
{"highlight": "+60 HOURS"}
```

This permits you to highlight long stays, as shown in the following figure:

... but long stays still remain constant

Since Q1, there is not any improvement for long stays, because no investment was done during this period for long stays.

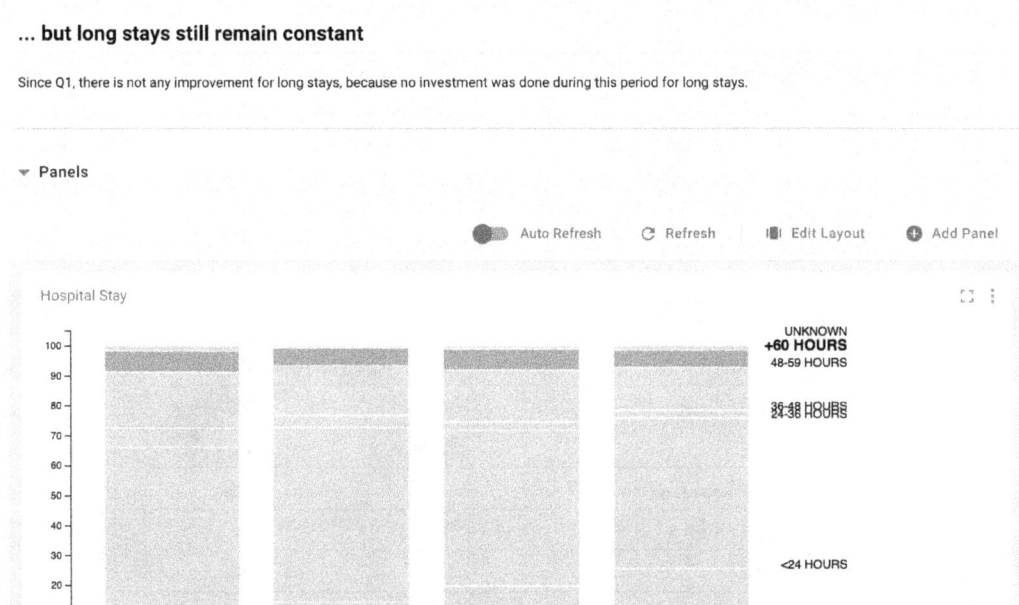

Figure 5.29 – The second section of the report

6. Similar to the previous section of the report, we can add some words that add context to the panel: **Since Q1, there is not any improvement for long stays, because no investment was done during this period for long stays**.

7. So far, you have built your report, which shows data turned into knowledge. You still need to add a wisdom section, which calls the audience to action. You can do it by adding a third section, which contains some questions, such as the following ones:

   - What can we do to reduce long stays?

   - Can we invest some money for long stays?

8. Now, your report is ready, and you can download it as a PDF file or share it with your audience.

You can continue to practice with Comet to build a narrative from data, so let's move on to the second example.

Example two

For the second example, you will solve another exercise provided by storytellingwithdata.com. The exercise is available at the following link: https://community.storytellingwithdata. com/exercises/visualize-the-insight. The dataset contains the output of a survey where customers expressed what they liked and what they did not like about a clothes retailer company, compared to all the other competitors. The following table shows the dataset:

Questions	Our store	All stores
The store is well-organized.	40%	38%
Fast and easy checkout.	33%	34%
Friendly and helpful employees.	45%	50%
Good promotions.	45%	65%
I can find what I'm looking for.	46%	55%
I can find the size I need.	39%	49%
A nice atmosphere.	80%	70%
Latest technology for easy shopping.	35%	34%
Lowest sales prices.	40%	60%
A wide selection.	49%	47%
Items I can't find elsewhere.	74%	54%
The latest styles.	65%	55%

Figure 5.30 – The survey dataset

For each question, the table shows the level of importance for **Our store** and for the competitors, named as **All stores**.

The objective of this example is to build a Comet report that calls the company's decision-makers to action. In other words, you should build a story with your data. The report will contain two panels built in Python, using the matplotlib library.

You can adopt the following strategy:

1. Firstly, you prepare the dataset by loading it as a `pandas` DataFrame and cleaning it, as shown in the following piece of code:

```python
import pandas as pd
df = pd.read_csv('source/data.csv')
columns = ['Our store', 'All stores']
for col in columns:
    df[col] = df[col].str.replace('%', '').astype(int)
```

    We loop over the two columns contained in the `columns` variable to remove the % symbol and convert strings to numbers.

2. Then, you can build two experiments in Comet, one for **Our store** and the other for **All stores**. For each experiment, you consider each question as a metric and you log it in Comet, as follows:

```python
from comet_ml import Experiment
def run_experiment(df, store):
    experiment = Experiment(project_name="data-
narrative-2")
    experiment.set_name(store)
    for i in range(len(df)):
        experiment.log_metric(df['Questions'].iloc[i],
df[store].iloc[i])

run_experiment(df,'Our store')
run_experiment(df,'All stores')
```

    We define a function, named `run_experiment()`, which receives as input the DataFrame and the column name to evaluate. The function builds an experiment and logs the metrics corresponding to the DataFrame column passed as an argument. Then, we call the function for both the columns, **Our store** and **All stores**.

3.   You build a custom panel using the Comet SDK, which compares the two experiments. Access the Comet SDK, as described in *Chapter 2, Exploratory Data Analysis in Comet*, and select Python as the programming language. The objective is to build the following panel:

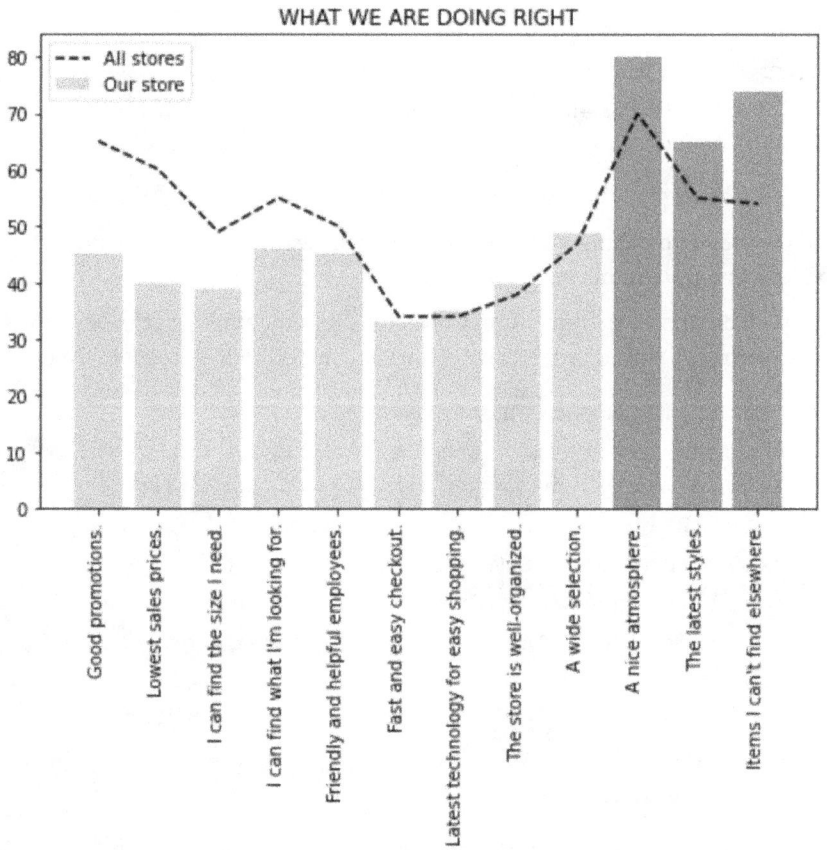

Figure 5.31 – A comparison between Our store and All stores

In the previous graph, we have calculated the difference between the values for **Our store** and **All stores**, and we have ordered questions by increasing the difference. Then, we have shown **Our store** as a bar chart and **All stores** as a line. These operations have turned data into information. In addition, we have highlighted in green what we are doing right.

4.   To build the previous graph, we first retrieve all the metrics' names, and we build a `pandas` DataFrame containing them, as follows:

```
from comet_ml import API, ui
import matplotlib.pyplot as plt
import pandas as pd
```

```
api = API()
options = api.get_panel_options()
metric_names = api.get_panel_metrics_names()
df = pd.DataFrame(metric_names, columns=['Questions'])
```

We use `get_panel_metrics_names()` provided by the `API` class to retrieve all the metrics' names.

5. Then, we get all the experiment keys and the objects containing the metric values, as follows:

```
experiment_keys = api.get_panel_experiment_keys()
metrics_obj = api.get_metrics_for_chart(experiment_
keys,metric_names)
```

The `get_metrics_for_chart()` method returns a dictionary that contains the metrics for each experiment.

6. We loop over each metric object to retrieve the single metric value for each experiment, as follows:

```
for experiment_key in metrics_obj:
    experiment = api.get_experiment_by_key(experiment_
key)
    column = []
    for metric in metrics_obj[experiment_key]["metrics"]:
        column.append(metric['values'][0])
    df[experiment.name] = column
```

We store the list of metrics for each experiment in a temporary variable called `column`, and then we append it to a new column of the DataFrame.

7. Now, the DataFrame is ready. We can calculate the difference between **Our store** and **All stores**, as follows:

```
df['diff'] = df['Our store'] - df['All stores']
df = df.sort_values(by=['diff'])
```

We also order the DataFrame by the new column, through the `sort_values()` method.

8. We define some color options, as follows:

```
cc=list(map(lambda x: '#508DED' if x < -1 else '#D9D9D9',
df['diff']))
if options['filter'] == 'greater':
    cc=list(map(lambda x: '#40B7AD' if x > 2 else
'#D9D9D9', df['diff']))
```

We use the `options` variable to make the plot customizable. In practice, depending on the filter option, a specific part of the bar chart will be highlighted, as shown in the following figure:

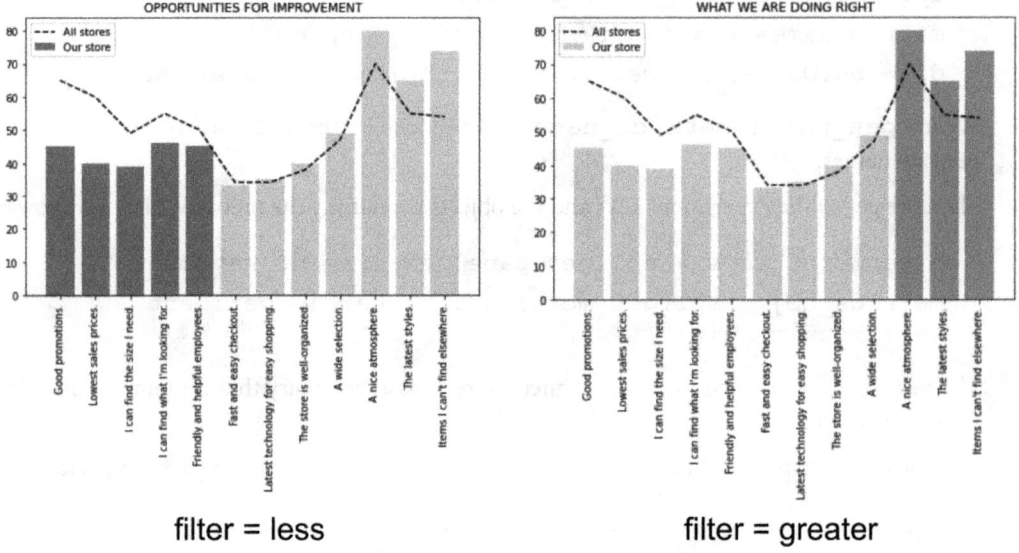

Figure 5.32 – Different highlights for different filters

If the filter is set to greater, then the greatest values are highlighted; otherwise, the lowest values are highlighted.

9.   We are now ready to plot the graph:

```
plt.figure(figsize=(8,4))
plt.bar(df['Questions'], df['Our store'], color = cc,
label='Our %store')
plt.plot(df['Questions'], df['All stores'],
color='#000000', ls='--', label='All stores')
plt.xticks(rotation=90)
plt.title(options['title'])
plt.legend()
plt.tight_layout()
ui.display(plt)
```

We set the figure size, as well as the title, extracted from `options`, and the legend, and finally, we show the graph through the `ui.display()` method.

10. Finally, you build a report that tells the story. You can create a new report, as described in *Chapter 2, Exploratory Data Analysis in Comet*. You can set the title and a short context, as shown in the following figure:

Our Customer Experience

In the previous period we did a questionnaire among our customers and we asked them what the important things are for them in choosing a store for back-to-school shopping. We then compared our results with those of the other stores.

Figure 5.33 – The report header

11. In the first section, you can add a text that calls to action, as well as the two panels, as shown in *Figure 5.19*. The following figure illustrates the resulting section:

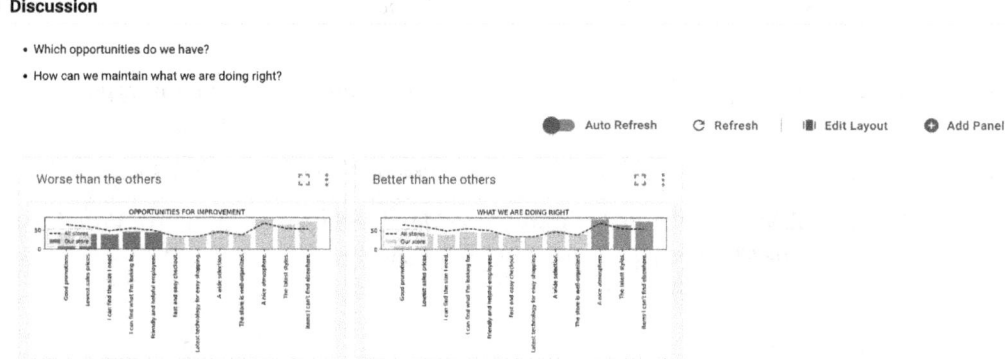

Figure 5.34 – The first section of the report

The discussion should focus on the following two questions: *Which opportunities do we have?* and *How can we maintain what we are doing right?*.

12. Now, your Comet report is ready. You can share it with your company or download it as a PDF document.

We have just completed the journey to build a narrative in Comet!

Summary

Throughout this chapter, you learned about the DIKW pyramid, with the related concepts of data, information, knowledge, and wisdom. You also learned how to turn your data into wisdom by building a story. You learned that to transform data into information, you should add meaning to your data, and to turn information into knowledge, you need to add context. Finally, to turn knowledge into wisdom, you should call your audience to action. In general, while building your story, you should always pay attention to your audience, who they are, and how they can understand your message.

In this chapter, you also learned how to build a story in Comet by using some concepts you already know – panels and reports. You learned how to share a report, either as a static PDF document or an interactive dashboard. You also learned how to build a panel in JavaScript.

Finally, you implemented two practical examples, which demonstrated how you can use Comet to build a narrative from data.

Data narrative is one of the final steps of a data science project life cycle. The other step is a deployed model. In the next chapter, you will review some basic concepts regarding DevOps, which will permit you to deploy your model.

Further reading

- Berengueres, A. F. J., and Sandell, M. (2019). *Introduction to Data Visualization & Storytelling: A Guide for the Data Scientist*

- Knaflic, C. N. (2015). *Storytelling with data: A data visualization guide for business professionals*. John Wiley and Sons

- Knaflic, C. N. (2019). *Storytelling with Data: Let's Practice!*. John Wiley and Sons

- Kriebel, A., and Murray, E. (2018). *# MakeoverMonday: Improving How We Visualize and Analyze Data, One Chart at a Time*. John Wiley & Sons

6

Integrating Comet
into DevOps

Model **deployment** and **monitoring** are the last two steps in a data science project life cycle. The former permits you to move your project from testing to production, and the latter provides with you all the strategies and tools to ensure that your project is running without errors, secure, and updated.

You can implement model deployment by adopting a particular philosophy, called **DevOps**. DevOps (short for **Development and Operations**) is a set of best practices that permit software developers and operations teams to collaborate during the whole software project life cycle to improve software development, speed, and efficiency through automatic techniques.

As you have already learned from the previous chapters, Comet permits you to track and monitor all your experiments, thus you can use it during the monitoring phase. So, you can easily integrate Comet into the DevOps strategy to monitor your model during the production phase.

In this chapter, you will review the basic concepts behind DevOps, including their principles and best practices. In addition, you will review the concept of **MLOps**, that is, DevOps applied to **machine learning (ML)**. Then, you will learn how to integrate Comet in the DevOps/MLOps philosophy, through the concept of a REST API. Next, you will analyze Docker and Kubernetes, two of the most common tools and frameworks used to implement DevOps. Finally, you will implement two practical examples, one in Docker and the other in Kubernetes, both integrating Comet.

The chapter is organized as follows:

- Exploring DevOps and MLOps principles and best practices
- Combining Comet and DevOps/MLOps
- Implementing Docker
- Implementing Kubernetes

Before moving toward the first step, let's install the software needed to run the code implemented in this chapter.

Technical requirements

The examples described in this chapter use the following software/tools:

- Python
- Docker
- Kubernetes

Python

We will use Python 3.8. You can download it from the official website (`https://www.python.org/downloads/`) and choose version *3.8*.

The examples described in this chapter use the following Python packages:

- `pandas 1.3.4`
- `scikit-learn 1.0`
- `comet-ml 3.23.0`
- `requests 2.27.1`
- `Flask 2.1.1`

We have already described the first two packages and how to install them in *Chapter 1, An Overview of Comet*, so please refer back to that for further details on installation.

`requests` is a Python package for managing HTTP requests simply. You can install it by running the following command:

```
pip install requests
```

For more details, you can read the `Requests` official documentation, available at the following link: `https://docs.python-requests.org/en/latest/`.

`Flask` is a Python package for creating an HTTP server. You can install it by running the following command:

```
pip install Flask
```

For more details, you can read the `Flask` official documentation, available at the following link: `https://flask.palletsprojects.com/en/2.1.x/`.

Docker

Docker is an open platform used to build and run applications. In this chapter, we will use Docker Desktop. You can download the Docker Desktop from its official website available at this link: `https://docs.docker.com/get-docker/`. Depending on your operating system, you should download a different file, but the procedure is always the same: you download the software, you install it, and then you run it. Once installed, you can run it as you usually do for the other applications.

If you are not able to install Docker Desktop, you can follow the alternative procedures described in the Docker official documentation, available at the following links: `https://docs.docker.com/desktop/`, `https://docs.docker.com/engine/install/`, and `https://docs.docker.com/desktop/windows/troubleshoot/`.

Kubernetes

Kubernetes is an open source platform used to manage containerized applications. Its official documentation is available at the following link: `https://kubernetes.io/docs/home/`.

In this chapter, we will use the local version of Kubernetes, provided by Docker Desktop.

Once you have installed Docker Desktop, you can enable Kubernetes as follows:

1. Launch Docker Desktop and access its main dashboard.
2. Select the **Settings** button, located in the top-right part of the dashboard.
3. Select **Kubernetes | Enable Kubernetes | Apply & Restart**.

Now, you should be able to run Kubernetes on your local computer.

Now that you have installed all the software required for this chapter, we can move to the next step, exploring DevOps and MLOps principles and best practices.

Exploring DevOps and MLOps principles and best practices

A traditional approach to moving software from testing to production includes two separate steps:

1. **Development** – Developers implement their code and test it in their local environment and under their local conditions.
2. **Operations** – When the code is ready, developers send the code to the operations teams, who are responsible for installing and maintaining the code in the production machine.

This approach requires a great effort from both teams – developers and operations teams – because, on the one hand, a little change in the code requires a new installation and configuration in the production machine. On the other hand, when the operations team discovers an anomaly in the code, they should communicate with the development team to run new tests, and then the process may become very slow.

If the number of tasks is small, a manual process could be acceptable, but if you have millions of tasks, this manual approach becomes infeasible.

And here is where DevOps comes in. DevOps is the acronym for software development (**Dev**) and IT operations (**Ops**). DevOps is a mentality, an approach, which tries to define a set of best practices to make development and operations teams work together, with a little effort, automatically and quickly, while preserving high quality in software delivery and guaranteeing software scalability.

In an ML environment, DevOps becomes MLOps, which aims at deploying and maintaining ML models in production.

In this section, you will review the following concepts:

- The DevOps life cycle
- Moving from DevOps to MLOps

Let's start with the first one, the DevOps life cycle.

The DevOps life cycle

The DevOps life cycle includes eight stages, as shown in the following figure:

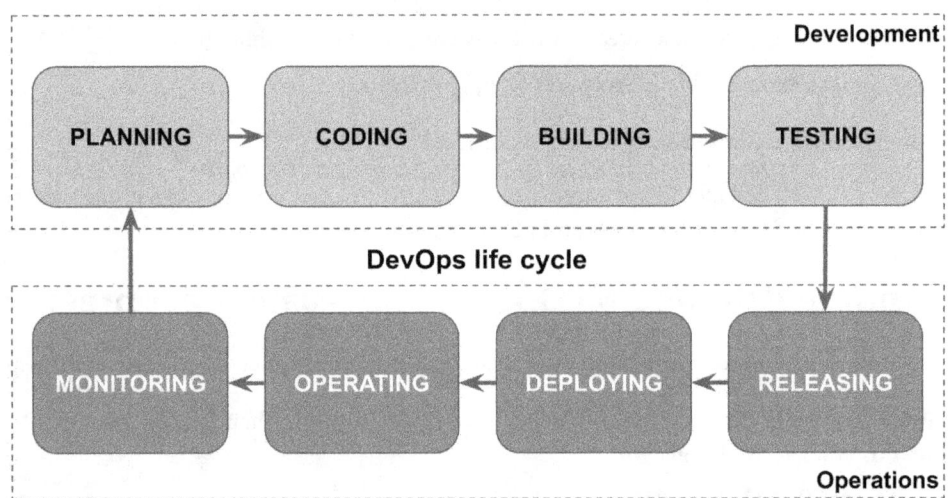

Figure 6.1 – The DevOps life cycle

The DevOps life cycle implements the eight stages as a continuous loop, where every iteration improves the previous one by adding new features or solving some problems. The eight stages are as follows:

1. **Planning** – Describing the work to do

2. **Coding** – Writing the code that does the work

3. **Building** – Compiling the code into runnable software

4. **Testing** – Verifying the correctness of the software by running some tests

5. **Releasing** – Publishing the runnable software into a shared and centralized repository

6. **Deploying** – Making the software available to users

7. **Operating** – Managing the infrastructure and software platforms

8. **Monitoring** – Controlling the infrastructure and software status

The first four steps (from planning to testing) refer to the **development** part, and the last four steps refer to the **operations** part (from releasing to monitoring). You have already learned how to implement the development part in the previous chapters. In this chapter, you will learn how to implement the deploying, operating, and monitoring stages, while in *Chapter 7, Extending the GitLab DevOps Platform with Comet*, you will learn about releasing. We have decided to discuss the releasing stage in a separate chapter, because it constitutes the connection between the development and operation parts, and requires particular attention by data scientists.

Typically, the deploying stage involves wrapping the generated software into an image that you can run in a standalone container. In this chapter, you will use Docker as the platform to build images and containers. The operating stage requires a platform that orchestrates all the possible containers. In this chapter, you will use Kubernetes as the orchestration platform. The monitoring stage involves all the tools to track critical issues and similar aspects. In this chapter, you will use Comet to track the issues related to ML models.

Now that you have learned the DevOps life cycle, we can move to the next step, moving from DevOps to MLOps.

Moving from DevOps to MLOps

Machine learning operations (**MLOps**) is a specialization of DevOps in the ML field. Thus, while DevOps is a generic name used for all the software development best practices, MLOps has a specific focus on ML.

You can consider the MLOps and DevOps life cycles quite similar. However, the MLOps life cycle also includes two additional stages: **data wrangling** and **modeling**, as shown in the following figure:

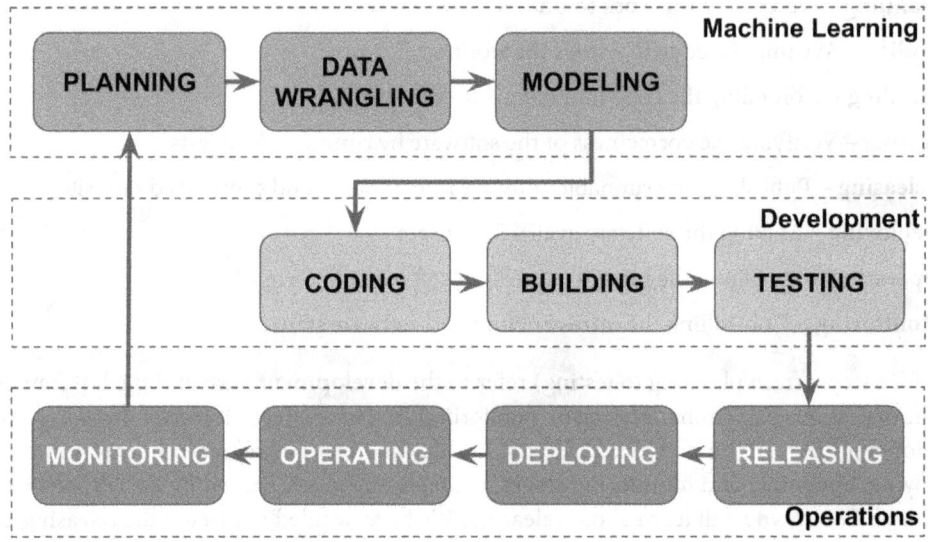

Figure 6.2 – The MLOps life cycle

Data wrangling involves data collection, cleaning, and preparation, while the modeling phase allows you to test different models and then choose the candidate model to send to production. You can provide the candidate model as input to a prediction application or service that will use it during the production phase. For example, if you want to make an image recognition system, then you need to define a model that classifies the images, and a prediction service that uses the model to recognize unknown images.

This means that you should not only monitor the prediction service, but you also have to ensure that the model behaves well and does not degrade over time. If the model degrades over time, you have to implement an automatic technique that updates both the model and the prediction service that uses it. And here is where Comet comes in.

In the next section, you will learn how to integrate Comet in the DevOps/MLOps life cycle.

Combining Comet and DevOps/MLOps

You can integrate Comet with the deployment process to monitor and keep track of your experiments. In addition to the techniques described in the previous chapters, Comet also provides a REST API service to access your projects, workspaces, experiments, and assets. You can easily integrate the REST API service with the other services provided by DevOps and MLOps platforms, as you will see later in this chapter.

The section is organized as follows:

- Comet in the DevOps life cycle
- Setting up the Comet REST API
- Using the Comet REST API

Let's start with the first point, Comet in the DevOps life cycle.

Comet in the DevOps life cycle

Let's suppose that you have already built a model and now you want to refine it, to improve its performance. If you do not use any tracking system, it becomes very easy to lose changes in your code, which are the hyperparameters for your model, and even the final artifacts.

A very low-level solution to keep order in your project evolution could be the use of a spreadsheet where you manually track all the changes in code, hyperparameters, and artifacts. However, this procedure could explode until it becomes very complicated and difficult to manage.

Here is where Comet comes to your aid. Comet provides the tools to track all changes in your code, experiments, and so on, in order to be easily integrated with the deployment phase. The following figure shows how you can integrate Comet with the deployment phase:

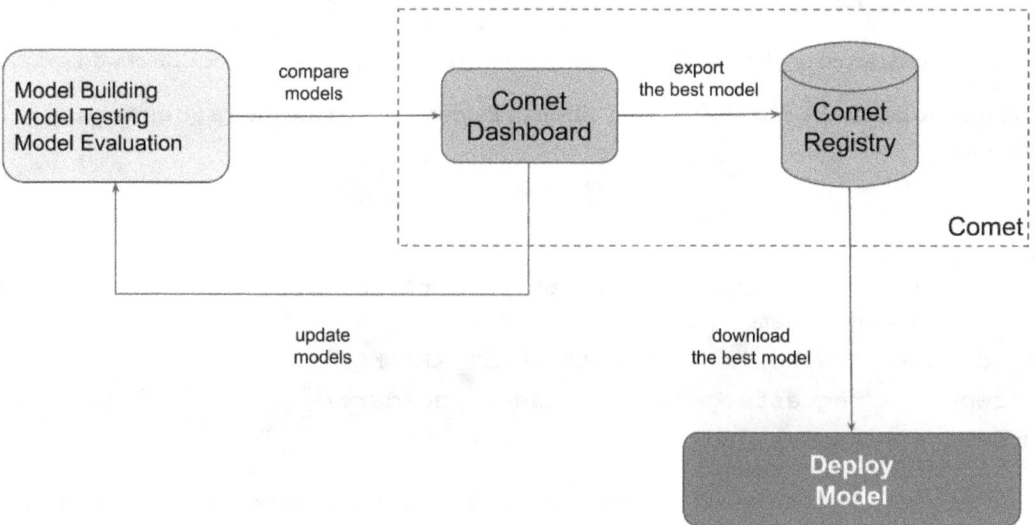

Figure 6.3 – Comet integration with deployment

You start by building, testing, and evaluating your candidate models. Then, you can save your experiments in Comet and compare them to choose the best model. You can save the best model in the Comet Registry, and, when it is ready, you can mark its stages as *production*. From the Comet Registry, you can download the best model, and deploy it as you want.

To download the best model, you can use the techniques you have already learned in the previous chapters or the Comet REST API service. In the next section, you will learn how to use the Comet REST API service.

Setting up the Comet REST API service

The Comet REST API service is available at the following URL: `https://www.comet.ml/api/rest/v2`.

To communicate with the service, you need to set up the HTTP Authorization header with your Comet API key. For example, if you want to access the REST API service in a terminal through `curl`, you can use the following command:

```
COMET_API_KEY='<MY_COMET_API>'
COMET_WORKSPACE='<MY_COMET_WORKSPACE>'
curl -s "https://www.comet.ml/api/rest/v2/registry-
model?workspaceName=$COMET_WORKSPACE" -H "Authorization:
$COMET_API_KEY"
```

The previous code accesses all the models contained in the registry related to a specific Comet workspace.

If you want to access the REST API service through Python, you can use the `requests` package, as follows:

```
import requests
COMET_API_KEY='<MY_COMET_API>'
url = f"https://www.comet.ml/api/rest/v2/registry-
model?workspaceName={COMET_WORKSPACE}"
headers = {"Authorization": f"{COMET_API_KEY}"}
response = requests.get(url, headers=headers)
models = response.json()
```

We use the `get()` function provided by the `requests` package, and then we parse the response as a JSON, through the `json()` function.

The Comet REST API service provides the following main methods:

- `/workspaces` – The list of all workspaces
- `/projects?workspaceName=myWorkspace` – The list of all projects within a workspace

- `/registry-model?workspaceName=myWorkspace` – The list of all registry models within a workspace

- `/registry-model/details?workspaceName=myWorkspace&modelName=myModelName` – The details related to a specific model within a specific workspace

- `/project?projectName=myProject&workspaceName=myWorkspace` – The details related to a specific project within a specific workspace

- `/experiments?projectId=myProjectId&archived=false` – The experiments within a project

- `/experiment/metadata?experimentKey=myExperimentAPIKey` – The metadata associated to a specific experiment

- `/registry-model/item/download?workspaceName=myWorkspace&modelName=myModel&version=myVersion` – To download a specific model

You can find the list of all the methods provided by the Comet REST API service in the Comet official documentation, available at this link: `https://www.comet.ml/docs/v2/api-and-sdk/rest-api/overview/`.

Now that you have learned some basic concepts about the Comet REST API service, let's move onto a practical example.

Using the Comet REST API

You will implement a scenario that uses the Comet REST API to download the *production* model from the Model Registry. The production model has a stage set for production. As an example, we will use the models saved in the Comet Registry, implemented in *Chapter 3*, *Model Evaluation in Comet*. You can download the code of this example from the official repository of this book, available at the following link: `https://github.com/PacktPublishing/Comet-for-Data-Science/tree/main/06/comet-rest-api`.

Let's begin:

1. First, we import all the needed libraries, as follows:

```
import os
import requests
import zipfile
```

2. Then, we import the Comet configuration parameters as environment variables:

```
COMET_API_KEY = os.environ.get("COMET_API_KEY")
COMET_WORKSPACE = os.environ.get("COMET_WORKSPACE")
```

We could have used the classic configuration file method, but using environment variables will come in handy later, as we will see when using Kubernetes.

3. We define the base name for each model, as well as the list of models to download:

```
base_project = "diamonds"
models = ["model", "color-feature-label-
encoder","clarity-feature-label-encoder","scaler",
"label-encoder"]
output_dir = "models"
```

We also set the output directory name to models.

4. For each model, we search for the version of the model associated to the stage production, then we download the compressed file related to that version, unzip the file, and save the model in output_dir. Eventually, we delete the original zipped file. The following piece of code implements the described steps:

```
for model_name in models:
    model_name = f"{base_project}-{model_name}"
    url = f"https://www.comet.ml/api/rest/v2/
registry-model/details?workspaceName={COMET_
WORKSPACE}&modelName={model_name}"
    headers = {"Authorization": f"{COMET_API_KEY}"}
    response = requests.get(url, headers=headers)
    model_details = response.json()
    for version in model_details['versions']:
        for stage in version['stages']:
            if stage == 'production':
                model_version = version['version']
                link = f"https://www.comet.ml/api/rest/
v2/registry-model/item/download?workspaceName={COMET_
WORKSPACE}&modelName={model_name}&version={model_
version}"
                path = f"{output_dir}/model.zip"
                model_resp = requests.get(link,
headers=headers)
                with open(path, 'wb') as f:
                    f.write(model_resp.content)
                with zipfile.ZipFile(path,"r") as zip_
ref:
```

```
                    zip_ref.extractall(output_dir)
            os.remove(path)
    break
```

Here are some notes on the previous code:

- We use the `https://www.comet.ml/api/rest/v2/registry-model/details?workspaceName={COMET_WORKSPACE}&modelName={model_name}` call to retrieve all the details related to a specific model. The objective is to retrieve the version of the model corresponding to the production stage.

- For a given model, we loop over all the available versions and all the available stages to check whether there is at least one stage corresponding to production. If so, we download it and put it in `output_dir`.

- Once we have downloaded a model, we unzip it through the `ZipFile()` class provided by the `zipfile` standard Python package, and we put it in `output_dir`. We also remove the original zipped file through the `os.remove()` function.

5. You can save the script as `get_models.py`. To run it, first, you need to configure the environment variables, as follows. Open a terminal and write the following code:

```
export COMET_API_KEY=<MY_API_KEY>
export COMET_WORKSPACE=<MY_WORKSPACE>
```

6. Now, you can create a directory named `models`, where your script will save the models. From your working directory, you can run the following command:

```
python get_models.py
```

7. In the `models` directory, you should see the downloaded models.

Now that you have learned how to use the Comet REST API service, we can move to the next step, implementing Docker.

Implementing Docker

Docker is one of the most popular containerization platforms, which constitutes the basic building block to run DevOps applications. **Podman** and **BuildKit** are other popular containerization platforms that are available as alternatives to Docker. In general, a containerization platform permits you to run a single application in an isolated environment. In Docker, this is achieved by wrapping your application in a **Docker container**, which contains your code, the operating system, and all required libraries used by your code. Within a Docker container, you can run a **Docker image**, which is the object in your filesystem that wraps your application.

As an alternative to Docker, you can use a virtual machine, which also provides an isolated environment and a ready-to-run code for your application. However, Docker is much smaller and easier to port than virtual machines.

As with any application, a Comet experiment can also be launched in a Docker container. In this section, you will implement a practical example that combines Comet and Docker.

This section is organized as follows:

- Overview of Docker
- Running Comet in a Docker container

Let's start from the first point, the overview of Docker.

Overview of Docker

The Docker platform is a client-server-based architecture, as shown in the following figure:

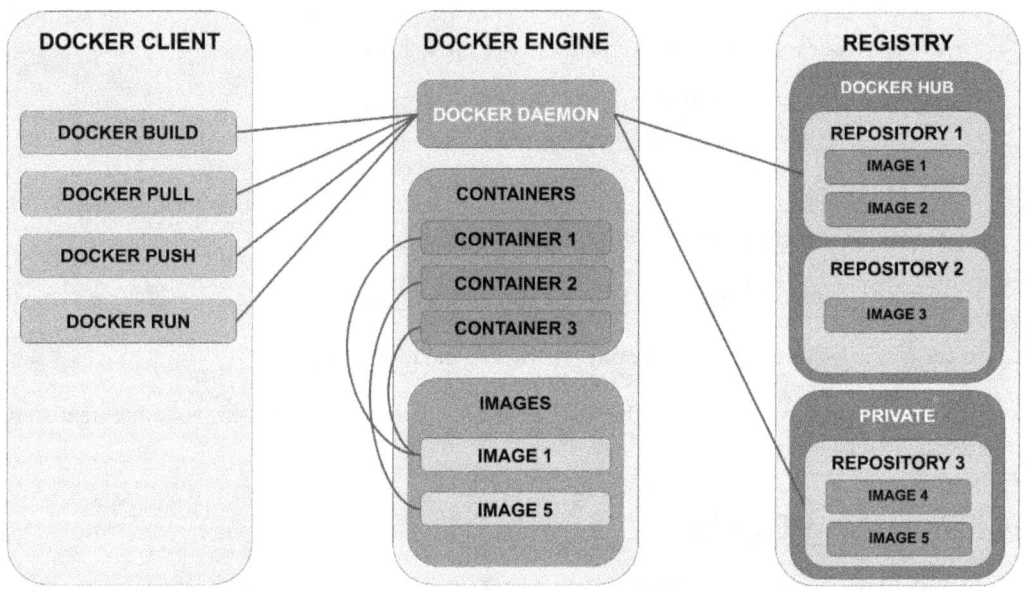

Figure 6.4 – The Docker architecture

The Docker platform is composed of the following main components:

- The **Docker client** – The client of the system, running either a command-line-based interface or using a Docker API. The Docker client runs four main commands:

  - `docker build` – To build a Docker image

  - `docker pull` – To pull an image from the registry

  - `docker push` – To push an image from the registry

  - `docker run` – To run an image as a container

- **Docker Engine**, which builds the images, and manages and runs the containers. It contains the **Docker daemon**, which is the running server, and the created containers and images. You can run more than one container with the same image. For example, image 1 is associated with both containers 1 and 3.

- The **Registry**, which contains public and private Docker images. The Registry is composed of **Docker Hub**, which contains public repositories and images, and private registries, which contain private repositories and images, which you can also use to share with your team.

Now that we have reviewed some basic concepts regarding Docker, we can implement a practical example, which wraps Comet in a Docker container.

Running Comet in Docker container

As a use case, we use a simplified version of the example implemented in *Chapter 3*, *Model Evaluation in Comet*. The objective is to build a standalone application that tests four different classification models (random forest, decision tree, Gaussian naive Bayes, and k-nearest neighbors), which classify cut diamonds into two categories (gold and silver) on the basis of some input parameters. In addition, the application logs all the models in Comet. As auxiliary functions, we use those described in *Chapter 3*, *Model Evaluation in Comet*.

You can download the code of this example from the GitHub repository of the book, available at the following link: `https://github.com/PacktPublishing/Comet-for-Data-Science/tree/main/06/docker-example`.

This example follows the workflow shown in the following figure:

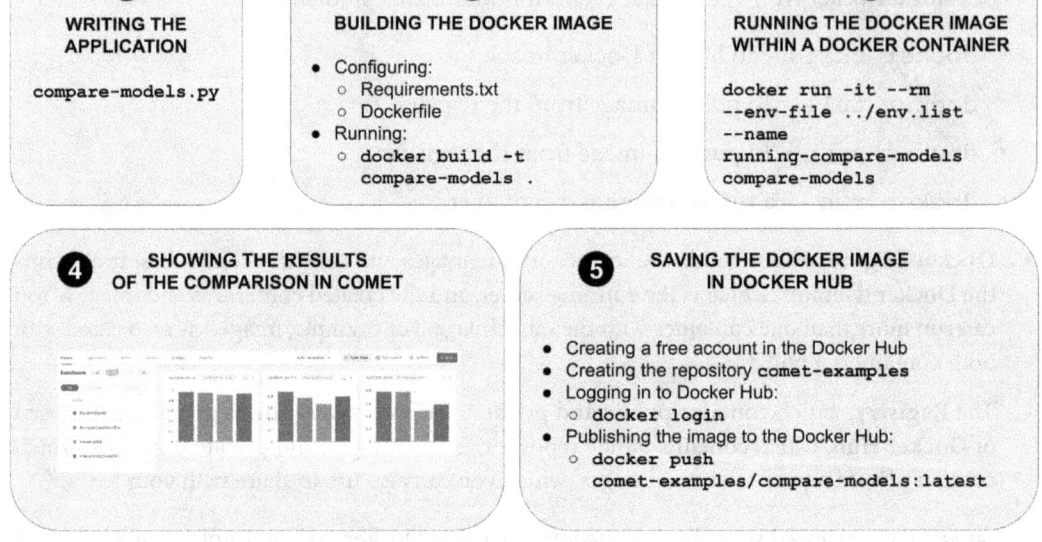

Figure 6.5 – The workflow followed by the use case

There are the following four steps:

- **WRITING THE APPLICATION**: Here is where you write your code that compares the four classification models and logs them in Comet:

1. Since you cannot encapsulate secrets in a Docker image, you should pass them to the container when you run it. In this example, we create an environment file, named `env.list`, containing all the configuration parameters for Comet, as follows:

    ```
    COMET_API_KEY=MY_COMET_API_KEY
    COMET_PROJECT=MY_COMET_PROJECT_NAME
    COMET_WORKSPACE=MY_COMET_WORKSPACE
    ```

2. Now, you are ready to write your application. We create a file, named `compare-models.py`, and we edit it as follows. First, we import all the required libraries:

    ```
    import os
    from comet_ml import Experiment
    from sklearn.model_selection import train_test_split
    # Import the scikit-learn models
    import pandas as pd
    ```

You should make sure to import `comet_ml` before any ML library.

3. Then, we retrieve the environment variables:

```
COMET_API_KEY = os.environ.get("COMET_API_KEY")
COMET_WORKSPACE = os.environ.get("COMET_WORKSPACE")
COMET_PROJECT = os.environ.get("COMET_PROJECT")
```

We use the `os.environ` object to retrieve the environment variables.

4. Now, we copy within the file all the functions described in *Chapter 3, Model Evaluation in Comet*. We only modify the `run_experiment()` function to make Comet read the environment variables:

```
experiment = Experiment(
        api_key=COMET_API_KEY,
        project_name=COMET_PROJECT,
        workspace=COMET_WORKSPACE,
    )
```

When we create the experiment, we pass it the configuration parameters and read from the environment variables.

5. Finally, we build the `main` function as follows:

```
if __name__ == "__main__":
    df = pd.read_csv('source/diamonds.csv')
    df = df.drop(["Unnamed: 0"], axis=1)
    df['target'] = df['cut'].apply(lambda x: set_
target(x))
    df.drop("cut", axis = 1,inplace=True)
    X = df.drop("target", axis = 1)
    y = df["target"]
    encode_labels(X)
    scale_numerical(X)
    label_encoder = LabelEncoder()
    y = label_encoder.fit_transform(y)
    X_train, X_test, y_train, y_test = train_test_
split(X, y, test_size=0.20, random_state=42)
    run_experiment(RandomForestClassifier,
'RandomForest')
    run_experiment(DecisionTreeClassifier,
```

```
'DecisionTreeClassifier')
    run_experiment(GaussianNB, 'GaussianNB')
    run_experiment(KNeighborsClassifier,
'KNeighborsClassifier')
```

We have already described the code in *Chapter 3, Model Evaluation in Comet*. The only difference is that we have added the line if __name__ == "__main__": to run the code *only* if it is called directly.

- **BUILDING THE DOCKER IMAGE**: You can create a new directory named docker-example and put the compare-models.py script in that directory. To transform your application into a Docker image, you need to define two files within the same directory: requirements. txt and the Dockerfile.

6. The requirements.txt file contains the list of Python packages to install. In our example, the file contains the following libraries:

```
comet_ml
numpy
pandas
scikit-learn
```

Notice that you do not need to install these libraries on your local machine because Docker will do it for you!

7. The Dockerfile file contains the instructions used by Docker to build the Docker image. In our example, the Dockerfile is very simple, as shown by the following code:

```
FROM python:slim
WORKDIR /code
COPY requirements.txt ./
RUN pip install --no-cache-dir -r requirements.txt
COPY . .
CMD ["python", "./compare-models.py"]
```

Here are some comments on the preceding code:

- The FROM directive specifies the base image upon which Docker should build the image. In our example, we use python:slim, which is downloaded from Docker Hub. You can check all the official Python images provided by Docker Hub at the following link: https://hub.docker.com/_/python.

- The WORKDIR directive specifies the directory where the application is executed within the container. This directive first creates the directory if it does not exist, and then moves it inside the created directory.

- The COPY directive copies the source into a destination. In our example, it copies the files contained in the current directory of our filesystem into the code directory of the Docker image.

- The RUN directive executes the specified commands during the building process. In our case, it installs the packages contained in the requirements.txt file.

- The CMD directive specifies the entry command to run when you run the image within a container. In our case, it runs the compare-models.py application.

8. Once you have configured the parameters, you can build the Docker image by running the following command within the docker-example directory:

```
docker build -t compare-models .
```

The -t part specifies the image name. Do not forget to also specify the directory containing the application, which is specified by the final . (dot) in our example. You can view the list of all images contained in your local Docker registry by running the following command:

```
docker images
```

The following piece of code shows the output of the command in our example:

```
REPOSITORY        IMAGE ID       CREATED        SIZE     TAG
compare-models df9200aae1d2   5 minutes ago   497MB latest
```

You can notice the size of the image, which is 497 MB. If we used another base image, such as the Python 3.8 image, the size of the image would have been much bigger.

- **RUNNING THE DOCKER IMAGE WITHIN A DOCKER CONTAINER**: Now, you can run the application from anywhere on your filesystem using the following command:

```
docker run --rm --env-file /path/to/env.list --name
running-compare-models compare-models
```

Here are some notes on the docker run command arguments:

- The --rm argument specifies that the container must be removed once it stops running.

- The --env-file argument specifies the path to the environment file, previously defined.

- The --name argument specifies the container name.

- **SHOWING THE RESULTS OF THE COMPARISON IN COMET**: Eventually, you can access the Comet dashboard as you usually do. Select the project and you will see the results, as shown in the following figure:

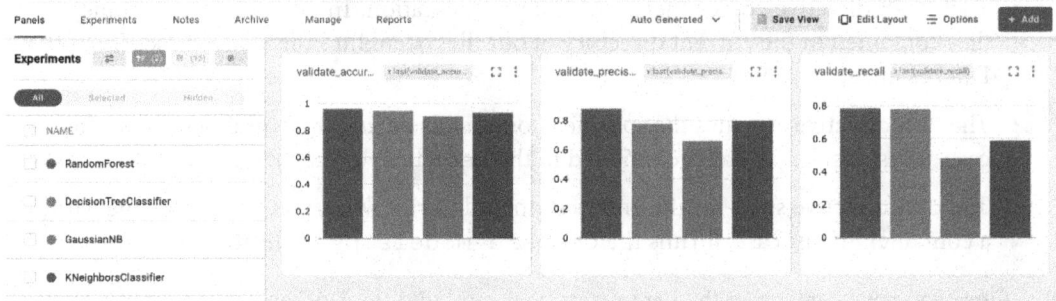

Figure 6.6 – The output of the application in Comet

You can see that using Docker does not affect the final result in Comet!

- **SAVING THE DOCKER IMAGE IN DOCKER HUB**: The last step involves pushing the Docker image into Docker Hub:

9. First, you need to create a free account at the following link: `https://hub.docker.com/`.

10. From the command line, log in to Docker Hub through the following command:

```
docker login
```

The command asks you for your username and password.

11. Add at least one tag to your image, as follows:

```
docker tag compare-models compare-models:v1
```

We have added the `v1` tag. The use of tags permits you to download a specific version of the image. If you do not specify any tag, the `latest` tag is automatically added to your image.

12. You should also tag the Docker image in Docker Hub, as follows:

```
docker tag compare-models:v1 my_account/compare-models:v1
```

13. Finally, you can save the Docker image in Docker Hub, as follows:

```
docker push my_account/compare-models:v1
```

14. You can download the Docker image through the following command:

```
docker pull my_account/compare-models
```

Now that you have integrated your Comet application with Docker, we can move to the next step, implementing Kubernetes.

Implementing Kubernetes

Kubernetes is an open source platform for managing containers and ensuring that they are running properly and consistently. It uses the principles of Docker to help you orchestrate your containers and workloads. Kubernetes can manage a cluster of containers, and make them communicate with each other. In addition, it can select the best place to run a container, ensuring that they are always in the right place so that they can be recovered if they fail.

This section is organized as follows:

- The Kubernetes architecture
- Configuring Kubernetes
- Deploying a local Kubernetes cluster

Let's start with the first step, the Kubernetes architecture.

The Kubernetes architecture

The following figure shows the Kubernetes basic architecture:

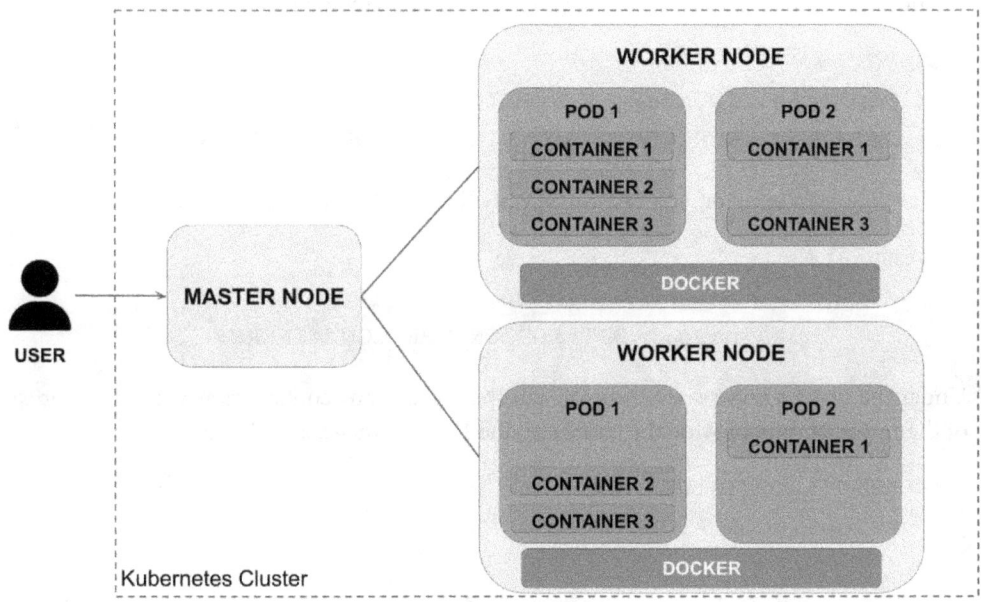

Figure 6.7 – The Kubernetes basic architecture

The Kubernetes cluster is composed of two main components: the **master Node**, which is responsible for the main orchestration and infrastructure management, and one or more **worker Nodes**, which run the applications and the services. Every worker Node contains a Docker engine, which runs applications as containers. Each worker Node runs one or more Pods. A **Pod** is the smallest deployable object in Kubernetes. It can run one or more containers.

Now that you have learned the basic Kubernetes architecture, we can move to the next step, configuring Kubernetes.

Configuring Kubernetes

In Kubernetes, you can configure different types of objects using the YAML language. In this chapter, we consider two types of objects: Deployment and Service.

Deployment

You can use a **Deployment** to configure how to update a Pod:

1. To define a Deployment in the YAML file, you should write the following code:

    ```
    apiVersion: apps/v1
    kind: Deployment
    ```

2. You can specify the number of replicas (Pods) for the current Deployment, as well as the criteria to find the Deployment in the cluster (for example, a matching label):

    ```
    spec:
      replicas: <NUMBER_OF_REPLICAS>
      selector: <CRITERION_TO_FIND_THE_DEPLOYMENT>
    ```

3. Under the `spec` selector, you can also specify the template for the Deployment, as follows:

    ```
    template:
      spec:
          containers: <DETAILS_ON_THE_CONTAINERS>
    ```

4. Under the `containers` selector, you can specify as many containers as you want. For each of them, you should provide the name and the Docker image, as follows:

    ```
    - name: <CONTAINER_NAME>
      image: <CONTAINER_IMAGE>
    ```

Next, we'll look at the Service option.

Service

A **Service** permits you to expose an application running on one or more containers as a network service:

1. To define a Service in the YAML file, you should write the following code:

    ```
    apiVersion: v1
    kind: Service
    ```

2. Usually, the YAML contains the spec selector, which specifies the ports to expose to the external world, as well as the type of Service:

    ```
    spec:
      ports:
        - port: <EXTERNAL_PORT>
          protocol: TCP
          targetPort: <CONTAINER_PORT>
      type: <TYPE_OF_SERVICE>
    ```

    The type selector specifies the type of Service to implement. Examples of types include LoadBalancer and ClusterIP. For more details, you can refer to the Kubernetes official documentation, available at the following link: https://kubernetes.io/docs/concepts/services-networking/service/.

Now that you have learned the basic concepts behind Kubernetes, we can move to a practical example, which deploys a local Kubernetes cluster.

Deploying a local Kubernetes cluster

As a use case, you will implement a web service that predicts the diamond cut (silver or gold) on the basis of some input features. The web service downloads the model contained in the Comet Registry, through the code implemented in this chapter, in the *Using the Comet REST API* section.

You can download the code used in this example from the official GitHub repository of this book, available at the following link: https://github.com/PacktPublishing/Comet-for-Data-Science/tree/main/06/kubernetes-example.

The following figure shows the architecture of the implemented use case:

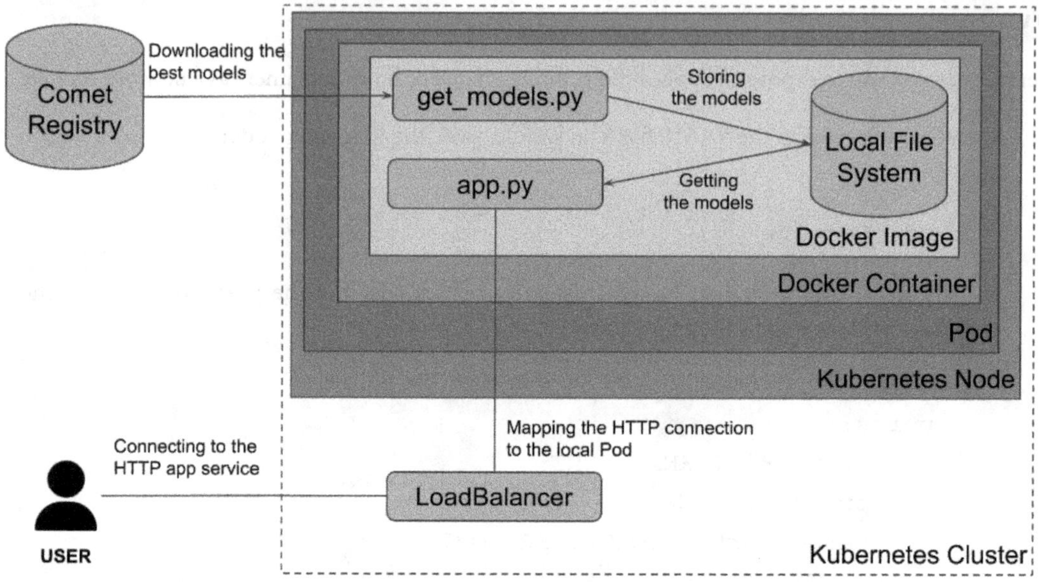

Figure 6.8 – The scenario architecture

Starting from the top-right corner of the figure, you can see the Comet Registry, which stores all the updated models. The `get_models.py` script downloads the best model (corresponding to the production stage) from the Comet Registry. You have already implemented the code contained in the `get_models.py` script in the previous section, *Using the Comet REST API*. This script stores the downloaded model in a local filesystem. The `app.py` script implements a web server that receives the diamond features from the web interface, then loads the models from the local filesystem, and predicts the diamond cut. The `get_models.py` and `get_models.py` scripts, as well as the local filesystem, are wrapped into a Docker image, which you can use to build a Docker container, which is then wrapped in a Pod. Finally, you can build a Kubernetes cluster with just one Pod. The load balancer, also contained in the Kubernetes cluster, forwards the external HTTP connections to local HTTP connections for the `app.py` script.

Now, you are ready to implement the described architecture. To deploy it, you can perform the following steps:

- Write the prediction service.

- Wrap the prediction service in a Docker image.

- Create the Kubernetes cluster.

Let's start with the first step, writing the prediction service.

Writing the prediction service

The prediction service is implemented by the `app.py` script:

1. First, we import all the required packages:

```
import pandas as pd
import pickle
import os
from flask import Flask, render_template, request
```

We import the `flask` package since you will implement the web server in `flask`.

2. Then, we define an auxiliary function, which loads a model from the filesystem, as follows:

```
def load_model(file_name):
    f = open(file_name, 'rb')
    unpickler = pickle.Unpickler(f)
    model = unpickler.load()
    f.close()
    return model
```

The function receives the filename as input and returns the model contained in it.

3. We initialize the `Flask` app:

```
app = Flask(__name__)
app.config["TEMPLATES_AUTO_RELOAD"] = True
```

We set the `TEMPLATES_AUT_RELOAD` configuration parameter to `True` to make sure that the app reloads the templates every time the page is loaded. This prevents template caching.

4. Now, we load all the models:

```
base_dir = os.path.abspath(os.path.dirname(__file__))
color_label_encoder = load_model(f"{base_dir}/models/
colorFeatureLabelEncoder.pkl")
clarity_label_encoder = load_model(f"{base_dir}/models/
clarityFeatureLabelEncoder.pkl")
scaler = load_model(f"{base_dir}/models/scaler.pkl")
model = load_model(f"{base_dir}/models/model.pkl")
label_encoder = load_model(f"{base_dir}/models/
labelEncoder.pkl")
```

We extract the project base directory, and then we access the `models` directory to retrieve all the models.

5. Now, we define the home page of our web server as follows:

```
@app.route('/')
def home():
    return render_template('index.html')
```

We set the route to `/` (the main page of the website) through the `@app.route('/')` statement. The `home()` function simply renders the template contained in the `index.html` page. This HTML page defines a form where the user can insert all the diamond features. You will see the content of this file later in this section.

6. The **Submit** button of the form contained on the `index.html` page triggers a call to the web server to calculate the diamond cut and show the result to the user. We define the handling function for the submit button as follows:

```
@app.route('/predict', methods=['GET', 'POST'])
def predict():
    params = request.form.to_dict()
    df = pd.DataFrame(params, index=[1])
    num_cols = df.columns.tolist()
    cat_cols = ['color', 'clarity']
    for col in cat_cols:
        num_cols.remove(col)
    df[num_cols] = df[num_cols].apply(pd.to_numeric,
errors='coerce')
    # prepare data
    df['color'] = color_label_encoder.
transform(df['color'])
    df['clarity'] = clarity_label_encoder.
transform(df['clarity'])
    df[df.columns] = scaler.transform(df[df.columns])
    target = model.predict(df)
    label = label_encoder.inverse_transform(target)
    return render_template('predict.html',
target=label[0] )
```

The `predict()` function extracts the form parameters through the `request.form.to_dict()` function, then it builds a DataFrame with all the parameters, converts numeric values to numbers, and transforms data by applying the encoders and the scalers.

Next, it calculates the target class through the `predict()` method provided by the model and extracts the original class by applying the `inverse_transform()` method of the label encoder. Finally, it sets the target variable in the `predict.html` template and returns it as result.

7. Finally, you can write the code to run the full app:

```
if __name__ == "__main__":
    app.run(host="0.0.0.0")
```

You make the app run on all the addresses available on the machine.

The `app.py` script renders two templates: `index.html` and `predict.html`. You can put both the `index.html` and `predict.html` files in a directory named `templates`, located at the same level of the `app.py` script. Let's describe both, starting with the first one, `index.html`.

8. The `index.html` file creates the HTML form with all the diamond features, as follows:

```
<div class="textbox">
        <form action="{{ url_for('predict') }}"
method='POST'>

            . . .


                <div class="btn-block">
            <button type="submit" href="/">CLASSIFY</
button>
            </div>
        </form>
    </div>
```

The `action` attribute of the form contains the following code: `"{{ url_for('predict') }}"`, which means that the `flask` server will automatically substitute it with the URL created for the `predict()` function.

9. You can create each numeric feature as an `input` tag within the form, as follows:

```
<div class="item">
    <label for="name">Carat<span>*</span></label><br/>
    <input id="carat" type="text" name="carat" required/>
</div>
```

We specify that the input tag is a text (`type="text"`) and it is required.

10. You can represent categorical features through a `select` tag, as shown in the following piece of code:

```
<div class="item">
    <label for="name">Color<span>*</span></label><br/>
    <select name="color">
        <option selected value="" disabled selected></
option>
        <option value="D" >D</option>
        <option value="E">E</option>
        <option value="F">F</option>
        <option value="G">G</option>
        <option value="H">H</option>
        <option value="I">I</option>
        <option value="J">J</option>
    </select>
</div>
```

For simplicity, we have wired in the code all the possible values of the category.

11. The `predict.html` file simply renders the predicted class, as follows:

```
<div class="textbox">
    <div class="mainbox">
        <h2>Diamond Classification</h2>
            <div id="target" class="target">
                {{ target }}
            </div>
    </div>
</div>
```

The `flask` server will substitute the `{{ target }}` variable with the actual value of prediction.

12. The app is ready. You can test whether it works by running the following command in a terminal in the same directory as your app:

```
flask run
```

The default web server should run at the following link: `http://localhost:5000`. If you point your browser at this link, you should see something like the following screenshot:

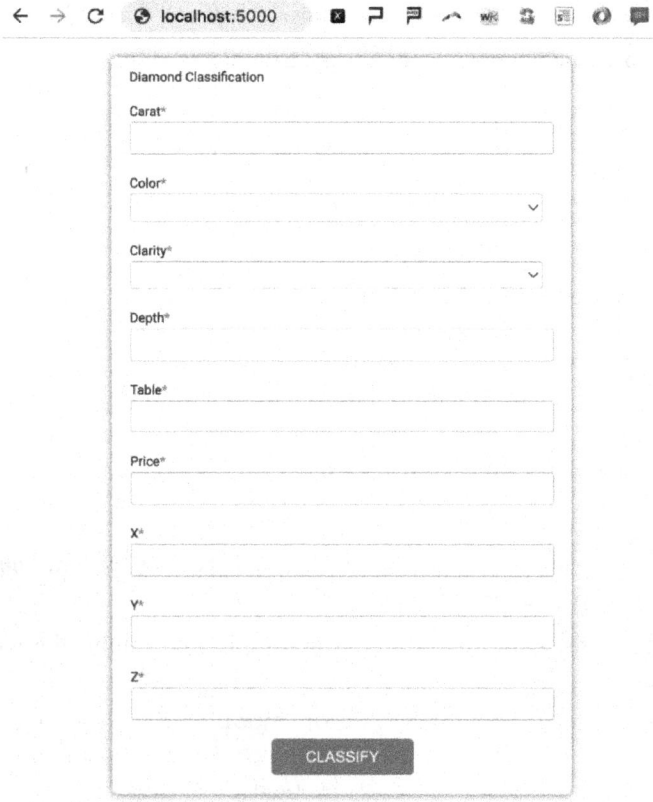

Figure 6.9 – The app running in the browser

> **Note**
>
> We have added some CSS style to the form, so do not worry if your form does not have exactly the style shown in the previous figure.

So far, you have built your app. Now it is time to wrap it into a Docker image.

Wrapping the prediction service in a Docker image

To wrap your app in a Docker image, you can proceed as follows:

1. To do it, we build an auxiliary script, named `run.sh`, which runs both the `get_models.py` and `app.py` scripts, as follows:

```
#!/bin/sh
python get_models.py
flask run --host=0.0.0.0
```

We have specified that the `flask` app will be available for all the available addresses.

2. Now, you need to define the `requirements.txt` file, as follows:

```
flask
pandas
requests
scikit-learn
```

3. You also need to define the Dockerfile, as follows:

```
FROM python:slim
WORKDIR /app
COPY . .
RUN pip install --no-cache-dir -r requirements.txt
CMD ["./run.sh"]
```

The Dockerfile is very similar to that described in the previous section, *Running Comet in a Docker container*, so you can refer to it for further details.

The following figure shows the final structure of the folder containing the app:

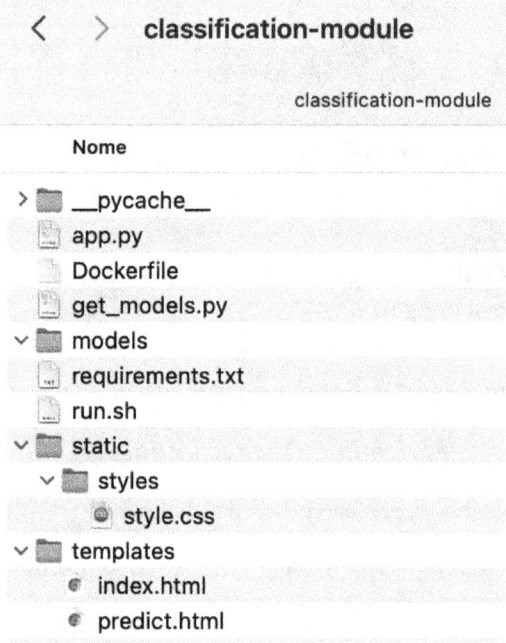

Figure 6.10 – The structure of the app folder

In addition to what you have already implemented, there is a static folder, which contains the CSS style file used to give a layout to the HTML templates.

4. You can run the `docker build` command to build your Docker image:

    ```
    docker build -t diamonds-prediction .
    ```

5. You can test whether the app works correctly by running the following command:

    ```
    docker run --rm --env-file ../../env.list -p 6002:5000
    --name cp diamonds-prediction
    ```

    The –p argument specifies a mapping between the local port in the container (5000) and the external port in the host (6002).

Finally, everything is ready to create the Kubernetes cluster! So let's move on to the final step, creating a Kubernetes cluster.

Creating the Kubernetes cluster

To make the app run in Kubernetes, you need to create three YAML configuration files:

*   `comet-secrets.yaml`
*   `deployment.yaml`
*   `app-service.yaml`

Let's analyze each configuration file separately, starting with the first one, `comet-secrets.yaml`.

comet-secrets.yaml

This configuration file stores the secrets to access Comet. We consider two secrets, `COMET_API_KEY` and `COMET_WORKSPACE`:

1. First, we encrypt them by running the following command in a terminal:

    ```
    echo -n 'MY_COMET_API_KEY' | base64
    ```

    You can copy the output of this command. You can run the same command for the Comet workspace:

    ```
    echo -n 'MY_COMET_WORKSPACE' | base64
    ```

2. Now, you can write the `comet-secrets.yaml` file, as follows:

    ```
    apiVersion: v1
    data:
      comet_api_key: MY_ENCRYPTED_COMET_API_KEY
    ```

```
        comet_workspace: MY_ENCRYPTED_COMET_WORKSPACE
    kind: Secret
    metadata:
      name: comet-secrets
    type: Opaque
```

Here are some comments on the previous code:

- You should specify all the secrets' names in lowercase.

- You will use the name, specified under the `metadata` selector, to retrieve the secrets from the other configuration files.

3. To create the secret, you can run the following command in a terminal:

```
    kubectl create -f comet-secrets.yaml
```

4. You can check whether the system has created the Deployment through the following command:

```
    kubectl get secrets
```

Let's move on.

Deployment.yaml

This configuration file defines the Pod and the associated containers. The following piece of code shows a possible implementation of the `deployment.yaml` file:

```
apiVersion: apps/v1
kind: Deployment
metadata:
  name: diamonds
  labels:
    app: dc
spec:
  replicas: 1
  selector:
    matchLabels:
      app: dc
  template:
    metadata:
      labels:
        app: dc
```

```
spec:
  containers:
    - name: prediction
      image: diamonds-prediction
      ports:
        - containerPort: 5000
      imagePullPolicy: Never
      env:
        - name : COMET_API_KEY
          valueFrom:
            secretKeyRef:
              key: comet_api_key
              name: comet-secrets
        - name : COMET_WORKSPACE
          valueFrom:
            secretKeyRef:
              key: comet_workspace
              name: comet-secrets
```

Here are some notes on the previous code:

- The deployment name is `diamonds` and its associated label is `dc` (diamond classification).

- Under the `spec` selector, we have set the number of replicas to 1. This means that we have just one Pod for this project. To select this Pod, we have used a matching criterion based on the label (`matchLabels:    app: dc`).

- The Pod contains just one container, called `prediction`. Its image is downloaded from the local registry (`imagePullPolicy: Never`).

- The container accepts connection at port 5000 (`containerPort:    5000`).

- The container receives as input three environment variables, whose value is taken from the Kubernetes secrets.

Let's go through it in steps.

1. To create the Deployment, you can run the following command in a terminal:

    ```
    kubectl create -f deployment.yaml
    ```

2. You can check whether the system has created the Deployment through the following command:

    ```
    kubectl get deployments
    ```

You should see the following output:

```
NAME        READY   UP-TO-DATE   AVAILABLE   AGE
diamonds    1/1     1            1           40h
```

Next, we'll look at the Service.

app-service.yaml

This configuration file defines the Service associated with the app. This file exposes the web server to the external world, thus it simply contains the mapping between the internal container port (to which the container listens) and the external port. The following piece of code shows an example of the app-service.yaml file:

```
apiVersion: v1
kind: Service
metadata:
  name: diamonds
  labels:
    app: dc
spec:
  ports:
    - port: 6002
      protocol: TCP
      targetPort: 5000
  type: LoadBalancer
  selector:
    app: dc
```

Here are some notes on the previous code:

- The kind selector specifies that we are defining a Service.
- Under the ports selector, we map the container port (targetPort) to the external port (port).
- We set the type of Service to LoadBalancer.

  Let's walk through it.

  I.   To create the Deployment, you can run the following command in a terminal:

  ```
  kubectl create -f app-service.yaml
  ```

II.   You can check whether the system has created the Deployment through the following command:

```
kubectl get services
```

You should see the following output:

```
NAME            TYPE          CLUSTER-IP      EXTERNAL-IP
PORT(S)         AGE
diamonds        LoadBalancer  10.108.230.69   localhost
6002:32590/TCP  41h
```

When you have created all the secrets, Deployments, and Services, you are ready to use your application. You can point your browser at the following address: `http://localhost:6002`.

As a further exercise, you can try to integrate the image deployed in the previous section, *Implementing Docker*, into Kubernetes.

Summary

We have just completed the journey to move your model from testing to production and use Comet to keep your model up to date!

Throughout this chapter, you have reviewed some basic concepts related to DevOps and MLOps, as well as the DevOps and MLOps life cycle. You have also learned how to integrate Comet in the DevOps/MLOps life cycle, and how to use it to keep track of the best model.

You have also reviewed the basic concepts behind Docker and Kubernetes, two popular platforms to build DevOps applications. Finally, you have implemented two examples: the first one integrated Comet in a Docker image, and the second one integrated Comet with an application deployed in Kubernetes.

In the next chapter, you will review some basic concepts behind releasing software, with a particular focus on GitLab and how to integrate it with Comet.

Further reading

- Arundel, J., and Domingus, J. (2019). *Cloud Native DevOps with Kubernetes: Building, Deploying, and Scaling Modern Applications in the Cloud*. O'Reilly Media.

- Gift N., Behrman, K., and Deza, A. (2019) *Python for DevOps: Learn Ruthlessly Effective Automation*. O'Reilly Media.

- Treveil, M., Omont, N., Stenac, C., Lefevre, K., Phan, D., Zentici, J., ... and Heidmann, L. (2020). *Introducing MLOps*. O'Reilly Media.

7

Extending the GitLab DevOps Platform with Comet

When you implement a data science project, you should consider that your model could age due to various factors, such as concept drift or data drift. For this reason, your project will probably need constant updates.

In the previous chapter, you learned the fundamental principles related to DevOps, which allow you to move the project from the test phase to the production phase. However, this is not enough to deal with constant updates efficiently. In fact, you may need to develop an automatic or semi-automatic procedure that allows you to pass from the build/test phase to the production phase easily without too many manual interventions.

In this chapter, you will review the basic concepts behind **Continuous Integration and Continuous Delivery (CI/CD)**, two strategies that permit you to easily and automatically update your code and move from building/testing to production efficiently. You will also learn how you can implement CI/CD using GitLab, a very popular platform for software management. Then, you will configure GitLab to work with Comet. Finally, you will see a practical example that will help you get familiar with the described concepts.

The chapter is organized as follows:

- Introducing the concept of CI/CD
- Implementing the CI/CD workflow in GitLab
- Integrating Comet with GitLab
- Integrating Docker with the CI/CD workflow

Before moving on to the first step, let's install the software needed to run the code implemented in this chapter.

Technical requirements

The examples described in this chapter use the following software/tools:

- Python
- Git client

Python

We will use Python 3.8. You can download it from the official website at `https://www.python.org/downloads/` and choose version 3.8.

The examples described in this chapter use the following Python packages:

- `comet_ml 3.23.0`
- `pandas 1.3.4`
- `scikit-learn 1.0`
- `requests 2.27.1`
- `Flask 2.1.1`

We described the first two packages and how to install them in *Chapter 1, An Overview of Comet*. Please refer back to that for further details on installation.

Git client

A Git client is a command-line tool that allows you to communicate with a **Source Control System** (**SCS**). In this chapter, you will use the Git client to interact with GitLab. You can install it as follows:

For macOS users, open a terminal and run the following commands:

1. Install the XCode Command Line Tools as follows:

    ```
    xcode-select --install
    ```

2. Then, install Homebrew as follows:

    ```
    /bin/bash -c "$(curl -fsSL https://raw.githubusercontent.
    com/Homebrew/install/HEAD/install.sh)"
    ```

    For more details, you can read the Homebrew official documentation, available at the following link: `https://brew.sh/`.

3. Finally, install Git as follows:

    ```
    brew install git
    ```

For Ubuntu Linux users, you can open a terminal and run the following commands:

```
sudo apt-add-repository ppa:git-core/ppa
sudo apt-get update
sudo apt-get install git
```

For Windows users, you can download it from the following link: https://git-scm.com/download/win. Once you have downloaded it, you can install it by following the guided procedure.

For all users, you can verify whether Git works by running the following command in a terminal:

```
git --version
```

For more details on how to install Git, you can read the Git official documentation, available at the following link: https://git-scm.com/book/en/v2/Getting-Started-Installing-Git.

Now that you have installed all of the software and tools needed to run the examples described in this chapter, we can move on to the first topic, which is introducing the concept of CI/CD.

Introducing the concept of CI/CD

DevOps best practices permit us to build a pipeline that connects the development phase with the operations phase through different steps. In the previous chapter, you learned how to deploy your Data Science project for the first time. You implemented all the steps manually by building a Docker image and then deploying it in Kubernetes. However, this described procedure does not scale if you perform daily updates to your software.

To automate the integration between the development and operation phases, we should introduce two new concepts, which are CI/CD and SCS.

This section is organized as follows:

- An overview of CI/CD
- The concept of an SCS
- The CI/CD workflow

Let's start from the first point, which is an overview of CI/CD.

An overview of CI/CD

When using software in production, either a generic app or a machine learning prediction service, you (or other users) may find some bugs in the code or may want to add new features. For this reason, you should be able to define an automatic procedure that permits you to update your software continuously. In this case, you could use the CI/CD strategy.

CI involves the automation of the building and testing phases every time there is a change in the code. CD deploys the software built during the CI phase to a production-like environment. Usually, this phase requires a manual approval step to move the software to production. A more advanced mechanism involves **Continuous Deployment**, which transforms the manual approval step into an automatic process.

To perform CI/CD, you need an SCS that keeps track of all software changes and makes the communication between the Development and Operation phases possible, as shown in the following figure:

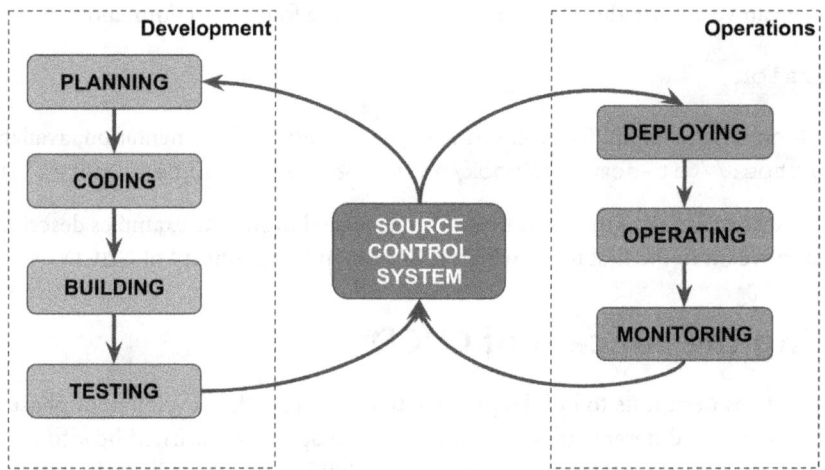

Figure 7.1 – The role of an SCS in DevOps

The preceding figure shows the DevOps life cycle, which was already described in *Chapter 6, Integrating Comet into DevOps*, where the releasing step has been substituted by the SCS. In practice, the SCS is a repository that stores the software and keeps track of its updates. The development team stores the software in an SCS and updates it regularly. The operations team downloads the software from the SCS, and when they approve it, the software moves to production. Thanks to the presence of the SCS, the CI/CD procedure is automatic.

Now that you are familiar with the concept of CI/CD, we can investigate the concept of an SCS in more detail and how it works.

The concept of an SCS

An SCS, also known as a **Version Control System** (**VCS**), enables you to store your code, keep track of the code history, merge code changes, and return to the previous code version if needed. In addition, an SCS permits you to share your code with your team and work on your code locally until it is ready. Since SCS is a centralized source for your code, you can use it to easily build the DevOps life cycle.

The following figure shows the basic architecture of an SCS:

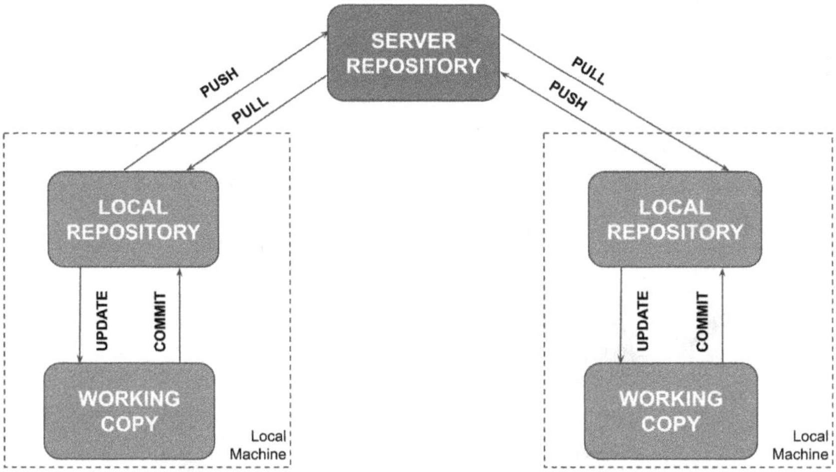

Figure 7.2 – The basic architecture of an SCS

An SCS is composed of a central server repository, which stores the code, and keeps track of all its history, as follows:

1. The first time you want to download the code, you need to clone it in your local machine where a local repository is built.

2. Then, you can work on your local working copy, which practically corresponds to a directory in your local filesystem. All of the changes you make to your code in your working copy do not affect either your local repository or the server repository.

3. To register your local changes to the local repository, you should run the COMMIT command.

4. To register the changes of your local repository to the server repository, you should run the PUSH command.

Before making any change to your code, you should always update your code from the server repository through the PULL command and from your local repository through the UPDATE command.

The described mechanism works well when multiple developers are working together because each developer works on their local machine, and when the code is ready, they save the changes to the server repository. Through an SCS, you can push your commits to other developers as well as pull their commits to your local repository.

Typically, when you want to propose some changes to your code, you do not push them to the main repository. Instead, you create a new branch that contains your changes. A **branch** is a version of the repository diverging from the main project because it contains some proposals for changes. Using a branch permits you to work on your code independently from its stable version available in the main repository. The operation of combining a branch with the main repository is called merging.

Now that you have learned the main concepts behind an SCS, we can move on to the next point, which is the CI/CD workflow.

The CI/CD workflow

The following figure shows the typical CI/CD workflow:

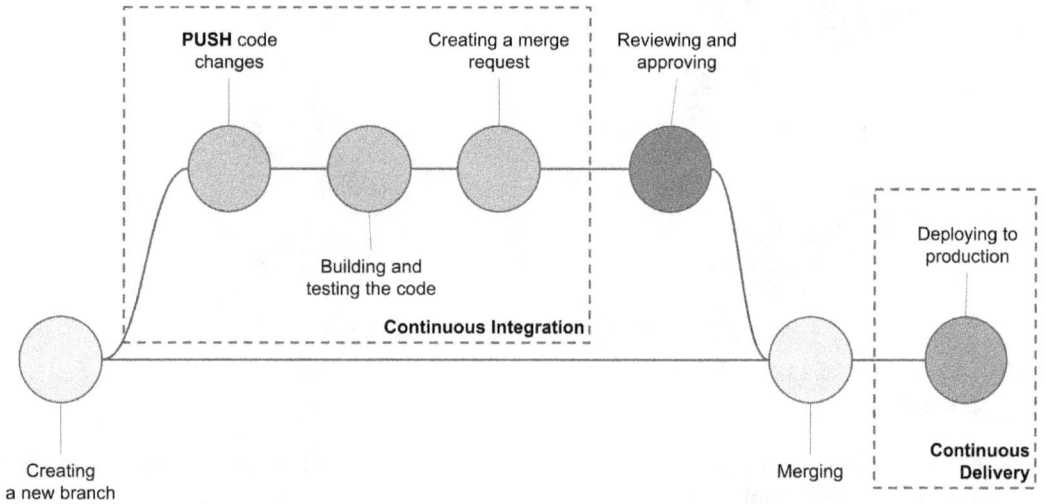

Figure 7.3 – A typical CI/CD workflow

Let's suppose that you want to make some changes to your code that is already hosted on an SCS. You should do the following:

1. Create a new branch for your code.

2. Next, push the new code to the server repository.

3. The CI strategy automatically builds and tests your code.

4. If your code passes the tests, you can create a merge request. Otherwise, you need to perform other changes to your code and create a new push, which triggers automatic building and testing.

5. Once your code has passed all of the tests, you need to manually approve it (in the case of CD) or an automatic procedure approves it (in the case of Continuous Deployment).

6. Upon approval, the branch is merged into the main repository.

7. Eventually, the new code is moved to production.

Many platforms implement the CI/CD workflow. In the next section, we will review GitLab, one of the most popular platforms in the industry.

Implementing the CI/CD workflow in GitLab

GitLab is an SCS that permits you to store your code through a versioning system. In addition, it provides you with all of the tools to implement a CI/CD workflow. The GitLab platform is available at the following link: https://gitlab.com/. You can get started with GitLab by creating an account at the following link: https://gitlab.com/users/sign_up.

The section is organized as follows:

- Creating/modifying a GitLab project
- Exploring the GitLab internal structure
- Exploring GitLab concepts for CI/CD
- Building the CI/CD pipeline
- Creating a release

Let's start with the first point, which is creating/modifying a GitLab project.

Creating/modifying a GitLab project

Basic operations on a GitLab project include the following steps:

1. Creating a new project
2. Adding a new branch to the project
3. Creating a merging request

Let's investigate each step separately, starting from the first one: creating a new project.

Creating a new project

You can create a new project in GitLab as follows:

1. Log in to the GitLab platform.
2. From the main dashboard, click on the **New Project** button located in the top-right corner of the page.

3. Select **Create blank project**, fill out the form, and then click on the **Create Project** button.

4. Now your project is on the GitLab platform. You can download it on your local machine by selecting **Repository** on the left menu, and then you can click on the **Clone** button. A popup window opens. You can copy the address defined by the **Clone with HTTPS** label as shown in the following figure:

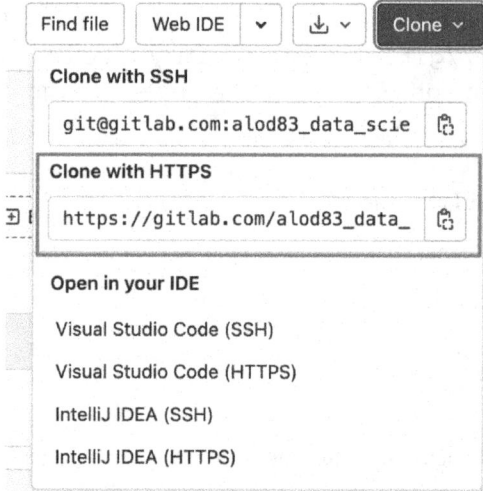

Figure 7.4 – The address to copy to clone a GitLab project in your local file system

The address starts with `https://gitlab.com/` and contains the path to your project group.

5. Open a terminal and run the following command:

```
git clone <PASTE_HERE_YOUR_COPIED_ADDRESS>
```

6. The terminal will ask you for your credentials. You can enter them. Once the procedure is completed, a new directory will appear in your filesystem. The name of the directory is the project name in GitLab.

At this point, you have a working copy of your GitLab empty project in your local filesystem. You can start editing it.

7. For example, you can add to the repository a simple file, named `helloWorld.py`, which simply prints the `'Hello World!'` string, as shown in the following piece of code:

```
if __name__ == "__main__":
    print('Hello World!')
```

8. You can add the script to the GitLab working copy as follows:

```
git add helloWorld.py
```

We use the add command to add a new file to the working copy.

9. You can add the script to the local GitLab repository as follows:

```
git commit -m "my message"
```

10. Finally, you can save changes to the remote server as follows:

```
git push origin main
```

The keyword main indicates that you are saving the changes to the main branch. The git command might prompt you to set the upstream, which is the default remote branch for the current local branch. You can set the upstream branch directly through the −u option as follows:

```
git push -u origin main
```

So far, you have worked with the main branch. However, when you want to add new features to a stable project, it is better to work on a separate branch. So, let's see how to create a new branch.

Adding a new branch

Let's suppose that now you want to change the original text printed by the script as follows:

```
if __name__ == "__main__":
    print('Hi!')
```

Now you want to save the changes to a new branch of the repository. You can proceed as follows:

1. First, create a new branch, named new_branch, as follows:

```
git checkout -b new_branch
```

You can use the checkout keyword to create a new branch as well as the −b option to set the branch name.

2. Then, add the script to the branch as follows:

```
git add helloWorld.py
```

3. Commit the changes to the local repository as follows:

```
git commit -m "modified greeting string"
```

4. Finally, push the changes to the new branch as follows:

```
git push origin new_branch
```

If you access the GitLab dashboard, you can see that the system has created a new branch as shown in the following figure:

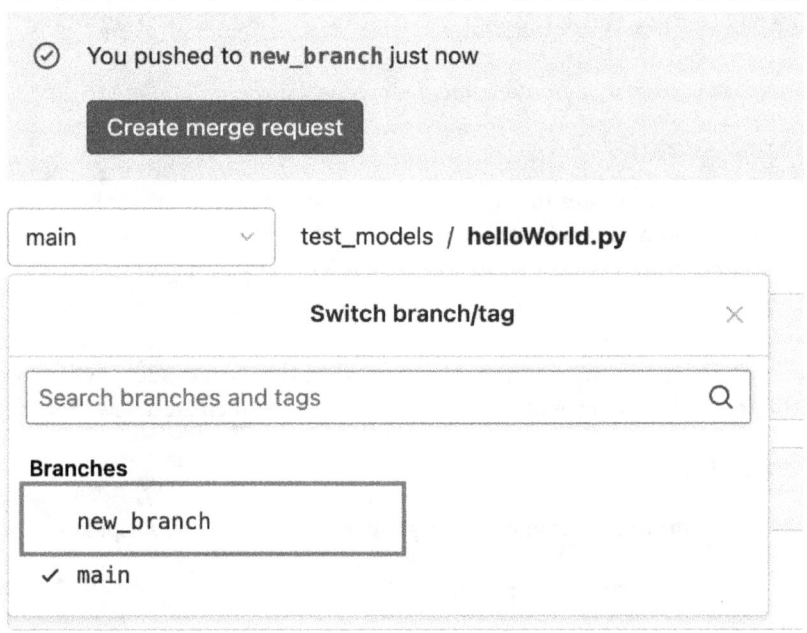

Figure 7.5 – The new branch created in GitLab after the git push command

The created branch remains pending until you create a merge request. Let's investigate how to perform it.

Creating a merge request

Figure 7.5 shows that the GitLab platform proposes to create a merge request of the branch (the green rectangle at the top of the figure, with the blue button). The following are the steps to creating a merge request:

1. If you click on the **Create merge request** button, the GitLab dashboard asks you to fill out some fields for the merge request, such as an associated message.

2. Then, you can click again on the button **Create merge request**.

3. The repository's maintainer will receive a notification, specifying that there is a merge request.

4. The maintainer (you, in this case) can approve the merge request simply by clicking on the **merge** button, as shown in the following figure:

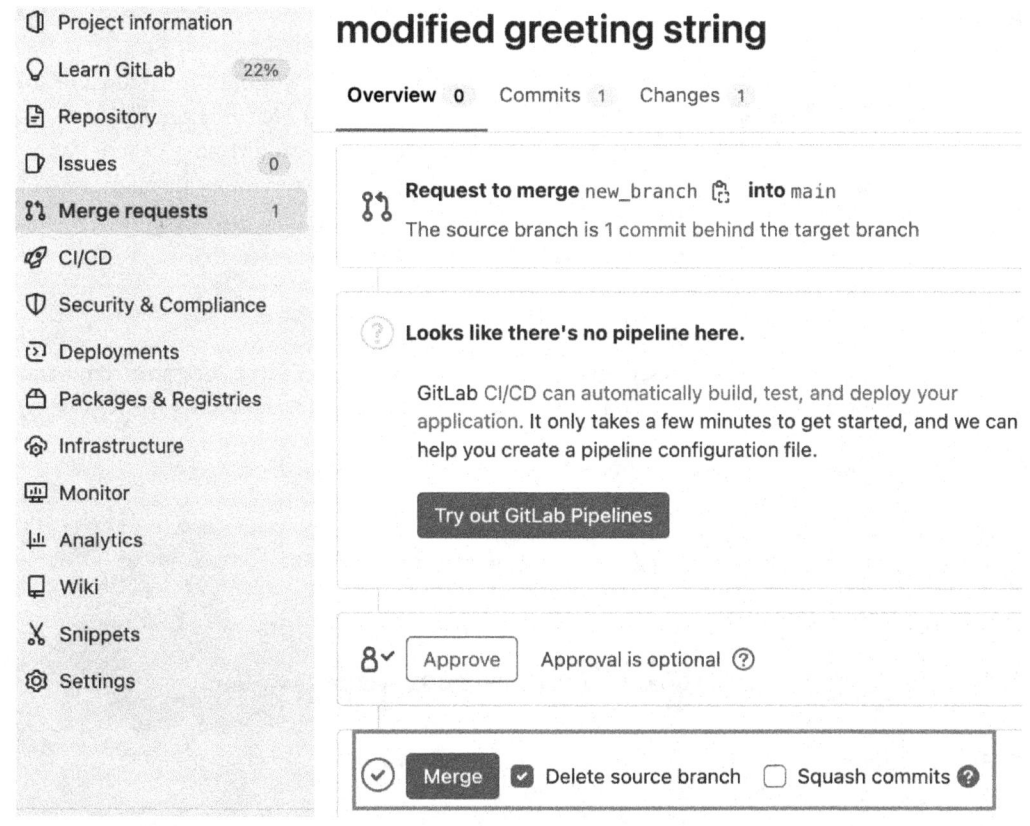

Figure 7.6 – Approving a merge request

Once the merge request is approved, all of the changes you performed in the branch are moved to the main branch.

Now that you have learned the basic concepts of creating and modifying a project in GitLab, we can move on to the next step, which is exploring GitLab's internal structure.

Exploring GitLab's internal structure

To store the content of a project, GitLab uses the following three main concepts:

- Blob
- Tree
- Commit

Let's investigate each concept separately.

Blob

A **blob** (short for **Binary Large Object**) represents the content of a file, without its metadata or the file name. In practice, a blob of a file is the associated SHA1 hash of that file. Different files with the same content have the same blob. GitLab stores all of the blobs in a local directory called `.git/objects` within the GitLab project. Whenever you add a new file to the repository through the `git add` command, a new blob object is added to the `.git/objects` directory, provided that the same content is not already available in the `.git/objects` directory.

Tree

Blobs are organized in trees, which represent directories. Every time you run a `git commit` command, GitLab creates a new tree in the `.git/objects` directory. The tree objects contain the following:

- The blob identifiers contained in the tree
- The paths of the files
- The metadata associated with all the files in that directory

The following piece of code shows an example of tree content:

```
100644 blob ce013625030ba8dba906f756967f9e9ca394464a
      file1.txt
100644 blob e69de29bb2d1d6434b8b29ae775ad8c2e48c5391
      file2.txt
```

The preceding tree contains two blobs named `file1.txt` and `file2.txt`.

Commit

When you run the `git commit` command, GitLab also adds a commit object in the `.git/objects` directory. The commit object points to the `tree` object, as shown in the following piece of code:

```
tree dca98923d43cd634f4359f8a1f897bf585100cfe
author Author Name <author's computer name> 1656927878 +0200
committer Author Name <author's computer name> 1656927878 +0200

My commit message
```

You can see the reference to the tree, the author who performed the commit, and the commit message (`My commit message` in the example).

If you access the `.git/objects` directory, you can check the type of the object as follows:

1. First, take the name of the directory containing the object you want to check (such as `7f`).

2. Next, append to the directory's name the first two characters of the object you want to check. For example, if the object name is `dca98923d43cd634f4359f8a1f897bf585100cfe`, you take `dc` and append it to `7f`. As a resulting string, you have `7fdc`.

3. From the `.git/objects` directory, run the following command to see the type of the object:

   ```
   git cat-file -t 7fdc
   ```

   You can output either `blob`, `tree`, or `commit` file types.

4. Alternatively, you can view the content of the object by running the following command:

   ```
   git cat-file -p 7fdc
   ```

   For the preceding command, you use the `-p` argument.

Now that you have learned the GitLab internal structure, we can move on to the next point, which is exploring GitLab concepts for CI/CD.

Exploring GitLab concepts for CI/CD

GitLab defines the following concepts, which implement the CI/CD workflow:

- **Job**: The smallest unit that you can run in a GitLab CI/CD workflow. For example, a job can be a compilation, building, or running task.
- **Stage**: A logical representation that defines when to run jobs.
- **Pipeline**: A collection of jobs organized in different stages.
- **Runner**: An agent that runs jobs.

To configure the CI/CD workflow in GitLab, you should add `.gitlab-ci.yml` to your project. The presence of a `.gitlab-ci.yml` file in your repository automatically triggers a GitLab runner that executes the scripts defined in the jobs.

To define the stages of your CI/CD workflow, you can use the `stages` keyword in the `.gitlab-ci.yml` file, as shown in the following piece of code:

```
stages:
  - build
  - test
  - deploy
```

In the preceding example, we defined three stages: build, test, and deploy. These stages are executed in sequential order, starting from the first stage to the last one.

To define a job, you can use a generic name, followed by the stage to which the job belongs, and the scripts to run, as shown in the following example:

```
my-job:
  stage: build
  script:
    - echo "Hello World!"
```

In the preceding example, we defined a job named my-job that belongs to the build stage and simply prints to screen the sentence "Hello World!".

By default, a GitLab runner runs all of the jobs in the same stage concurrently. This process is also called a basic pipeline. However, you can customize the order of jobs, regardless of the stage they belong to, by using the needs keyword within the job definition. For example, if you want to specify that job2 must be run after job1, you can write the following code:

```
job1:
  stage: build
  script:
    - echo "Hello World from Job1!"
job2:
  needs:
    - job1
  stage: build
  script:
    - echo "Hello World from Job2!"
```

We use the needs keyword for job2 to specify its dependence on job1.

Now that you have learned the basic GitLab concepts for CI/CD, we can implement a practical example.

Building the CI/CD pipeline

Let's suppose that now you want to implement a CI/CD workflow that is triggered every time you push a change to a branch of your repository. The idea is to implement the CI/CD workflow illustrated in *Figure 7.3*.

As a use case, we can extend the example described in the *Creating/modifying a GitLab project* section by adding a new script to the test_models repository. The script simply builds a linear regression model on the well-known diabetes dataset and calculates the mean squared error metric.

To configure the CI/CD workflow for this project, you should perform the following steps:

- Writing the main script
- Configuring a runner
- Configuring the `.gitlab-ci.yml` file
- Creating a merge request

Let's start with the first step, which is writing the main script.

Writing the main script

The script named `linear_regression.py` loads the well-known diabetes dataset provided by the scikit-learn library, trains it with a training set, and then calculates the **Mean Squared Error (MSE)** metric.

The following piece of code shows the code that implements the preceding steps:

```python
from sklearn.datasets import load_diabetes
from sklearn.model_selection import train_test_split
from sklearn.linear_model import LinearRegression
from sklearn.metrics import mean_squared_error
diabetes = load_diabetes()
X = diabetes.data
y = diabetes.target
X_train, X_test, y_train, y_test = train_test_split(X, y, test_
size=0.20, random_state=42)
model = LinearRegression()
model.fit(X_train,y_train)
y_pred = model.predict(X_test)
mse = mean_squared_error(y_test,y_pred)
print(f"MSE: {mse}")
```

The script simply prints the MSE value. Since the script uses the `scikit-learn` library, we also include a `requirements.txt` file in the project repository. This file simply contains the `scikit-learn` library name.

Now you can perform the following steps:

1. Add the files to the GitLab repository as explained in the preceding section:

    ```
    git add linear_regression.py
    git add requirements.txt
    ```

2. Then, you can commit changes to your local repository first and then to the GitLab remote server as follows:

```
git commit -m "added linear regression "
git push origin main
```

Once the push operation terminates, you should be able to see the files on your GitLab dashboard.

Now that you have written the main script, we can implement the CI/CD workflow by following the next step, which is configuring a runner.

Configuring a runner

First, we need to set up a runner, which is a process that executes jobs. In this example, we will use the shared runners provided by GitLab. However, you can implement your own custom runner as described in the GitLab official documentation, available at the following link: `https://docs.gitlab.com/runner/`.

To configure a runner, you can proceed as follows:

1. On the GitLab dashboard, access your project by selecting **Menu | Projects | Your Projects**. Then, select your project name, which is `test_models` in our case.

2. In the left menu, select **Settings** and then **CI/CD**. In the **Runners** section, click on the **Expand** button.

3. Make sure that the slider under **Enable shared runners for this project** is active. If not, enable it. The following figure shows where you should click to enable runners:

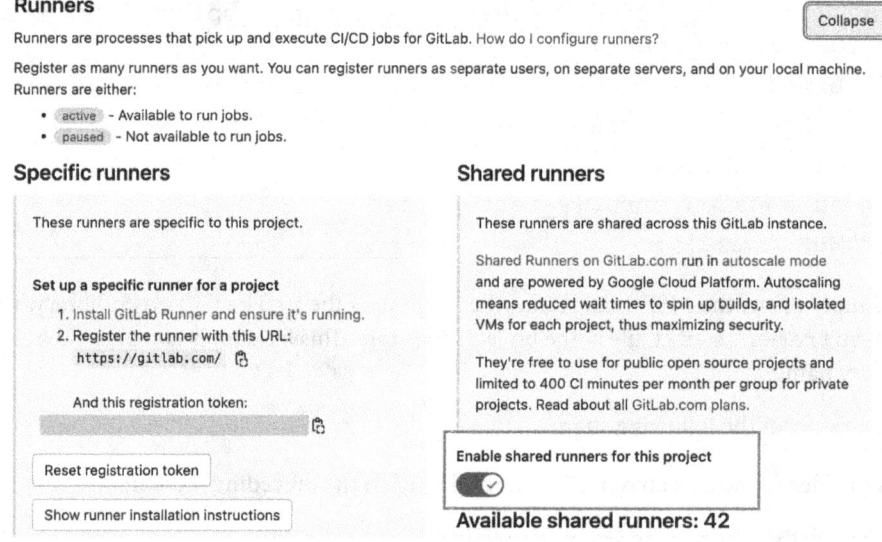

Figure 7.7 – How to enable GitLab runners

Now that you have configured the runners, you can configure the `.gitlab-ci.yml` file.

Configuring the .gitlab-ci.yml file

We need to configure `.gitlab-ci.yml` as follows:

1. In the left menu, select **CI/CD** and then **Editor**. An **Editor** opens with a basic template of the `.gitlab-ci.yml` file. You can modify the basic template through the following steps.

2. First, you can import the basic Docker image as follows:

    ```
    image: python:3.8
    ```

    Since our script is written in Python, we only need to import the Python interpreter.

3. Next, you can define just one stage as follows:

    ```
    stages:
      - run
    ```

4. To make the installed packages also accessible during the subsequent run stages, we cache them. First, we define a caching working directory as follows:

    ```
    variables:
      PIP_CACHE_DIR: "$CI_PROJECT_DIR/.cache/pip"
    ```

    `$CI_PROJECT_DIR` contains the current directory of the project.

5. To cache the installed packages, you should install them in a virtual environment and cache them as well:

    ```
    cache:
      paths:
        - .cache/pip
        - venv/
    ```

    The `cache` keyword specifies the list of files that should be cached. The `paths` keyword determines which files to add to the cache.

6. Now you define a special job, named `before_script`, that specifies all of the operations that should be run before each script. In our case, we create and activate a virtual environment as well as install the required packages as follows:

    ```
    before_script:
      - pip install virtualenv
      - virtualenv venv
      - source venv/bin/activate
      - pip install -r requirements.txt
    ```

7. Next, you can define the `run` job as follows:

```
run-job:
  stage: run
  script:
    - python linear-regression.py
```

Simply, the job runs the `linear-regression.py` script.

8. Finally, you can click on the **Commit changes** button. This operation will trigger the CI/CD pipeline, as defined in your `.gitlab-ci.yml` file.

If everything is okay, you should see an output similar to that shown in the following figure:

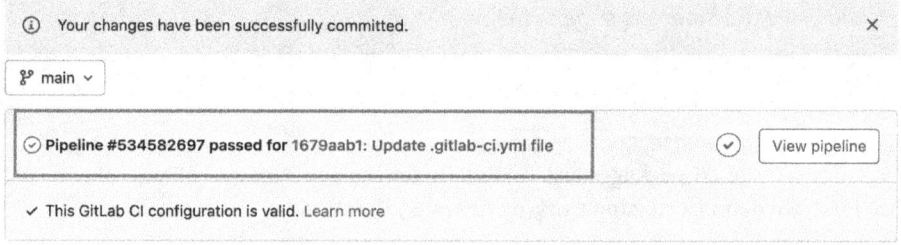

Figure 7.8 – The output of a successful pipeline

If you click on the **View pipeline** button, you can view all of the details related to the built pipeline, as shown in the following figure:

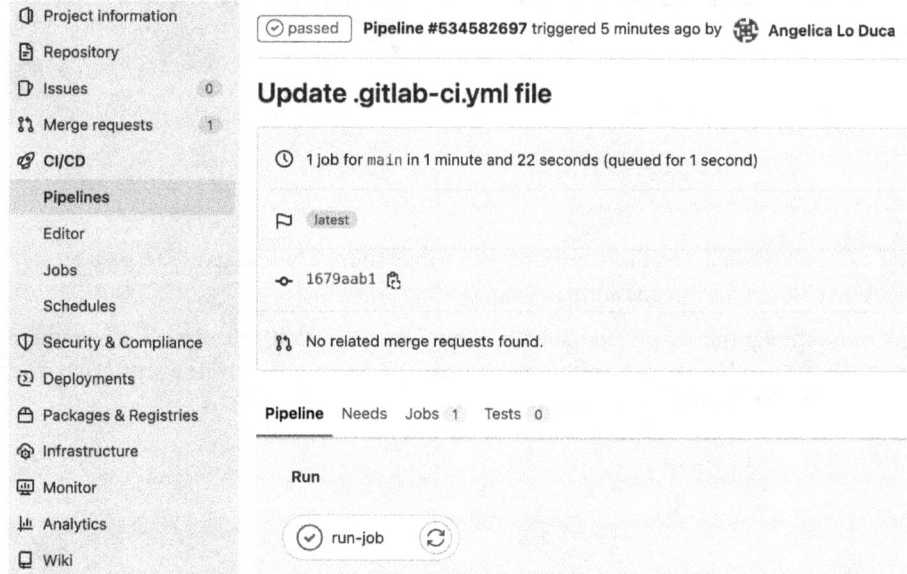

Figure 7.9 – Details related to a successful pipeline

At the bottom part of the screen, you can see the jobs contained in the pipeline. In our case, there is just one job named **run-job**. You can click on the **run-job** button to see the output of the console, as shown in the following figure:

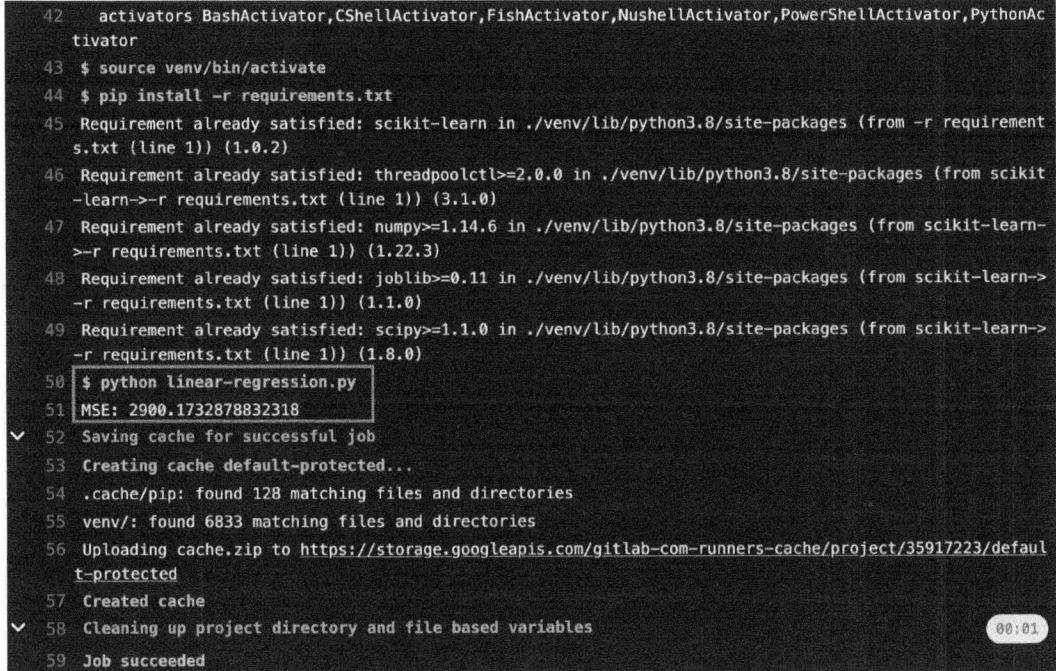

```
42    activators BashActivator,CShellActivator,FishActivator,NushellActivator,PowerShellActivator,PythonAc
      tivator
43  $ source venv/bin/activate
44  $ pip install -r requirements.txt
45  Requirement already satisfied: scikit-learn in ./venv/lib/python3.8/site-packages (from -r requirement
    s.txt (line 1)) (1.0.2)
46  Requirement already satisfied: threadpoolctl>=2.0.0 in ./venv/lib/python3.8/site-packages (from scikit
    -learn->-r requirements.txt (line 1)) (3.1.0)
47  Requirement already satisfied: numpy>=1.14.6 in ./venv/lib/python3.8/site-packages (from scikit-learn-
    >-r requirements.txt (line 1)) (1.22.3)
48  Requirement already satisfied: joblib>=0.11 in ./venv/lib/python3.8/site-packages (from scikit-learn->
    -r requirements.txt (line 1)) (1.1.0)
49  Requirement already satisfied: scipy>=1.1.0 in ./venv/lib/python3.8/site-packages (from scikit-learn->
    -r requirements.txt (line 1)) (1.8.0)
50  $ python linear-regression.py
51  MSE: 2900.1732878832318
52  Saving cache for successful job
53  Creating cache default-protected...
54  .cache/pip: found 128 matching files and directories
55  venv/: found 6833 matching files and directories
56  Uploading cache.zip to https://storage.googleapis.com/gitlab-com-runners-cache/project/35917223/defaul
    t-protected
57  Created cache
58  Cleaning up project directory and file based variables                                      00:01
59  Job succeeded
```

Figure 7.10 – The output of run-job

You can clearly see the output of the `python linear-regression.py` command, which shows the calculated **Root Mean Squared Error (RMSE)**.

> **Note**
>
> It is possible that your pipeline fails. In fact, the first time you run a CI/CD pipeline in GitLab, you need to validate your account by providing a valid credit card. Once you fill in the required information, you should be able to run the CI/CD pipeline.

Now that you have learned how to configure the `.gitlab-ci.yml` file, we can move on to the next step, which is creating a merge request.

Creating a merge request

The objective of this step involves modifying the `linear-regression.py` script, saving the changes to a new branch, and creating a merge request.

1. First, we modify `linear-regression.py` by simply printing a new metric, which is the RMSE. We can write the following code:

    ```
    from math import sqrt
    # train the model and calculate MSE
    rmse = sqrt(mse)
    print(f"RMSE: {rmse}")
    ```

2. Next, we create a new branch named `test` as follows:

    ```
    git checkout -b test
    ```

3. We commit changes to our local repository and the remote GitLab repository as follows:

    ```
    git commit -m "modified linear regression"
    git push origin test
    ```

    You can see that we have pushed changes to the `test` branch.

4. If you access the GitLab dashboard, you can see that a new branch has been created. A new message appears, asking you to create a merge request, as shown in the following figure:

Figure 7.11 – The message appearing after pushing to a new branch

You can click on the **Create merge request** button to trigger a merge request. This operation triggers the CI/CD pipeline.

5. In the **Merge requests** menu, you can see the merge request we just created, as shown in the following figure:

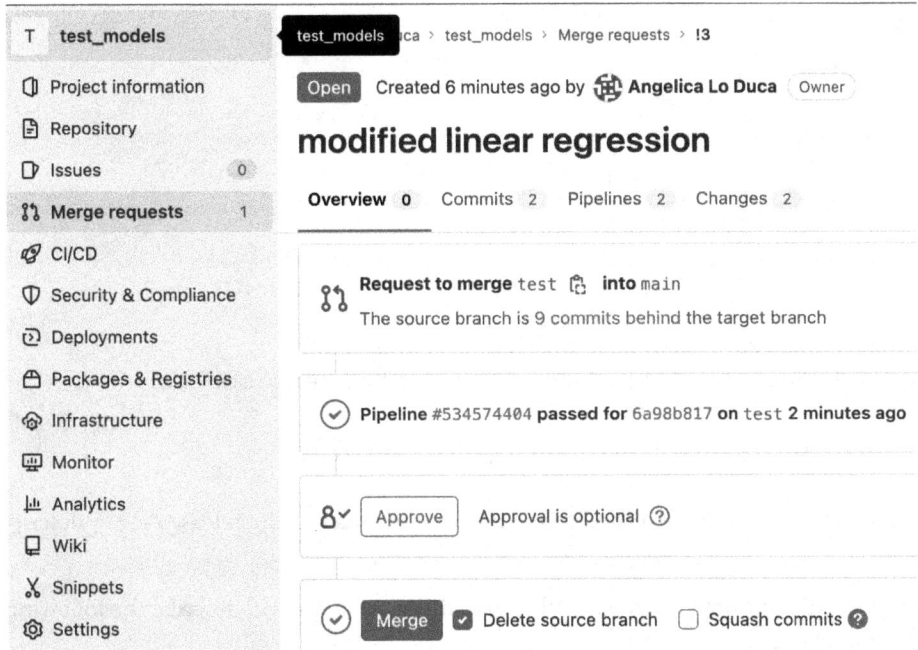

Figure 7.12 – An overview of the merge request

You can see details on the pipeline, and you can decide to approve and merge the test branch with the main branch.

6. Click on the **Merge** button to merge the changes in the main branch.

Now that you have learned how to build the CI/CD pipeline, we can move to the last step, which is creating a release.

Creating a release

When your project is ready to be published and downloaded by other people, you can create a release. A release is a snapshot of your project, which also includes the installation packages and the release notes. To create a release, your project must have at least one tag, which can be, for example, the version of your project.

A release includes four zipped versions of your code, including four different formats: .zip, .tar. gz, .tar.bz2, and .tar. The release also includes a JSON file, which lists the release content, as shown in the following example:

```
{"release":{
    "id": 5269390,
    "name": "Test Models V1",
```

```json
        "project":{
            "id": 35917223,
            "name": "test_models",
            "created_at": "2022-05-05T16:33:04.210Z",
            "description": ""
        },
        "tag_name": "v1",
        "created_at": "2022-07-04T19:45:04.219Z",
        "milestones": [],
        "description": "The first release of test models"
        }
    }
```

The JSON file includes metadata related to the name of the release, the release date, a description, and so on.

GitLab provides different ways to create a release. In this section, we will describe the following two ways to create a release:

- **Using the GitLab dashboard**: To create a new release, you can click on the left tab of your dashboard and select **Deployments | Releases**. Then, you can add a new tag, such as v1, and follow the guided procedure.

- **Using the .gitlab-ci.yml file**: You can add a new stage, called release, to your configuration file, as shown in the following piece of code:

```yaml
release_job:
  stage: release
  image: registry.gitlab.com/gitlab-org/release-
cli:latest
  rules:
    - if: $CI_COMMIT_TAG
  release:
    tag_name: '$CI_COMMIT_TAG'
```

In this example, you build a `release` only if the commit operation includes a tag. You should also make sure that your `release` job has access to the `release-cli`, so you should add it to the `PATH` or use the image provided by the GitLab registry, as shown in the preceding example.

To add a tag to your commit operation, you can proceed as follows:

A.   From the command line, you can add a tag before committing your code by using the `tag` keyword as follows:

```
git tag my_tag
```

B.   Then, you can save your tag to the remote repository as follows:

```
git push origin my_tag
```

C.   You add your tag name at the end of the `git push` command.

For more details on the other ways to create a release, you can refer to the GitLab official documentation, available at the following link: `https://docs.gitlab.com/ee/user/project/releases/`.

Now that you have learned the basic concepts behind GitLab, we can investigate how to integrate Comet with GitLab.

Integrating Comet with GitLab

Thanks to a collaboration between Comet and GitLab, Comet experiments are fully integrated with GitLab. You can integrate Comet and GitLab in two ways as follows:

- Running Comet in the CI/CD workflow
- Using Webhooks

Let's investigate the two ways separately, starting with the first one.

Running Comet in the CI/CD workflow

The following figure shows how Comet can be integrated with the CI/CD pipeline:

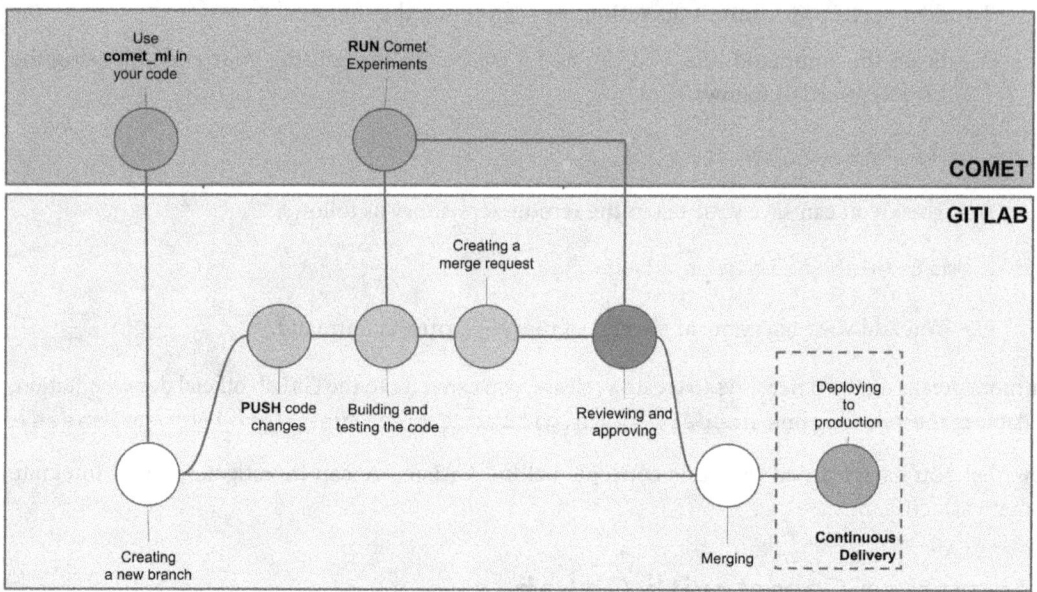

Figure 7.13 – Integration of Comet in the CI/CD workflow

Let's suppose that you have changed your code to support Comet experiments. If your code is written in Python, then you have imported the comet_ml library and used it to track your experiments. You can start the CI/CD workflow by creating a new branch for your project. As usual, you push code changes, and you build and run the code. This process also triggers a connection with the Comet platform. Then, the CI/CD workflow continues as described in the preceding section. When you review and approve the code, you should also consider the output of the Comet experiments to make sure that the proposed changes improve the model.

To make the integration between Comet and GitLab work, you should configure the Comet secrets in the GitLab project. You can proceed as follows:

1. In GitLab, select your project. Then, from the left menu, select **Settings** | **CI/CD** | **Variables** | **Expand**.

2. Add the following two protected variables: COMET_API_KEY, which stores your Comet API key, and COMET_WORKSPACE, which stores your Comet workspace, as shown in the following figure:

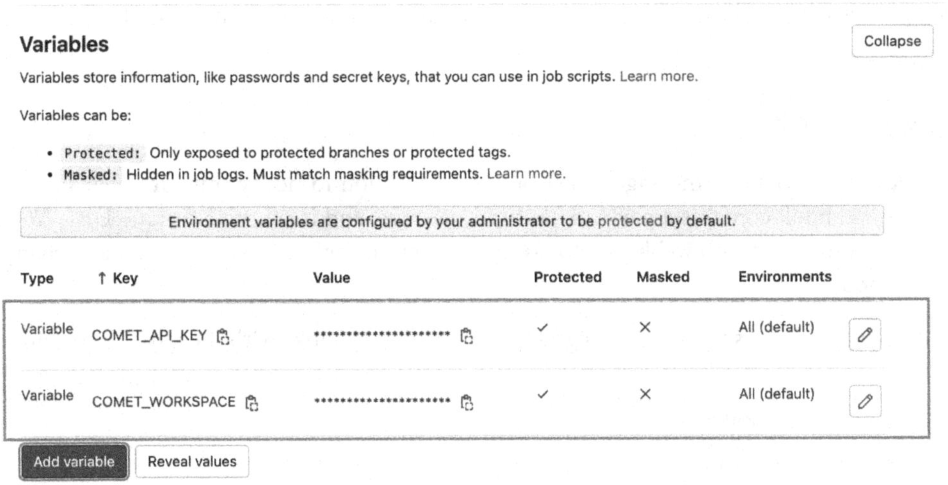

Variables   Collapse

Variables store information, like passwords and secret keys, that you can use in job scripts. Learn more.

Variables can be:

- `Protected:` Only exposed to protected branches or protected tags.
- `Masked:` Hidden in job logs. Must match masking requirements. Learn more.

Environment variables are configured by your administrator to be protected by default.

Type	↑ Key	Value	Protected	Masked	Environments	
Variable	COMET_API_KEY	••••••••••••••••••••	✓	✕	All (default)	✐
Variable	COMET_WORKSPACE	••••••••••••••••••••	✓	✕	All (default)	✐

Add variable   Reveal values

Figure 7.14 – The configured variables in GitLab

We have configured both variables as **protected**. This means that only protected branches can access them. By default, the main branch is protected. You can set a branch as **protected** in the **Protected Branches** menu, which is available by following this path from the project menu: **Settings →Repository →Protected Branches →Expand**. When you want to protect a branch, you should select who can modify the branch. For example, you can choose the **maintainers**, as shown in the following figure:

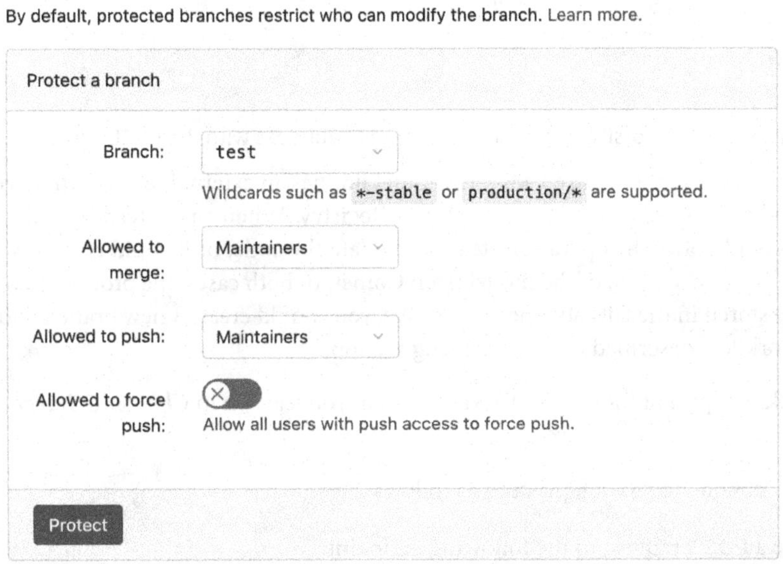

By default, protected branches restrict who can modify the branch. Learn more.

Protect a branch

Branch: test

Wildcards such as `*-stable` or `production/*` are supported.

Allowed to merge: Maintainers

Allowed to push: Maintainers

Allowed to force push: ⊗   Allow all users with push access to force push.

Protect

Figure 7.15 – How to protect a branch in GitLab

Now that you have learned how to run Comet in a CI/CD pipeline, we can move on to the next step, which is using webhooks.

Using webhooks

A **webhook** is an automated message sent from an application to notify someone that something happened. A webhook contains a payload, that is a message, and is sent to a unique URL. You can configure Comet to work with webhooks to notify an external application that there was a change in a registered model.

The following figure shows a possible integration of Comet webhooks with the CI/CD workflow:

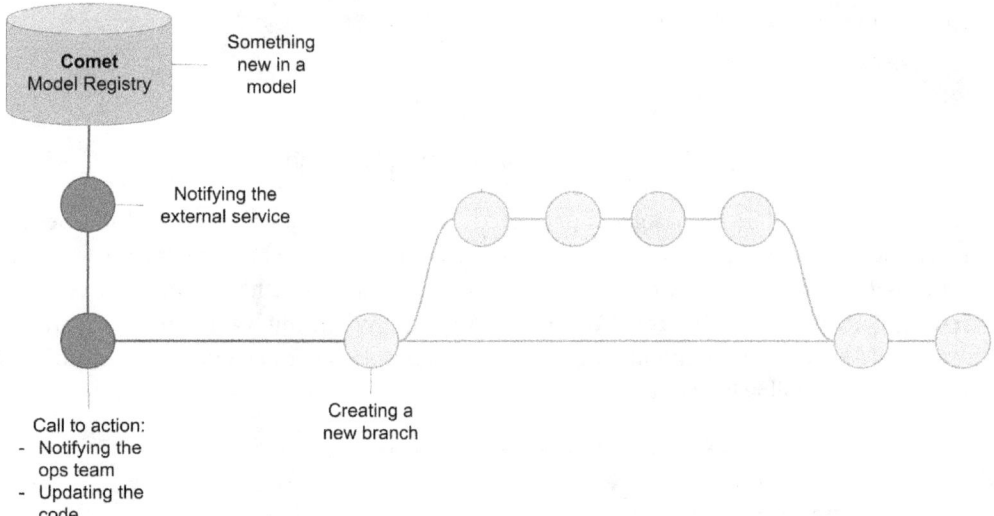

Figure 7.16 – A possible integration of Comet webhooks with the CI/CD pipeline

Let's suppose that you have configured a webhook that notifies an external service whenever a change occurs in a model registered in the Comet Model Registry. When the external service receives a notification, it could notify the operations team to update the code, or it could trigger an automatic downloading of the new version of the model from Comet. In both cases, the produced action should update the code stored in the GitLab repository. Thus, you should create a new branch and follow the usual CI/CD workflow described in the preceding section.

Comet webhooks are part of the Comet REST APIs that you learned in *Chapter 6, Integrating Comet into DevOps*.

To configure a webhook in Comet, proceed as follows:

1. Generate a POST request to the following endpoint:

    ```
    https://www.comet.ml/api/rest/v2/webhooks/config
    ```

2. The POST request should contain the following payload:

```
payload = { "workspaceName": <YOUR_COMET_WORKSPACE>,
    "modelName": <YOUR MODEL NAME>,
    "webhookUrls": [{ "url": <URL to the external
service>,
        "header": { "Authorization": <secret token>,
            "Other": "other_info"
        }
    }]
}
```

You should define the workspace name and the model name in the Comet Registry that you want to notify. Under the webhooksUrls keyword, you can specify as many external services as you want. For each service, you should specify the secret token used to access it.

3. In Python, you use the requests package to send the webhook configuration to the Comet REST API service as follows:

```
import requests
headers = {'Content-Type':'application/json',
    "Authorization": f"{<YOUR_COMET_API_KEY>}"
}
 response = requests.post(url, headers=headers,
json=payload)
```

Once you have configured the Comet webhook, every time there is a change in the stage of your model, a notification will be sent to your external service.

Now that you have learned how to integrate Comet with GitLab, let's move on to a practical example that shows how to integrate Docker with the CI/CD workflow.

Integrating Docker with the CI/CD workflow

Let's see how to integrate Docker with the GitLab CI/CD workflow through a practical example. As a use case, let's use the example implemented in *Chapter 6, Integrating Comet into DevOps*, in the *Implementing Docker* section. The example built an application that tested four different classification models (random forest, decision tree, Gaussian Naive Bayes, and k-nearest neighbors), which classified diamonds cut into two categories (Gold and Silver) based on some input parameters. In addition, the application logged all of the models in Comet.

You can download the full code of the example from the GitHub repository of the book available at the following link: `https://github.com/PacktPublishing/Comet-for-Data-Science/tree/main/07`.

In that example, we built a Docker image with the code and we ran it. In this section, the idea is to wrap that example in a CI/CD workflow and run it in GitLab. In other words, the idea is to build and run the Docker image automatically within the CI/CD workflow, without writing all the commands manually. To do so, proceed with the following steps:

1. First, we log in to the GitLab platform and we create a new project as described in the preceding section.

2. Then, we clone the project in our local filesystem, and we copy all of the files implemented in *Chapter 6*, *Integrating Comet into DevOps*, in the *Implementing Docker* section, in the created directory.

3. We add all of the files to the GitLab repository through the following command:

    ```
    git add <file_name>
    ```

    The copied files include the following:

    *   `Dockefile`
    *   `compare-models.py`
    *   `requirements.txt`
    *   `source/diamonds.csv`

4. We save changes to our local repository through the `commit` command and to the remote server through the `push` command as follows:

    ```
    git commit -m "initial import"
    git push origin main
    ```

5. Now you should be able to see the files on your GitLab dashboard.

6. The next step involves creating the environment variables related to Comet. We add the following environment variables, through the GitLab dashboard, as described in the preceding section:

    *   `COMET_API_KEY`
    *   `COMET_PROJECT`
    *   `COMET_WORKSPACE`

7. Now you are ready to create a new `.gitlab-ci.yml` file. You can follow the procedure described in the preceding section to open the file editor. Once you have opened the editor, you are ready to configure it. We will configure the CI/CD workflow to work with Docker images as follows:

```
image: docker:latest
  services:
    - docker:dind
```

8. Then, we build two stages, `build` and `run` as follows:

```
stages:
  - build
  - run
```

During the `build` stage, we will build the Docker image, and during the `run` stage, we will wrap it in a container, and we will run the code in the container.

9. We then define some auxiliary variables that will be used in the next steps as follows:

```
variables:
  CONTAINER_TEST_IMAGE: $CI_REGISTRY_IMAGE:$CI_COMMIT_
REF_SLUG
  CONTAINER_RELEASE_IMAGE: $CI_REGISTRY_IMAGE:latest
```

`CONTAINER_TEST_IMAGE` contains the name of the produced image during the `build` stage. We build it from some system variables available in GitLab. For more details on these system variables, you can check the GitLab official documentation available at the following link: `https://docs.gitlab.com/ee/ci/variables/predefined_variables.html`.

10. To make the `run` job use the image built during the `build` stage, we need to save the image in a registry. GitLab provides a registry to store images. So, before every job, we log in to the GitLab registry as follows:

```
before_script:
  - docker login -u $CI_REGISTRY_USER -p $CI_REGISTRY_
PASSWORD $CI_REGISTRY
```

We use the `docker login` command described in *Chapter 6, Integrating Comet into DevOps* but the difference is that now we log in to the GitLab registry. We use the username and the password provided by the GitLab variables.

11. We define the `build` job as follows:

```
job-build:
  stage: build
  script:
    - docker build --pull -t $CONTAINER_TEST_IMAGE .
    - docker push $CONTAINER_TEST_IMAGE
```

Simply, the job builds the Docker image and saves it to the GitLab registry.

12. Now we define the `run` job as follows:

```
job-run:
  stage: run
  script:
    - docker pull $CONTAINER_TEST_IMAGE
    - docker run --rm -e COMET_API_KEY -e COMET_PROJECT
-e COMET_WORKSPACE $CONTAINER_TEST_IMAGE
```

The job pulls the image from the registry and runs it in a container. As for the `run` command defined in *Chapter 6, Integrating Comet into DevOps*, we pass each environment variable separately.

13. We can save the `.gitlab-ci.yml` file by clicking on the button **Commit changes**. Now the pipeline is triggered. Let's wait for the results.

If everything works as expected, you should be able to see your results in Comet. In addition, you can access the produced image by selecting your project, then **Packages & Registries | Container Registry**, as shown in the following figure:

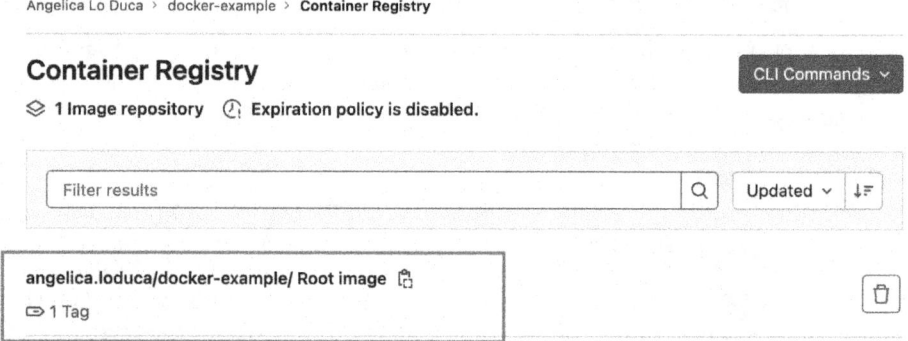

Figure 7.17 – The Container Registry in GitLab with the saved image

Now that the CI/CD pipeline is configured, every time you make a change in your code and you push it to GitLab, your code is automatically deployed as a Docker container.

Summary

You have just learned how to build and run a CI/CD workflow in GitLab!

Throughout this chapter, you have reviewed some advanced concepts related to DevOps, with a focus on the CI/CD workflow. You have also learned some basic concepts related to the GitLab platform and how to use them to implement the CI/CD workflow. Then, you have learned how to integrate Comet with GitLab. Finally, you have implemented a practical example that wraps the Docker building and running processes in the CI/CD pipeline. The CI/CD workflow is very important when deploying an application because it permits you to make the software releasing process easy, quick, and automatic.

In the next chapter, we will review some basic concepts related to machine learning in general, and how to use Comet to build and run a complete machine learning example.

Further reading

- Evertse, J. (2019). *Mastering GitLab 12: Implement DevOps Culture and Repository Management Solutions*. Packt Publishing.

- O'Grady, A. (2018). *GitLab Quick Start Guide: Migrate to GitLab for All Your Repository Management Solutions*. Packt Publishing Ltd.

- Umali, R. (2015). *Learn Git in a Month of Lunches*. Simon and Schuster.

Section 3 – Examples and Use Cases

In this final section, you will learn how to use Comet for model building. You will focus on four different types of models, depending on either the specific technology you are using or the different tasks you want to solve. You will learn how to use Comet to build models for machine learning (*Chapter 8, Comet for Machine Learning*), natural language processing (*Chapter 9, Comet for Natural Language Processing*), deep learning (*Chapter 10, Comet for Deep Learning*), and time series analysis (*Chapter 11, Comet for Time Series Analysis*).

In each chapter of this section, you will see an overview of the considered technology, a description of a Python library that implements that technology, and finally, a practical example, which describes step by step how to combine Comet with a specific technology.

The main focus of this section is to provide you with practical examples that you can use as guidelines for your future data science projects.

This section includes the following chapters:

- *Chapter 8, Comet for Machine Learning*
- *Chapter 9, Comet for Natural Language Processing*
- *Chapter 10, Comet for Deep Learning*
- *Chapter 11, Comet for Time Series Analysis*

8

Comet for Machine Learning

Artificial intelligence (**AI**) is the ability of a computer to perform operations and tasks that are usually done by humans. AI includes different subfields, such as machine learning, natural language processing, deep learning, and time series analysis. In this chapter, we will focus on machine learning, and in the following ones, you will review other subfields of AI, including natural language processing (*Chapter 9, Comet for Natural Language Processing*), deep learning (*Chapter 10, Comet for Deep Learning*), and time series analysis (*Chapter 11, Comet for Time Series Analysis*).

Machine learning aims at using computational algorithms to transform data into usable models. In other words, machine learning tries to build models that learn from data. You can use machine learning algorithms for different purposes and in different domains, such as describing a phenomenon, predicting future values, or detecting anomalies in a phenomenon under investigation. You have already learned some concepts about machine learning in previous chapters, including exploratory data analysis (*Chapter 2, Exploratory Data Analysis in Comet*) and model evaluation (*Chapter 3, Model Evaluation in Comet*). In this chapter, we will focus on model training, which involves building the correct model to represent a given phenomenon.

In recent years, different open source libraries and tools were available to build machine learning models. In this chapter, we will focus on scikit-learn and **XG-Boost**. You should already be familiar with `scikit-learn`, since you have already used it in the examples described in previous chapters. You have also learned how to integrate Comet with `scikit-learn`. In this chapter, you will implement a complete use case, which will permit you to implement a complete machine learning pipeline in Comet.

The chapter is organized as follows:

- Introducing machine learning
- Reviewing the main machine learning models
- Reviewing the scikit-learn package
- Building a machine learning project from setup to report

Before moving on to the first step, let's see the technical requirements to run the software used in this chapter.

Technical requirements

We will run all the experiments and code in this chapter using Python 3.8. You can download it from the official website at `https://www.python.org/downloads/` and choose version 3.8.

The examples described in this chapter use the following Python packages:

- `comet-ml 3.23.0`
- `matplotlib 3.4.3`
- `numpy 1.19.5`
- `pandas 1.3.4`
- `scikit-learn 1.0`
- `shap 0.40.0`

We have already described the first five packages and how to install them in *Chapter 1, An Overview of Comet*. So please refer back to that for further details on installation.

shap

shap is a Python package that permits you to calculate the Shapley value and plot some related graphs. To install the `shap` package, you can run the following command in a terminal:

```
pip install shap
```

For more details on the `shap` package, you can refer to its official documentation, available at the following link: `https://shap-lrjball.readthedocs.io/en/latest/index.html`.

Now that you have installed all of the software needed in this chapter, let's move on to how to use Comet for machine learning, starting with reviewing some basic concepts on machine learning.

Introducing machine learning

Machine learning is a subfield of AI that aims to build models that automatically learn from data. You can use these models for different purposes, such as describing a particular phenomenon, predicting future values, or detecting anomalies in an observed phenomenon. Machine learning has become very popular in recent years thanks to the spread of huge quantities of data that derive from different sources, such as social media, open data, sensors, and so on.

The section is organized as follows:

- Exploring the machine learning workflow
- Classifying machine learning systems

- Exploring machine learning challenges

- Explaining machine learning models

Let's start with the first step: exploring the machine learning workflow.

Exploring the machine learning workflow

The following figure shows the simplest machine learning workflow:

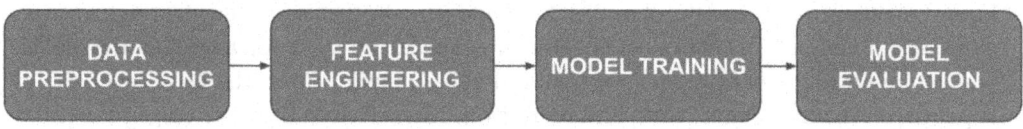

Figure 8.1 – The simplest machine learning workflow

Provided that you already know the problem you want to solve, there are four steps:

1. **Data preprocessing**: You prepare your data by performing all the cleaning operations, including dealing with missing values and anomalies, normalization, standardization, and dropping duplicates. In this phase, you also split your data into training, dev, and test sets.

2. **Feature engineering**: You choose the set of features in your data to send as input to the model.

3. **Model training**: You train your model on the training set, with a focus on tuning model parameters (hyperparameter tuning). In this phase, usually, you apply cross-validation.

4. **Model evaluation**: You evaluate the performance of your model on the dev set by choosing the set of evaluation metrics. You have already learned how to perform the model evaluation in *Chapter 3, Model Evaluation in Comet.*

We will review how `scikit-learn` permits you to implement the preceding steps later in this chapter in the *Reviewing the scikit-learn package* section.

Now that you have reviewed the simplest machine learning workflow, we can move on to the next step: classifying machine learning systems.

Classifying machine learning systems

You can classify machine learning systems based on the following three main criteria:

- *The nature of the problem to solve* (supervised, unsupervised, semi-supervised, and reinforcement learning)

- *The learning technique used* (batch and online learning)

- *The internal nature of the algorithm* (instance-based and model-based learning)

Let's investigate each criterion separately, starting with the first one: the nature of the problem to solve.

The nature of the problem to solve

In *Chapter 3, Model Evaluation in Comet*, you already encountered two types of machine learning models: supervised learning and unsupervised learning. There is an additional technique, called semi-supervised learning. The following are the main objectives of each technique:

- The main objective of **supervised learning** is to learn the mapping function between the input and the output values. In supervised learning, for each input value in the training set, you also know the output value, and the objective is to build a model that learns the mapping function from input to output. Output values are also known as labels. Once you have built the model, you can use it to predict the labels of new and unseen input values. There are two types of supervised learning: classification and regression. You reviewed the basic concepts behind classification and regression in *Chapter 3, Model Evaluation in Comet*.

- The main objective of **unsupervised learning** is to group input values based on some criteria of similarity. The main types of unsupervised learning include clustering, anomaly detection, and dimensionality reduction. You reviewed the basic concepts behind clustering in *Chapter 3*. We will review the other two types of unsupervised learning in the *Reviewing the scikit-learn package* section.

- **Semi-supervised learning** has the same objective as supervised learning. However, in semi-supervised learning, only a subset of input values is labeled, so the model should combine both supervised and unsupervised learning techniques to predict the output value.

Now that you have learned how to classify machine learning models based on the nature of the problem to solve, we can move on to the next criterion: the learning technique used.

The learning technique used

If you consider the learning technique used, you can classify machine learning systems into the following two types:

- **Batch learning systems**: You perform the training process offline, just once. You cannot update the model on the fly. Usually, this technique requires a lot of computational resources.

- **Online learning systems**: You can train the system incrementally. Usually, in each step, you feed the system with a small batch of data, also called *mini-batches*. You can use this technique when you have a continuous flow of data that you can feed to the model.

Now that you have learned how to classify machine learning models based on the learning technique used, we can move on to the final criterion: the internal nature of the algorithm.

The internal nature of the algorithm

Depending on the internal nature of the algorithm, you can have the following types of learning:

- **Instance-based learning**: The algorithm learns from data in the training set to find a data pattern that can be used for future predictions. In practice, the algorithm predicts the output for new samples based on similarity with the data in the training set. This type of algorithm preserves the original training set that is used at each prediction. The main drawback of this category of algorithms is that the model size could be huge if the training set size is huge. The k-nearest neighbor classifier, for example, falls in this category of algorithms.

- **Model-based learning**: The algorithm builds a mathematical model that approximates the data in the training set. Once the algorithm has built the model, the original dataset can be deleted. The decision tree classifier is an example of model-based learning. The main drawback of this category of algorithms is that you can hardly apply online learning in this case.

Now that you have briefly reviewed how you can classify machine learning systems, we can move on to the next step: exploring machine learning challenges.

Exploring machine learning challenges

When you build a machine learning model, you may encounter different challenges and issues that can be grouped into the following two big families:

- Data challenges
- Model challenges

Let's investigate each family separately, starting with the first: data challenges.

Data challenges

The following table shows the most common data challenges:

Data challenge	Description	Possible countermeasures
Insufficient quantity of data	The dataset is too small to represent the problem.	Searching for new data Enriching the dataset with synthetic samples
Poor quality of data	Presence of duplicates or outliers, missing values, and other similar issues.	Performing data cleaning
Non-representative data	The dataset does not represent the problem or partially represents the problem.	Searching for new data

Figure 8.2 – The most common data challenges

The table also describes some possible countermeasures against the described challenges. For example, if you have an insufficient quantity of data, you may search for new data, or enrich the dataset with new data or even with synthetic data.

Model challenges

The following table shows the most common model challenges:

Model challenge	Description	Possible countermeasures
Overfitted/underfitted model	The model depends on the training set too much and is not able to generalize the problem.	Balancing the dataset
Low performance	The model does not behave as expected.	Hyperparameter tuning Changing the model
Time-consuming/process-consuming training	Training the model requires a lot of time or computational resources.	Parallelization techniques
Concept drift	Over time, the relationship between input and output variables changes and is not represented by the model anymore.	Updating/changing the model

Figure 8.3 – The most common model challenges

When compared to data challenges, model challenges are more complicated to solve because they depend on the specific model you are using. The best way to deal with model challenges is to try different models and select the one with the best performance.

Now that you have briefly reviewed the main machine learning challenges, we can move on to the next step: explaining machine learning models.

Explaining machine learning models

Usually, you see a machine learning model as a black box that takes some features as input and produces an output (also called a target). What happens inside the black box depends on the specific algorithm you are using. To understand how each feature contributes to the output in the model, you can use different techniques. In this section, you will learn about SHAP.

The **SHapley Additive exPlanations (SHAP)** algorithm uses the concept of the Shapley value, which derives from game theory where you have a game and many players. In machine learning, the game is the output of the model and the players are the input features. The Shapley value calculates the contribution of each player to the game. In other words, it calculates the contribution of each input feature to build the final prediction of the model.

For each observation in the training set, you have a different game, thus you can use the Shapley value to analyze a single output each time.

Python provides a package, named `shap`, to calculate the Shapley value and to plot some useful graphs, which help you understand the contribution of each input feature to the output. The `shap` library is fully integrated with Comet.

In the remainder of this section, you will learn the following topics:

- Using the `shap` library
- Integrating the `shap` library in Comet

Let's start with the first topic: using the `shap` library.

Using the shap library

To calculate the Shapley value, you need to perform the following steps:

1. First, you should create an `Explainer` object. The `shap` library provides different types of `Explainer` objects, including, but not limited to, `TreeExplainer`, `GradientExplainer`, `DeepExplainer`, and so on. Each `Explainer` is related to the specific implemented algorithm. The `Explainer` object receives as input the trained model, as shown in the following piece of code:

    ```
    import shap
    shap.initjs()
    explainer = TreeExplainer(model)
    ```

    You import the library, then you need to call the `initjs()` function, and, finally, you can build the `Explainer` object. In this case, we have created a `TreeExplainer`.

2. Once you have created the object, you can calculate the Shapley value for a single observation or many observations as follows:

    ```
    shap_values = explainer.shap_values(X)
    ```

    If the model represents a classification task, the `shap_values()` method returns a list containing the Shapley values for each target class.

3. Using the calculated Shapley values, you can plot different graphs, including bar plots, decision plots, summary plots, and so on. For a complete list of descriptions of the available plots, you can refer to the `shap` official documentation, available at the following link: https://shap.readthedocs.io/en/latest/api_examples.html#plots.

Now that you have seen an overview of the `shap` package, we can investigate how to integrate the graphs produced with `shap` in Comet.

Integrating the shap library in Comet

To integrate one or more graphs produced with the shap library in Comet, you can perform the following steps:

1. Import the Comet library before importing shap, as shown in the following piece of code:

    ```
    from comet_ml import Experiment
    import shap
    shap.initjs()
    ```

2. Create a Comet Experiment and then an Explainer object as follows:

    ```
    experiment = Experiment()
    explainer = shap.Explainer()
    ```

3. Use any of the functions provided by shap to plot a graph as follows:

    ```
    shap_values = explainer.shap_values(X_test)
    shap.summary_plot(shap_values, X_test)
    ```

    The preceding code plots a summary plot for the Shapley values passed as input argument. The graph will be automatically logged in Comet under the **Graphics** section.

Now that you have learned some basic concepts regarding the shap library and how to integrate it in Comet, we can move on to the next step: a review of the main machine learning models.

Reviewing the main machine learning models

A machine learning model is an algorithm that can make predictions for some unseen data based on what it has learned from some training data. As already discussed in the preceding section, you can distinguish machine learning models into two categories, which depend on the specific task you want to solve: supervised models and unsupervised models.

Many machine learning models exist in the literature. In this section, you will review the most popular models used to perform supervised learning and unsupervised learning. We will focus on the following models in detail:

*   Supervised learning
*   Unsupervised learning

In the remainder of the section, you will review an introduction to the most popular machine learning models. For more details, you can read the books proposed in the *Further reading* section. Let's start with the first category of models: supervised learning.

Supervised learning

A supervised algorithm receives a sample as input, performs some computations, and returns a predicted value as output. If the output is a continuous variable, you will have a regression problem, but if the output is a discrete variable, such as a class label, you will have a classification problem.

The following supervised algorithms are some of the most popular ones:

- **Linear regression**: The algorithm searches for a line that best fits the input samples.

  - **Logistic regression**: The algorithm uses the logistic unit to model the input samples. Contrary to what the name might suggest, logistic regression is used to solve classification problems.

  - **Support vector machines (SVM)**: The algorithm searches for a hyperplane that groups training data by common features.

  - **Naive Bayes**: The algorithm assumes that the input features are independent. It uses Bayes's theorem to produce results.

  - **Decision trees**: The algorithm uses a tree to predict the output. Each node of the tree represents an input feature, whereas the leaves of the tree represent the possible outputs.

  - **K-nearest neighbors**: The algorithm uses proximity to group the input samples and predict the output.

  - **Random forest**: The algorithm uses multiple trees to predict the output.

After a brief overview of the most popular algorithms for supervised learning, we can move on to reviewing the algorithms for unsupervised learning.

Unsupervised learning

An unsupervised algorithm aims at grouping similar objects. A typical example of unsupervised learning is clustering.

One of the most popular algorithms for unsupervised learning is **k-means**, which tries to split the dataset into k non-overlapping sub-groups.

Now that you have learned the main machine learning models, we can move on to the next topic: a review of the `scikit-learn` package.

Reviewing the scikit-learn package

`scikit-learn` is a very popular Python package for machine learning. You have already encountered this package in previous chapters. In particular, you have focused on some examples using supervised learning and model selection. However, the `scikit-learn` package also provides other classes and methods, as shown in the following figure:

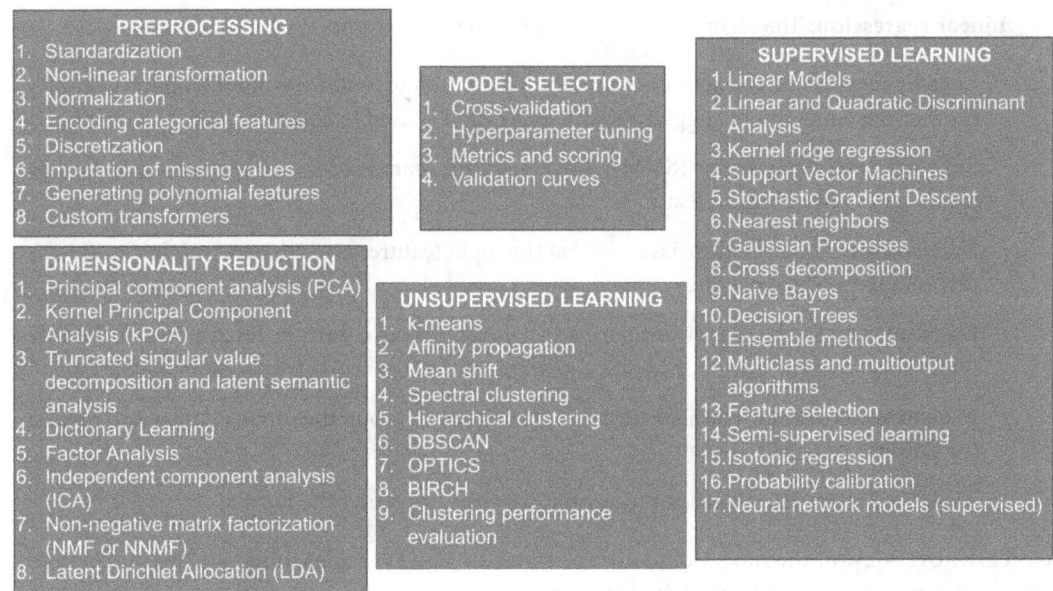

Figure 8.4 – An overview of the scikit-learn package

The package is divided into the following subpackages:

- Preprocessing

- Dimensionality reduction

- Model selection

- Supervised learning

- Unsupervised learning

Let's investigate each subpackage briefly, starting with the first one: preprocessing. For a more in-depth analysis of each subpackage, you can refer to the *Further reading* section at the end of this chapter.

Preprocessing

Preprocessing contains all of the classes and methods that permit us to manipulate the dataset before giving it as input to a machine learning model. You can also use the methods provided by pandas as an alternative to almost all of the classes and methods provided in the preprocessing subpackage.

Preprocessing includes classes for the following:

- **Feature scaling** to perform data standardization and normalization
- **Feature binarization** to convert features into binary values
- **Feature encoding** to convert categorical features into numerical values
- **Non-linear transformations** to apply non-linear transformations to features, such as power transform
- **Other classes** to perform other specific transformations

Now that you have briefly reviewed the most important classes provided by the preprocessing package, we can analyze the next subpackage: dimensionality reduction.

Dimensionality reduction

Dimensionality reduction is a technique that reduces the number of input features from a high-dimensional space to a low-dimensional space.

Dimensionality reduction is especially useful when you have millions of input features, which could slow down the model training process. However, although the use of dimensionality reduction techniques speeds up the training process, it still leads to the loss of information.

Dimensionality reduction is also useful for data visualization because if you reduce the number of features down to two or three, you can easily plot them and perform a visual exploratory data analysis.

Figure 8.4 shows the most important techniques provided by scikit-learn to perform dimensionality reduction. Among them, the most popular technique is **principal component analysis (PCA)**. The following piece of code shows how to implement PCA in scikit-learn:

```
from sklearn.decomposition import PCA
pca = PCA(n_components=2)
X_reduced = pca.fit_transform(X)
```

We have used the PCA class, which receives as input the final dimensionality of features (n_components). To get the reduced feature dataset, you should call the fit_transform() method on the input features (X).

Now that you have seen an overview of the dimensionality reduction package, we can analyze the next subpackage: model selection.

Model selection

Model selection includes the following techniques that help you select the best model for your problem:

- Cross-validation
- Hyperparameter tuning
- Metrics and curves

Let's investigate the first two techniques, starting with cross-validation.

> **Metrics and curves**
>
> You remember that model selection also includes the study of metrics and curves for model evaluation. We reviewed this aspect in *Chapter 3, Model Evaluation in Comet*, so you can refer to that for further details.

Cross-validation

Cross-validation is a technique that permits you to calculate the performance of a machine learning algorithm on a dataset. k-fold cross-validation is one of the most popular techniques to perform cross-validation. It divides the datasets in k non-overlapping folds, then it fits k models and calculates the performance of each one. At each iteration, it uses k-1 folds as the training set and the remaining fold as the test set, as shown in the following figure:

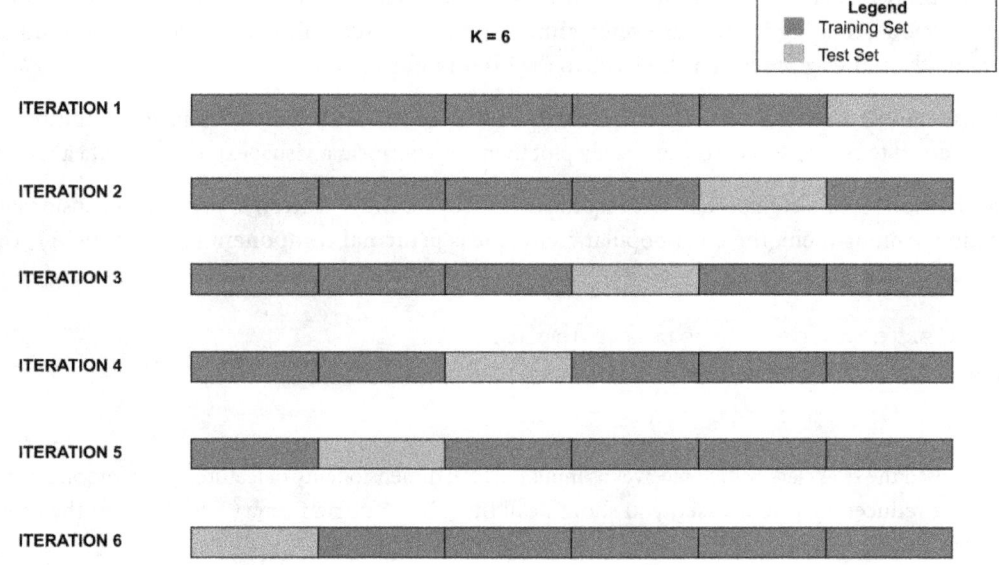

Figure 8.5 – k-fold cross-validation with k = 6

The figure shows how the k-fold cross-validation algorithm builds the training and test sets at each iteration in the case of k = 6. A common value for k is 10, but, usually, you should run different tests to calculate the most appropriate value for k. In the last section of this chapter, you will learn a practical strategy to calculate the best value for k in a practical example.

The performance of the model is calculated as the average value of the results of the k models.

`scikit-learn` provides the `KFold` class to perform basic k-fold cross-validation. This class receives the following parameters as input:

- `n_splits`: The number of folds, representing the parameter k.
- `shuffle`: A Boolean representing whether or not to shuffle the dataset before splitting it into folds.
- `random_state`: If `shuffle` is `True`, `random_state` affects how `shuffle` is performed.

`scikit-learn` also provides other classes to perform cross-validation, such as `StratifiedKFold` and `GroupKFold`. For more details on them, you can refer to the `scikit-learn` official documentation, available at the following link: `https://scikit-learn.org/stable/modules/cross_validation.html`.

Now that you have reviewed the basic concepts behind cross-validation, we can move on to the next point: hyperparameter tuning.

Hyperparameter tuning

Hyperparameter tuning permits you to search for the best parameters of a specific machine learning algorithm. Usually, you perform hyperparameter tuning in combination with cross-validation. You have already learned how to perform hyperparameter tuning in Comet in *Chapter 4, Workspaces, Projects, Experiments, and Models*. In this chapter, you will review the classes and methods provided by `scikit-learn` to perform hyperparameter tuning.

In general, you define a grid of parameters to test and you pass it as input of an algorithm, which fits as many models as the different combinations of parameters. Finally, the algorithm calculates the model with the best performance and returns it as the best model.

`scikit-learn` implements different algorithms to perform hyperparameter tuning, including, but not limited to, the following ones:

- `GridSearchCV`: The algorithm tests all of the possible combinations of parameters.
- `RandomizedSearchCV`: The algorithm performs a randomized search over parameters.
- `HalvingGridSearchCV`: The algorithm first tests all of the parameters on a small dataset, then iteratively discards the parameters with the lowest performance and continues the tests on the remaining parameters by adding new samples to the dataset.

In the last section of this chapter, you will see a practical example that uses the GridSearchCV algorithm. For more details on how to use the other algorithms, you can refer to the scikit-learn official documentation, available at the following link: https://scikit-learn.org/stable/modules/classes.html#module-sklearn.model_selection.

Now that you have reviewed the basic concepts behind hyperparameter tuning, we can move on to the next point: supervised and unsupervised learning.

Supervised and unsupervised learning

scikit-learn implements the most popular algorithms to perform both supervised and unsupervised learning. In all cases, to implement a model, you should perform the following steps:

1. Create the model as follows:

```
model = MyModel()
```

You should replace the MyModel() class with the specific class provided by scikit-learn, such as KNeighborsClassifier or LinearRegression.

2. Fit the model with the training set as follows:

```
model.fit(X_train, y_train)
```

3. Use the trained model to predict the output for unseen inputs as follows:

```
y_pred = model.predict(X_test)
```

The preceding steps describe the very basic operations you should perform to build a machine learning model. In the following section of this chapter, you will implement a more complex example that also considers cross-validation and hyperparameter tuning as well as the integration with Comet. So, let's move on to this practical example.

Building a machine learning project from setup to report

In this section, you will further improve the practical example of diamond cuts described in *Chapter 3, Model Evaluation in Comet*, and deployed in *Chapter 6, Integrating Comet into DevOps*. In this chapter, you will focus on the following aspects:

- Reviewing the scenario
- Selecting the best model
- Calculating the SHAP value
- Building the final report

Let's start with the first step: reviewing the scenario.

Reviewing the scenario

As our use case, we will use the diamonds dataset provided by ggplot2 under the MIT licenses (https://ggplot2.tidyverse.org/reference/diamonds.html) and available on Kaggle as a CSV file (https://www.kaggle.com/shivam2503/diamonds). With respect to the original version, already described in *Figure 3.3* in *Chapter 3*, we use the cleaned version produced in the same chapter and shown in the following figure:

	carat	color	clarity	depth	Table	price	x	y	z	target
0	0.23	E	SI2	61.5	55.0	326	3.95	3.98	2.43	Gold
1	0.21	E	SI1	59.8	61.0	326	3.89	3.84	2.31	Gold
2	0.23	E	VS1	56.9	65.0	327	4.05	4.07	2.31	Silver
3	0.29	I	VS2	62.4	58.0	334	4.2	4.23	2.63	Gold
4	0.31	J	SI2	63.3	58.0	335	4.34	4.35	2.75	Silver
5	0.24	J	VVS2	62.8	57.0	336	3.94	3.96	2.48	Gold
6	0.24	I	VVS1	62.3	57.0	336	3.95	3.98	2.47	Gold
7	0.26	H	SI1	61.9	55.0	337	4.07	4.11	2.53	Gold
8	0.22	E	VS2	65.1	61.0	337	3.87	3.78	2.49	Silver

Figure 8.6 – The cleaned version of the diamonds dataset

The dataset contains 53,940 rows and 10 columns. The objective of this scenario is to build a classification model that, when given a set of input features, predicts the target category (Gold, Silver). We suppose that the input features are stored in a variable called X and the target in a variable called y, as shown in the following piece of code:

```
X = df.drop("target", axis = 1)
y = df["target"]
```

We have supposed that the dataset is loaded as a pandas DataFrame. We preprocess the dataset by encoding labels and scaling numeric values. For more details on how to perform these operations, you can refer to *Chapter 3*.

Now that you have reviewed the employed dataset, we can move on to the next step: selecting the best model.

Selecting the best model

To select the best model, we compare the performance of the following four classification algorithms, as already described in *Chapter 3, Model Evaluation in Comet*: random forest, decision tree, Gaussian Naive Bayes, and k-nearest neighbors. For each algorithm, we search for the optimal model by performing the following steps:

- Searching for the best number of folds for cross-validation
- Performing hyperparameters tuning with cross-validation

We will build a Comet experiment for each test performed. We use accuracy as a metric to select the best model.

Let's start with the first step: searching for the best number of folds for cross-validation.

Searching for the best number of folds for cross-validation

To get familiar with cross-validation, we first run a simple experiment, which compares the performance of each algorithm with and without the use of cross-validation, with a fixed number of folds (10). When you build the model without cross-validation, you should use the train/test splitting operation to fit the model on the training set and evaluate it on the test set. On the other hand, if you build the model with cross-validation, you can use the whole dataset because cross-validation already performs train/test splitting. Perform the following steps to find the best number of folds for cross-validation:

1. We split the dataset into training and test sets as follows:

    ```
    from sklearn.model_selection import train_test_split
    X_train, X_test, y_train, y_test = train_test_split(X, y,
    test_size=0.10, random_state=42)
    ```

    We reserve 10% of records for the test set and the remaining for the training set.

2. We define a function, named `run_experiment()`, that builds an `Experiment` in Comet and then logs in Comet the accuracy of the model passed as an argument, either when using cross-validation or not, as follows:

    ```
    from sklearn.model_selection import KFold
    from sklearn.model_selection import cross_val_score
    from sklearn.metrics import accuracy_score
    import numpy as np
    def run_experiment(ModelClass, name, n_splits):
        experiment = Experiment()
    ```

```
        experiment.set_name(name)
        experiment.add_tag(name)

        cv = KFold(n_splits=n_splits, random_state=1,
    shuffle=True)
        model = ModelClass()
        # calculating accuracy with KFold
        scores = cross_val_score(model, X, y,
    scoring='accuracy', cv=cv)
        experiment.log_metric('accuracy-cv', np.mean(scores))

        # calculating accuracy without KFold
        model.fit(X_train, y_train)
        y_pred = model.predict(X_test)
        experiment.log_metric('accuracy', accuracy_score(y_
    test, y_pred))
```

We use KFold() with the shuffle=True parameter to perform an initial shuffle of the dataset. To calculate the accuracy in the case of k-fold, we use the cross_val_score() function. We log the accuracy obtained from cross-validation as accuracy-csv and the accuracy obtained without cross-validation as accuracy.

3. Now we can run the experiments simply by calling the run_experiment() function for the four algorithms as follows:

```
from sklearn.ensemble import RandomForestClassifier
from sklearn.tree import DecisionTreeClassifier
from sklearn.naive_bayes import GaussianNB
from sklearn.neighbors import KNeighborsClassifier

n_splits = 10
run_experiment(RandomForestClassifier, 'RandomForest',n_
splits)
run_experiment(DecisionTreeClassifier,
'DecisionTreeClassifier',n_splits)
run_experiment(GaussianNB, 'GaussianNB',n_splits)
run_experiment(KNeighborsClassifier,
'KNeighborsClassifier',n_splits)
```

We set the number of splits to 10, so for each iteration of K-Fold, 90% of the dataset is reserved for the training set and 10% for the test set.

4. Now we can see the results in Comet. We access the Comet dashboard and we click on the **Experiments** tab. Then, we select the following columns to show the results by clicking the **Columns** button, then **DURATION, ACCURACY-CV**, and **ACCURACY**. The following figure shows the results of this operation:

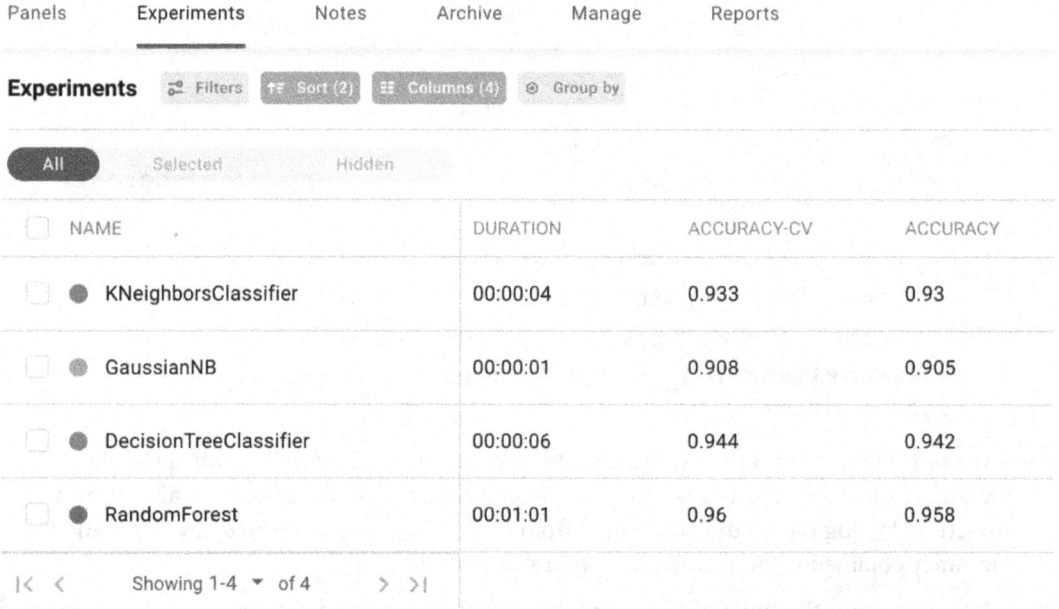

Figure 8.7 – The results of experiments in Comet

We note that, in general, all of the algorithms perform better with cross-validation. Gaussian Naive Bayes is the fastest algorithm while random forest is the slowest.

5. Now we build a **parallel coordinates chart**, which shows the behavior of each algorithm across the two metrics, **accuracy-cv** and **accuracy**. Comet provides the parallel coordinates chart as a built-in panel, so you can simply add it by clicking on **Add | New Panel | BUILT-IN | Parallel Coordinates Chart**. When the popup window opens, you can select **accuracy** as the target variable and add **accuracy-cv** on the **Y-axis**. The following figure shows the produced panel:

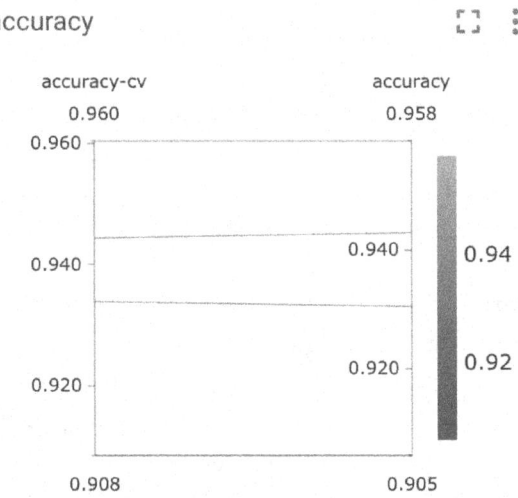

Figure 8.8 – The parallel coordinates chart for accuracy with and without cross-validation

So far, we have set the number of folds to 10. However, this might not be the best choice. To choose the best value for the number of folds, we need to improve the preceding code in the following step.

6. We define a function, called `run_experiment_kfold_numbers()`, that receives the maximum number of folds as input and calculates the accuracy of the model passed as an argument as the number of folds changes. Then, the function returns the number of folds with the highest score. The following piece of code implements the described operations:

```
def run_experiment_kfold_numbers(ModelClass, name, max_n_
splits):
    experiment = Experiment()
    experiment.set_name(name + '-kfold')
    experiment.add_tag(name + '-kfold')
    experiment.add_tag('kfold')


    scores_list = []
    min_n_splits = 2
    for n_splits in range(min_n_splits, max_n_splits):
        model = ModelClass()
        # calculating accuracy with KFold
        cv = KFold(n_splits=n_splits, random_state=1,
shuffle=True)
```

```
        scores = cross_val_score(model, X, y,
scoring='accuracy', cv=cv)
        mean_scores = np.mean(scores)
        scores_list.append(mean_scores)
        experiment.log_metric('accuracy-cv', mean_scores,
step = n_splits)


    # get the best number of folds
    best_score_value = np.max(scores_list)
    return scores_list.index(best_score_value) + min_n_
splits
```

We have created a Comet experiment and we have added the kfold tag to it. In addition, we have set the minimum number of folds to 2 to make cross-validation work in the minimal condition.

7. Now we can call the function for each model as follows:

```
max_n_splits = 20
random_forest_kfold_value = run_experiment_kfold_
numbers(RandomForestClassifier, 'RandomForest',max_n_
splits)
decision_tree_kfold_value = run_experiment_
kfold_numbers(DecisionTreeClassifier,
'DecisionTreeClassifier',max_n_splits)
gaussian_nb_kfold_value = run_experiment_kfold_
numbers(GaussianNB, 'GaussianNB',max_n_splits)
knn_kfold_value = run_experiment_
kfold_numbers(KNeighborsClassifier,
'KNeighborsClassifier',max_n_splits)
```

We have set the maximum number of folds (max_n_splits) to 20.

8. After running the experiment, we are ready to see the results in Comet. For example, you can select KNeighborsClassifier-kfold and analyze the trend in accuracy while varying the number of folds, as shown in the following figure:

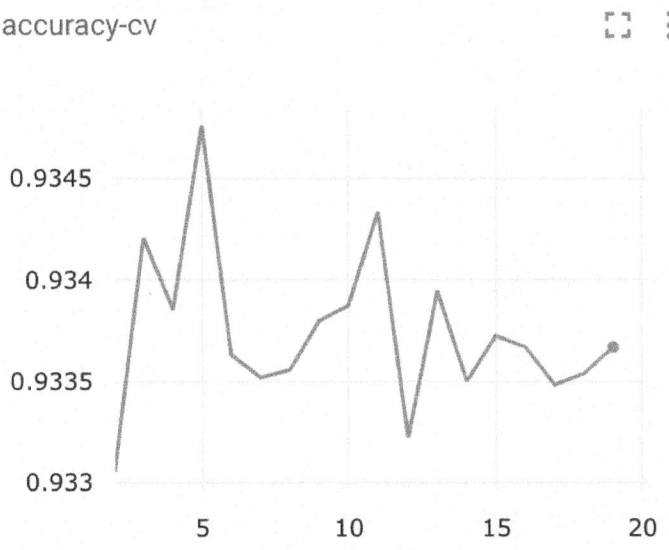

Figure 8.9 – Accuracy versus the number of folds for the k-nearest neighbor classifier

9. Now we can compare all of the produced models. Since the current Comet project does not contain only the experiments just completed, but also those executed at the beginning of this section, we need to build a filter that shows only the latest experiments. These experiments have a tag equal to kfold. We can build a filter in the Comet dashboard, as explained in *Chapter 4*, to select only those experiments. You can select the **Filters** button and then **Add filter | tag | is | kfold | Done**. You should be able to see only the experiments associated with a variable number of folds, as shown in the following figure:

Experiments ⇄ Filters (1) ↑↓ Sort (2) ☰ Columns (5) ⊘ Group by

	NAME	DURATION	ACCURACY-CV	ACCURACY	TAGS	
📌 ●	KNeighborsClassifier-kfold	00:00:45	0.933		KNeighborsClassifier-kfold	kfold
☐ ●	GaussianNB-kfold	00:00:06	0.908		GaussianNB-kfold	kfold
☐ ●	DecisionTreeClassifier-kfold	00:00:51	0.944		DecisionTreeClassifier-kfold	kfold
☐ ●	RandomForest-kfold	00:16:18	0.96		RandomForest-kfold	kfold

|< < Showing 1-4 ▼ of 4 > >|

Figure 8.10 – The experiments with a tag equal to kfold shown in the Comet dashboard

10. In Comet, you can add a new panel that shows the accuracy compared to the number of folds for all the experiments. You can click on **Add | New Panel | BUILT-IN | Line Chart**. On the **Y-axis**, you can select **accuracy-cv** and then leave the other fields to the default values. Finally, click on **Done**. The following figure shows the resulting panel:

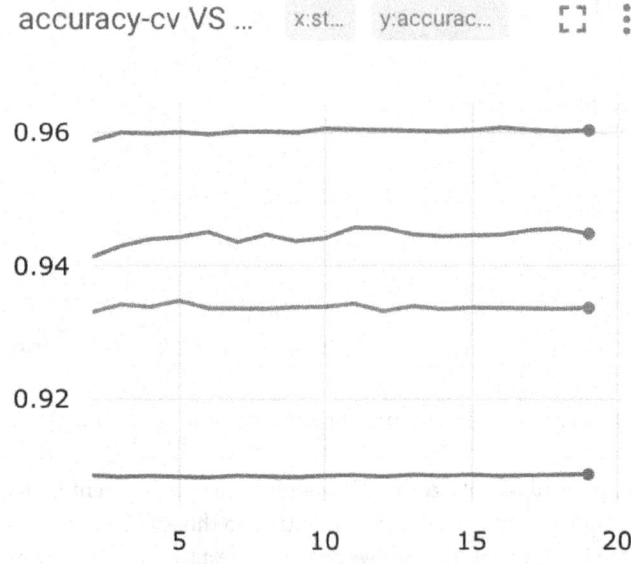

Figure 8.11 – Accuracy versus the number of folds for all of the algorithms

The panel is interactive, so if you click on a line, you should see the name of the associated model.

The implemented code returns the best number of folds for each model; thus, we can use the obtained values to perform model optimization. So, you are ready to move on to the next step: performing hyperparameters tuning with cross-validation.

Performing hyperparameters tuning with cross-validation

In *Chapter 4, Workspaces, Projects, Experiments and Models*, you have already learned the concept of Comet Optimizer, which permits you to perform model optimization. In this section, you will use the GridSearchCV() class provided by scikit-learn to tune the hyperparameters of each model.

We will perform the following steps:

1. We define the following function named `run_experiment_grid_search()`:

```python
from sklearn.model_selection import GridSearchCV
def run_experiment_grid_search(ModelClass, name, n_
splits, params):
    model = ModelClass()
    clf = GridSearchCV(model, params, cv = n_splits)
    clf.fit(X_train, y_train)

    for i in range(len(clf.cv_results_['params'])):
        experiment = Experiment()
        experiment.add_tag(name + '-gs')
    experiment.add_tag('gs')
        for k,v in clf.cv_results_.items():
            if k == "params":
                experiment.log_parameters(v[i])
            else:
                experiment.log_metric(k,v[i])
```

    The function receives the same parameters as the `run_experiment()` function as well as the dictionary of parameters to tune. Then, it builds a `GridSearchCV()` object with the model passed as an argument and fits it. Finally, for each tested model, the function builds a Comet Experiment and logs its parameters and its metrics.

2. We run the experiments and then we access the associated project in Comet. Since we have saved the experiments in the same project as the cross-validation example, we need to build a filter that shows only the experiments related to hyperparameter tuning. In the Comet dashboard, you can select the **Filters** button and then **Add filter | tag | is | gs | Done**.

3. Now we group experiments by model. To perform this operation, you can click on the **Group by** button and then **Tags | Done**.

4. You can select the columns to show by clicking on the columns button and adding the columns you prefer. For example, you could choose the parameters you have tuned.

5.   Finally, you can sort results by decreasing the mean test score by clicking on the **Sort** button and then selecting the mean test score as the criterion. You can also click on the bottom arrow to set the decreasing order. You should also delete all of the other parameters appearing in the popup window by simply clicking on the **X** button. The following figure shows a piece of the resulting screen for the decision tree classifier:

	NAME	DURATION	MEAN_TEST_SCORE ↓	CRITERION
	Tags: DecisionTreeClassifier-gs			
	managing_university_2092	00:00:01	0.946	gini
	random_carp_9145	00:00:03	0.946	gini
	stale_water_2279	00:00:04	0.945	gini
	sticky_gulf_2120	00:00:02	0.945	entropy
	greasy_wolf_9654	00:00:02	0.944	entropy
	colonial_jig_527	00:00:02	0.944	entropy

Experiments   Filters (1)   Sort (1)   Columns (6)   Group by (1)

All   Selected   Hidden

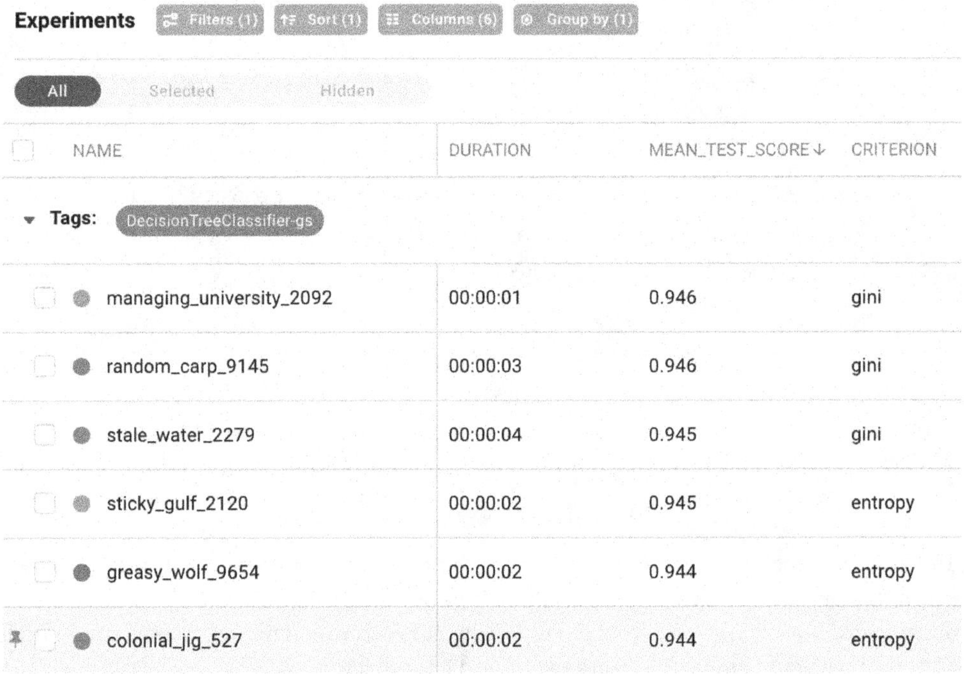

Figure 8.12 – The results of experiments in Comet for the decision tree classifier

The previous figure shows the duration, the main test score (accuracy), and the tested criterion for each experiment with the decision tree classifier. Experiments are ordered by decreasing the mean test scores, thus making it easy to recognize the best model.

6.   Eventually, we can select the best model for our purpose from the different algorithms we tested. For example, we could select the best model that has a mean test score greater than 0.94 and a duration less than or equal to 1 second. You can perform this task easily in Comet as follows:

I.   You can select the **Filters** button, remove the current filter, and then select **Add filter | mean_test_score | greater or equal | 0.9465660707805703 | Done**.

II.   Then, you can add another filter remaining on the same popup window and click on **Add filter | duration | less or equal | 1 sec | Done**. You should see the best model, as shown in the following figure:

Figure 8.13 – The best model of the experiments

The best model is the decision tree with a mean test score equal to 0.946 and a duration of 1 second.

Now that you have learned how to select the best model using Comet, we can move on to the next step: calculating the SHAP value.

Calculating the SHAP value

In the Comet dashboard, you have seen that the best model is a decision tree with the following parameters:

- `min_samples_split=4`
- `criterion='gini'`

We can build a new experiment for this model and calculate the Shapley values for a single observation.

We proceed as follows:

1. First, we import all the needed libraries and build a Comet experiment as follows:

```
from comet_ml import Experiment
import shap
shap.initjs()
experiment = Experiment()
experiment.set_name('BestModel')
```

2. Then, we build and train the model as follows:

```
from sklearn.tree import DecisionTreeClassifier
import numpy as np
model = DecisionTreeClassifier(min_samples_split=4,
criterion='gini')
model.fit(X_train,y_train)
```

3. Since the model is a decision tree, we create a `TreeExplainer` object as follows:

```
explainer = shap.TreeExplainer(model)
```

4. For simplicity, we decide to calculate the Shapley value related to the first sample in the test set, so we extract it as follows:

```
sample = X_test.iloc[0]
```

5. Now we generate the decision plot. Since our model is a classifier, we should calculate the Shapley value for each target class. Thus, we define a function, which receives as input the target class (either 0 or 1), and plots the decision plot, as follows:

```
def decision_plot(target):
    print(f"target: {target}")
    shap_values = explainer.shap_values(sample)[target]
    print(f"expected value: {explainer.expected_
value[target]}")
    shap.decision_plot(explainer.expected_value[target],
shap_values, sample)
```

The function calculates the Shapley values by calling the shap_values() method of the explainer. Then, it calls the decision_plot() function, which receives as input the expected value, the Shapley values, and the original sample.

6. We call the decision_plot() function for each target class as follows:

```
decision_plot(0)
decision_plot(1)
```

The following figure shows the resulting plots:

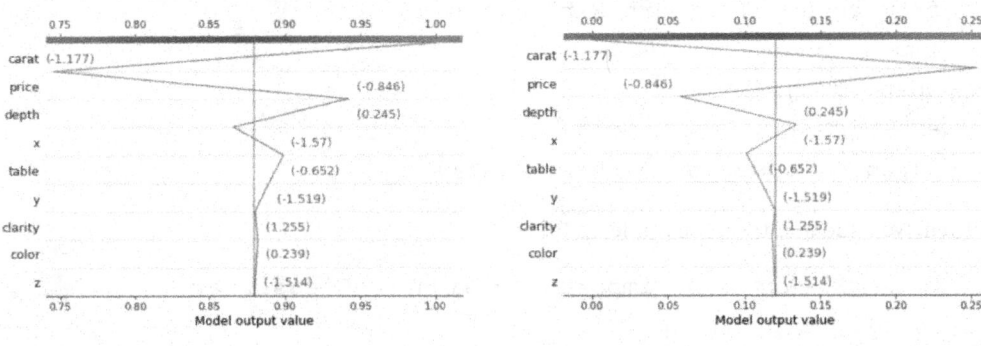

Figure 8.14 – The decision plot

On the left, there is the decision plot related to target class 0, while on the right we see the plot for target class 1. The vertical line represents the predicted value for each class. You can see that the predicted value for target class 0 is greater than that obtained for target class 1. This corresponds to real data in the test set where the actual class is 0. The decision plot shows the contribution of each feature (available on the left of the figure) to the final result. The red lines indicate a positive contribution, whereas the blue lines indicate a negative contribution.

7. Now, we build a summary plot as follows:

```
shap_values = explainer.shap_values(X_test)
shap.summary_plot(shap_values, X_test)
```

First, we extract the Shapley values from the explainer through the shap_values() method, then we call the summary_plot() function for all of the observations in the test set. The following figure shows the resulting plot:

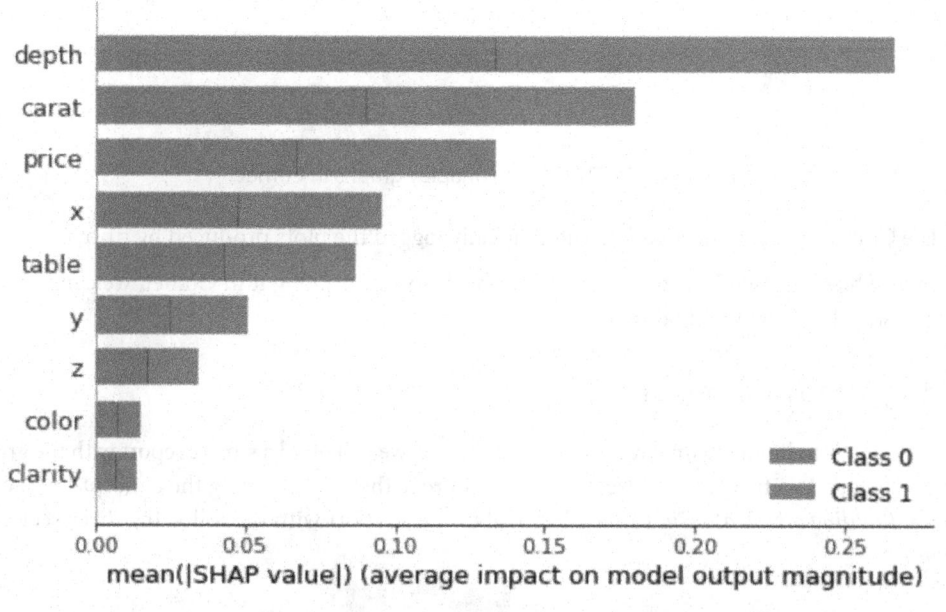

Figure 8.15 – The summary plot

The preceding figure shows how the average Shapley value over all of the observations in the test set depends on the input features. You can see that **depth** is the most influential feature, followed by **carat** and **price**. The less important feature is **clarity**.

8.   We run the experiment and we can see the results in Comet, in the **Graphics** section, as shown in the following figure:

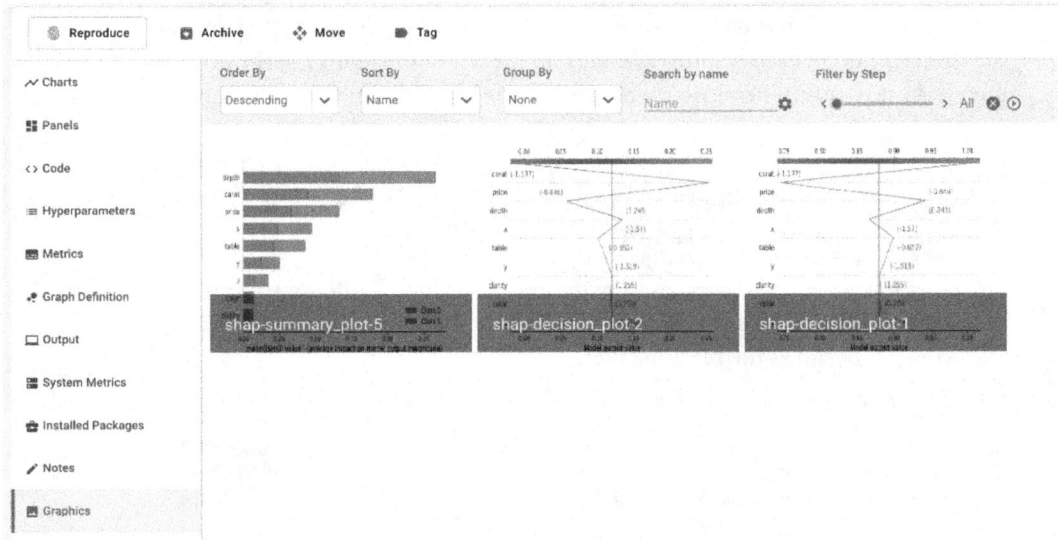

Figure 8.16 – The logged Shapley graphs in Comet

The Comet experiment class has automatically logged the plots produced by shap.

Now that you have learned how to calculate the Shapley value and log it in Comet, we can move on to the final step: building the final report.

Building the final report

Now you are ready to build the final report. In this example, we will build a simple report with the graphs we just produced. As a further exercise, you could improve them by applying the concepts learned in *Chapter 5, Building a Narrative in Comet*. You will create a report with the following three sections:

- Cross-validation
- Hyperparameter tuning
- The Shapley value

Let's start with the first section: cross-validation.

Cross-validation

In the Comet dashboard, you can perform the following operations:

1. Click on the **Reports** tab, then on **New Report** add a report name, such as *Selecting the best model for diamond cut prediction*. Finally, click on **Save**.

2. Come back to your main project by simply clicking on the project name on the path bar located in the top-left corner of the screen, as shown in the following figure:

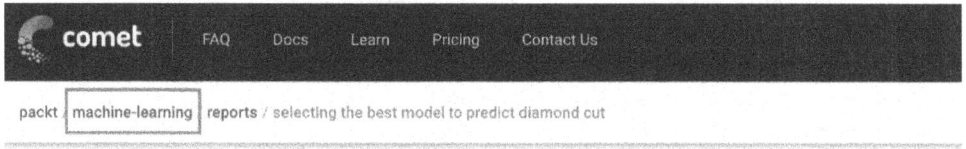

Selecting the Best model to Predict Diamond Cut

Figure 8.17 – A quick way to come back to the main project through the path bar

3. In the **Panels** section, you should be able to see the graphs described in *Figure 8.8* and *Figure 8.11*. If not, you can view them by removing all of the filters.

4. Click on the **Add** button, then **Add to report**, then your report name. A new section is added to your report containing the graphs. If you access the report, you can customize the section we just created.

Hyperparameter tuning

To add a new section in the report on hyperparameter tuning, you can perform the following operations:

1. You can create a new section in the report related to hyperparameter tuning. To add a new section, simply click at the end of the preceding section. A new button appears called **Add section here**. You can click on it to add a new section.

2. Under the **Experiments** part of the new section, you can select **Add filter | tag | is | gs**. We are simply repeating the steps performed for cross-validation in the preceding section, this time for hyperparameter tuning. If you want to add all of the experiments to the panel, you should click on the arrow near the **Show 1-25** text button, located at the bottom right part of the section, and select 50.

3. Then, click on **Add Panel | BUILT-IN | Bar Chart | Add**. Now you can customize your graph by selecting the **mean_test_score** for the **Y-Axis** as well as selecting the **Y-Axis ranges**. Finally, click on the **Done** button.

The Shapley value

You can add a new section to the report, as described in the preceding section. Then, within this new section, you can perform the following operations:

1. Under the **Experiments** part of the new section, you can select **Add filter | Name | is | BestModel**.

2. Now you can add a panel by clicking on the **Add Panel** button, then **FEATURED |Show images | Add | Done**.

3. A new panel is added to the report containing the figures described in *Figure 8.16*.

The report is finally ready. You can further improve it by applying the techniques described in the previous chapters.

Summary

We just completed the journey to building a machine learning model in `scikit-learn` and tracking it in Comet!

Throughout this chapter, we described some general concepts regarding machine learning as well as the main structure of the `scikit-learn` package. We also illustrated some important concepts, such as cross-validation, hyperparameter tuning, and the Shapley value.

In the last part of the chapter, you implemented a practical use case that showed you how to track some machine learning experiments in Comet as well as how to build a report with the results of the experiments.

In the next chapter, we will review the basic concepts related to natural language processing and how to perform it in Comet.

Further reading

- Géron, A. (2019). *Hands-on Machine Learning with Scikit-Learn, Keras, and TensorFlow: Concepts, Tools, and Techniques to Build Intelligent Systems*. O'Reilly Media, Inc.

- Raschka, S. Liu, Y.H. and Mirjalili V. (2022). *P Machine Learning with PyTorch and Scikit-Learn*. Packt Publishing Ltd.

9
Comet for Natural Language Processing

Natural Language Processing (**NLP**) is a subfield of artificial intelligence, aimed at making computers capable of understanding human natural language in the form of both text and spoken words. You can use NLP applications to build virtual assistants, such as Alexa or Siri, sentiment analyzers, document translators, chatbots, and document classifiers. In this chapter, you will review the main concepts behind NLP, including the basic NLP pipeline and how to transform texts into data structures.

Over the last year, different open source tools and libraries have been implemented to perform NLP, including Spark NLP, spaCy, and **Natural Language Toolkit** (**NLTK**). In this chapter, you will review the Spark NLP library and see how to integrate it with Comet. We will focus on how to perform NLP on texts, although you can also apply NLP to audio documents.

Training a good NLP model can be very time- and process-consuming because it usually requires a huge quantity of texts as input. For this reason, you can find many pretrained models available on the web, such as those provided by Hugging Face, a very popular website, used by about 5,000 organizations, including Google, Facebook, Microsoft, and Amazon Web Services. In this chapter, you will review the most popular hubs for pretrained models.

In the last part of this chapter, you will implement a practical use case that uses the Spark NLP library and track the results in Comet.

This chapter is organized as follows:

- Introducing basic NLP concepts
- Exploring the Spark NLP package
- Setting up an environment for Spark NLP
- Using NLP, from project setup to report building

Before starting to review the basic NLP concepts, let's install the required software needed to run the examples described in this chapter.

Technical requirements

We will run all the experiments and codes in this chapter using Python 3.8. You can download it from the official website, https://www.python.org/downloads/, choosing the 3.8 version.

The examples described in this chapter use the following Python packages:

- `comet-ml 3.23.0`
- `pandas 1.3.4`
- `scikit-learn 1.0`

We already described these packages and how to install them in *Chapter 1, An Overview of Comet*, so please refer back to that for further details on installation.

In addition, the running examples will need other specific requirements, which will be described in the *Setting up the environment for Spark NLP* section of this chapter.

Now that you have installed all the software needed in this chapter, let's look at how to use Comet for NLP, starting by reviewing some basic concepts.

Introducing basic NLP concepts

NLP is a subfield of artificial intelligence, aimed at analyzing and synthesizing human language and speech. You can use these models for different purposes, such as translation, chatbots, spam filters, grammar correction software, and voice assistants. NLP has become very popular in the last few years, thanks to the spread of huge quantities of text that can be analyzed to build very domain-specific tools.

The section is organized as follows:

- Exploring the NLP workflow
- Classifying NLP systems
- Exploring NLP challenges
- Reviewing the most popular models' hubs

Let's start from the first step, exploring the NLP workflow.

Exploring the NLP workflow

The following figure shows the simplest NLP workflow:

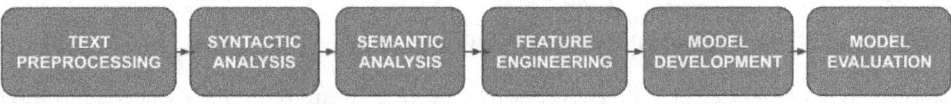

Figure 9.1 – The simplest NLP workflow

The workflow involves the following steps:

- **Text preprocessing** – This step involves preparing text for further analysis. Typically, during this phase, the text is split into sentences, and stop words are removed. Then, sentences are split into tokens and, finally, normalized.

- **Syntactic analysis** – This step involves sentence parsing and **Part of Speech** (**PoS**) tagging.

- **Semantic analysis** – This step involves **Named Entity Recognition** (**NER**) and concept recognition.

- **Feature engineering** – In this step, you extract features to build a machine learning model or a rule-based system, which performs a specific NLP task, such as text classification or translation.

- **Model development** – This step involves the development of the model, which can be either a machine learning or a rule-based model. Usually, you can use a machine learning model for a classification task, while you can use a rule-based model to extract some patterns from text.

- **Model evaluation** – This final step calculates the performance of the developed model.

The following table shows an example of workflow, which includes sentence splitting, stop word removal, tokenization, PoS tagging, and NER, for the following text:

Yesterday I went to Rome. I visited the Colosseum.

Sentence splitting	Stop words removal	Tokenization	PoS tagging	NER
Yesterday I went to Rome.	Yesterday I went to Rome	Yesterday	Noun	Rome
		I	Pronoun	
		went	Verb	
		to	Preposition	
		Rome	Noun	
I visited the Colosseum.	I visited the Colosseum	I	Pronoun	Colosseum
		visited	Verb	
		the	Determiner	
		Colosseum	Noun	

Figure 9.2 – A part of the NLP workflow for the text "Yesterday I went to Rome. I visited the Colosseum."

The text contains two sentences, separated by the intermediate dot (full stop). The stop word removal task simply removes the final full stop from each sentence. Tokenization splits each sentence into tokens, while PoS tagging recognizes the parts of speech for each token. Finally, NER recognizes two entities: *Rome* and *Colosseum*.

Now that we have reviewed the basic NLP workflow, we can move to the next step, classifying NLP systems.

Classifying NLP systems

Generally speaking, you can classify NLP systems into two big families:

- **Functionality libraries** – A functionality library is a collection of functions aiming at implementing some specific NLP tasks. Usually, these functions are independent of each other, and if, for example, you perform a PoS tagging task, the relative function also performs the operations that must be performed upstream, such as sentence splitting and tokenization. These libraries are good for research purposes, but they perform less well than the annotation libraries, since they do not provide any combination among the different tasks.

- **Annotation libraries** – An annotation library ensures that each NLP task is synchronized with the previous and subsequent ones, thanks to the concept of the document-annotation model. Thus, if you have to perform a PoS tagging task, you will first have to perform separately all the tasks that prepare the text for PoS tagging.

In this chapter, we will focus on Spark NLP, which is an annotation library.

Now that we have reviewed the basic classification of NLP systems, we can move to the next step, exploring NLP challenges.

Exploring NLP challenges

Although NLP has evolved enormously in recent years, thanks to the dissemination of large quantities of texts to analyze, many challenges still appear, including the following:

- **Contextual words** – A word may have multiple meanings, which depend on the context where it is inserted. Although a human can recognize the specific meaning easily, this process is not so trivial for a machine.

- **Synonyms** – Different words may have the same meaning. This issue is complementary to the previous one, but it has the same effects on machines, which may encounter difficulties associating the same meaning with different words.

- **Irony and sarcasm** – Usually, a word will be either positive or negative. However, when you use irony and sarcasm, you invert the original polarity of a word; thus, a negative word is used to express a positive concept, and vice versa. For a machine, recognizing irony and sarcasm can be very problematic because it associates a single polarity with each word.

- **Ambiguity** – This aspect refers to the possibility that a sentence has multiple interpretations. There are different levels of ambiguity, including lexical, semantic, and syntactic ambiguity.

- **Errors in text or speech** – It can happen that a word is not written correctly because it includes a typo or is misspelled. For a machine, recognizing this kind of error can be difficult.

- **Colloquialisms and slang** – Recognizing colloquialisms and slang in spoken language can be problematic for a machine, especially when the documents to analyze come in the form of audio.

- **Domain-specific language** – Some pretrained models work very well on generic texts, but they need to be adapted to specific domains. In this case, it may be difficult to find texts to train the models.

Now that we have reviewed some popular challenges of NLP, we can move toward the next step, an overview of the most popular model hubs for NLP.

Reviewing the most popular models' hubs

Training an NLP model on large amounts of data can be time-consuming as well as resource-consuming. For this reason, over time, communities, companies, and researchers have combined their efforts and produced pretrained models that are ready to use. You can use these models directly as they are, or you can adapt them to a specific domain. There are several websites, or **model hubs**, that collect pretrained models that can be downloaded and used for free.

The most popular aforementioned model hubs include the following:

- Model Zoo
- TensorFlow Hub
- Hugging Face

Let's briefly describe each hub, starting from the first one, Model Zoo.

Model Zoo

Model Zoo is a model hub for deep learning models, available at the following link: `https://modelzoo.co/`. Models are organized by category and framework. Categories include computer vision, NLP, generative models, reinforcement learning, unsupervised learning, and audio and speech.

Frameworks include TensorFlow, Caffe, Caffe2, PyTorch, and Keras. For each model, there is a detailed description that shows how to install and use it.

TensorFlow Hub

TensorFlow Hub contains pretrained models using TensorFlow, related to different categories, which include NLP for texts and audio, as well as computer vision tasks. For each model, you can find the description and a snippet of code on how to use it.

TensorFlow hub is available at the following link: `https://tfhub.dev/`.

Hugging Face

Hugging Face is a Python library, which also provides a model and dataset hub, available at the following link: `https://huggingface.co/models`. You can find pretrained models for NLP, image classification, reinforcement learning, and much more.

To use a specific pretrained model, you need to install the Hugging Face library and load the model from the library.

Comet is fully integrated with Hugging Face. For more details on how to use Comet with Hugging Face, you can read the Comet official documentation, available at the following link: `https://www.comet.ml/docs/v2/guides/getting-started/tutorials/nlp-data/`.

Now that we have reviewed some popular model hubs of NLP, we can move on to the next step, a review of the Spark NLP package.

Exploring the Spark NLP package

Spark NLP is an open source library for NLP released by John Snow Labs. It supports different programming languages, including Python, Java, and Scala. Spark NLP is widely used in production, since it is natively integrated with Apache Spark, a multi-language engine for large-scale analytics.

Spark NLP provides more than 50 features, including tokenization, NER, and sentiment analysis.

In this section, you will investigate the following aspects:

- Introducing the Spark NLP package
- Integrating Spark NLP with Comet

Let's start from the first point, introducing the Spark NLP package.

Introducing the Spark NLP package

Spark NLP is an open source library built on top of Apache Spark and Spark ML (a machine learning library implemented on top of Apache Spark). The Spark NLP library provides almost all the NLP tasks, including tokenization, stemming, lemmatization, PoS tagging, sentiment analysis, spellchecking, and NER. The full list of implemented tasks is available in the Spark NLP official documentation, available at the following link: `https://nlp.johnsnowlabs.com/docs/en/quickstart`.

Spark NLP defines the following main concepts:

- Annotator
- Pipeline

Let's investigate each concept separately, starting from the first one, the annotator.

Annotator

An **annotator** is either a Spark ML **estimator** or **transformer**. In Spark NLP, an estimator is an algorithm that you can train on a given DataFrame to produce a model, called a transformer. A transformer is a trained algorithm that can transform one input DataFrame into an output DataFrame. The difference between input and output DataFrames is that the input DataFrame contains features, while the output DataFrame contains predictions. The following figure shows how estimators and transformers are related to each other:

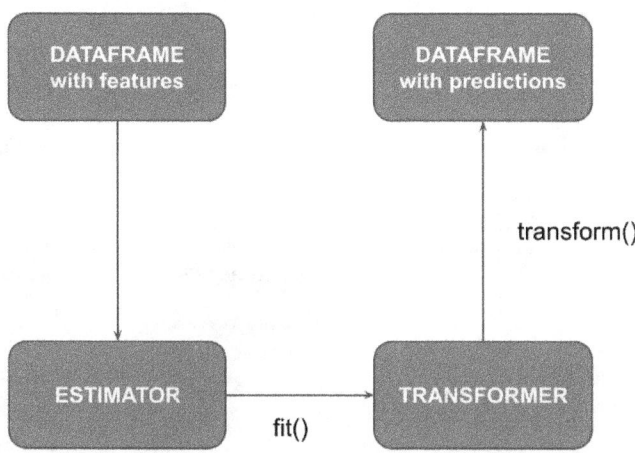

Figure 9.3 – The relationship between estimators and transformers in Spark ML

Spark NLP provides two types of annotators:

- **An annotator approach**, which extends the Spark ML estimator to implement the `fit()` method
- **An annotator model**, which extends the Spark ML transformer to implement the `transform()` method

Spark NLP also provides many pretrained annotator models, which have been already trained by someone else and are ready to use. You can check the NLP Models Hub website for the complete list of all the pretrained models, available at the following link: `https://nlp.johnsnowlabs.com/models`. All the pretrained models also provide the `pretrained()` method, which permits you to select the specific model to load, as shown in the following piece of code:

```
lemmatizer = LemmatizerModel.pretrained('lemma_antbnc', 'en')
```

The previous code loads the pretrained `lemma_antbnc` model.

All the annotators, whether an annotator approach or an annotator model, implement the following two methods:

- `setInputCols(column_names)`: The list of input column names required by this annotator

- `setOutputCol(column_name)`: The name of the output column containing the result of the current annotator

Now that we have reviewed the basic concepts behind annotators, we can move on to the next concept defined by Spark NLP, a pipeline.

Pipeline

A pipeline is an ordered sequence of stages, and each stage is either a transformer or an estimator. Each stage updates the DataFrame by adding new annotations, as shown in the following figure:

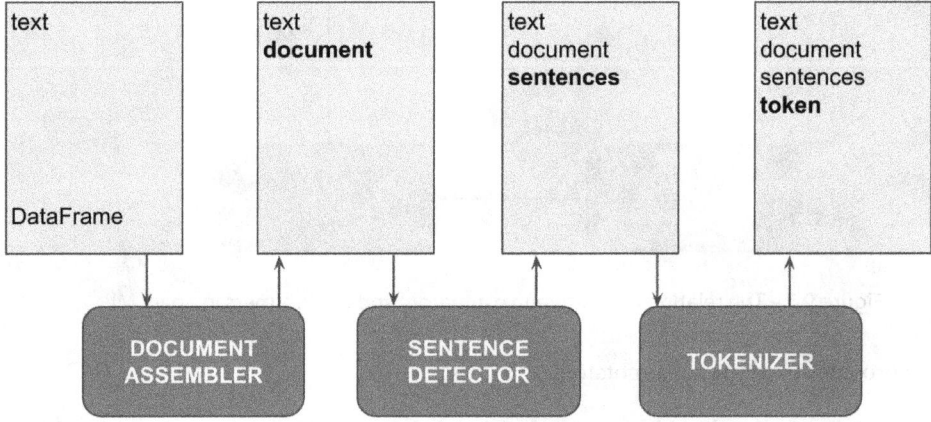

Figure 9.4 – A simple NLP pipeline

The previous figure implements a simple pipeline, composed of the following three stages:

- **Document assembler** – This Annotator is the initial stage of each Spark NLP pipeline. It creates the first annotation on the DataFrame, which is read by the next stage in the pipeline. The output column of this annotator is called a **document**.

- **Sentence Detector** – This annotator splits the document into sentences and adds a new annotation, called **sentences**, to the DataFrame.

- **Tokenizer** – This annotator tokenizes each sentence, and adds a new annotation, called **token**, to the DataFrame.

Similar to pretrained models, Spark NLP also provides pretrained pipelines, such as `BasicPipeline`, which implements sentence splitting, tokenization, lemmatization, stemming, and PoS tagging.

For more information on pipelines, you can read the Spark NLP official documentation, available at the following link: https://nlp.johnsnowlabs.com/docs/en/concepts.

Now that we have reviewed the basic concepts behind Spark NLP, we are ready to learn how to integrate Spark NLP with Comet.

Integrating Spark NLP with Comet

You can monitor your NLP experiments directly in Comet, either during the training process or at the end of an experiment. To integrate Spark NLP with Comet, you should create a CometLogger object, as follows:

```
from sparknlp.logging.comet import CometLogger
logger = CometLogger()
```

The CometLogger class provides the following main methods:

- log_pipeline_parameters(): Logging the parameters of each stage of the pipeline in Comet

- log_visualization(): Uploading the NER visualization provided by Spark NLP Display to Comet

- log_metrics(): Logging evaluation metrics in Comet

- log_parameters(): Logging multiple parameters in Comet

- log_completed_run(): Submitting the results of training metrics once the pipeline training process is complete

- log_asset(): Uploading an asset to Comet

- monitor(): Monitoring the training of the model and submitting logs to Comet while the pipeline training process is running

For the complete list of methods, you can read the documentation, available at the following link: https://nlp.johnsnowlabs.com/api/python/modules/sparknlp/logging/comet.html.

You can access the experiment class defined in Comet through the logger.experiment field.

You can only use the monitor() and log_completed_run() methods for some annotator approach objects, such as a MultiClassifierDLApproach object. When you configure the pipeline to work with one of the supported annotators, you should enable logs, as well as set the path where to save them, as shown in the following piece of code:

```
from sparknlp.annotator import MultiClassifierDLApproach()
PATH=/path/to/my/directory
```

```
model = MultiClassifierDLApproach()
      # add other configuration parameters
.setEnableOutputLogs(True) \
      .setOutputLogsPath(PATH)
```

We have used the `setEnableOutputLogs(True)` method to enable logs, and the `setOutputLogsPath()` method to set the path where to save logs. `CometLogger` will read the output of the log during the training process, as shown in the following piece of code:

```
logger.monitor(PATH, model)
```

Alternatively, `CometLogger` can read the log when the training process finishes:

```
logger.log_completed_run('/Path/To/Log/File.log')
```

The following figure shows an example of the output in Comet produced by the `monitor()` method:

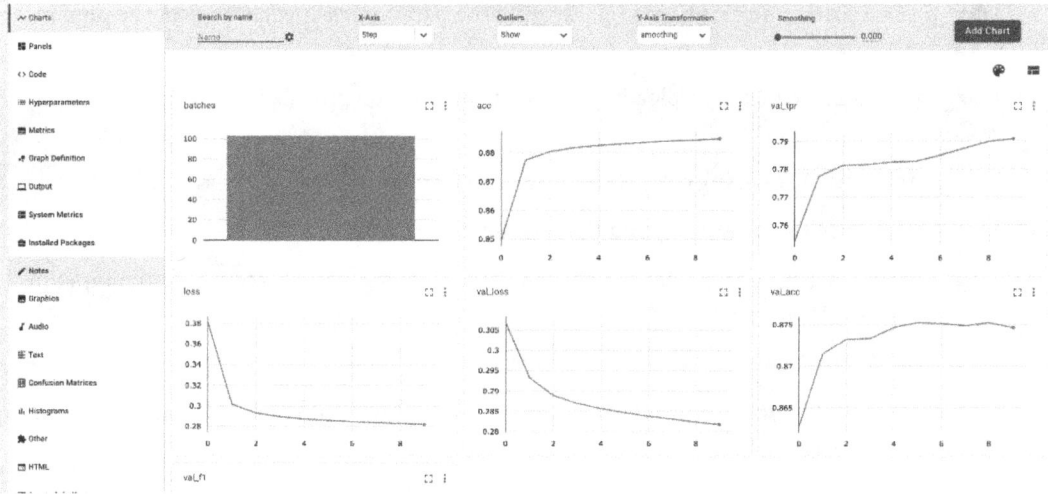

Figure 9.5 – An example of the output of the monitor() method

Each graph shows how a single metric evolves during the training phase.

We have just reviewed the main concepts behind Spark NLP and how to integrate it with Comet. Now, it is time to practice. Let's start by setting up the environment to work with Spark NLP.

Setting up the environment for Spark NLP

Since Spark NLP runs on Apache Spark, you need to first set up Apache Spark to make Spark NLP work properly. Apache Spark requires Java to run properly; thus, you also need to install Java. Optionally, you can install Scala, another programming language.

To install Spark NLP, you should install the following frameworks and libraries:

- Python
- Java
- Scala (optional)
- Apache Spark
- PySpark and Spark NLP

We have already installed Python following the procedure described in the *Technical requirements* section, so we can start installing the software from the second step, Java.

Installing Java

Spark NLP is built at the top of Apache Spark, which can be installed on any operating system supporting Java. Apache Spark requires **Java 8** to work properly:

1. To verify whether Java is already installed on your computer, as well as its current version, you can open a terminal and run the following command:

    ```
    Java -version
    ```

    If Java is already installed, the command should return something similar to the following output:

    ```
    openjdk version "1.8.0_322"
    OpenJDK Runtime Environment (build 1.8.0_322-bre_2022_02_
    28_15_01-b00)
    OpenJDK 64-Bit Server VM (build 25.322-b00, mixed mode)
    ```

    The second digit after the version keyword indicates the current Java version, which is 8 in this case.

2. If Java is not installed, you can download it either from the official Oracle website or from other open source websites, such as OpenJDK. In Ubuntu, you can install openjdk-8 as follows:

    ```
    sudo apt-get install openjdk-8-jre
    ```

    In macOS, you can run the following command, provided that you have brew installed:

    ```
    brew install openjdk@8
    ```

    In all cases, including Windows, you can download Java 8 from the following link, https://java.com/en/download/, and then follow the wizard.

3. If Java is already installed but you do not have version 8, you can install it following the procedure described in *step 2*, and then you set the JAVA_HOME environment variable to the path to the Java 8 directory.

4. Run *step 1* again to make sure that Java 8 is installed properly.

Once you have installed Java, you can move on to the next step, installing Scala.

Installing Scala (optional)

Scala is a programming language used in data processing, distributed computing, and web development. You can find more information about Scala on its official website, available at the following link: https://www.scala-lang.org/.

Apache Spark requires Scala 2.12 or 2.13 to work properly:

1. You can install Scala 2.12.15 following the procedure described at the following link: https://www.scala-lang.org/download/2.12.15.html.

2. To verify that you have installed Scala correctly, you can run the following command:

    ```
    scala -version
    ```

    The command should return version 2.12.

Once you have installed Scala, you are ready to install Apache Spark.

Installing Apache Spark

You can download Apache Spark from its official website, available at the following link: https://spark.apache.org/downloads.html.

1. For the examples described in this chapter, we will use version 3.1.2, available at the following link https://archive.apache.org/dist/spark/spark-3.1.2/.

2. Once you have downloaded the package, you can extract it and place it wherever you want in your filesystem. Then, you should add the path to the bin directory contained in your spark directory to the PATH environment variable. In Unix systems, you can export the PATH variable by adding the following line of code to your .bashrc or .zprofile (or other similar files, depending on your shell) file:

    ```
    export PATH=$PATH:/path/to/spark/bin
    ```

    The exact filename you should modify depends on the specific shell you are using.

3. You also need to export the SPARK_HOME environment variable with the path to your spark directory. In Unix systems, you can export the SPARK_HOME variable as follows:

    ```
    export SPARK_HOME="/path/to/spark"
    ```

You should add the previous line of code to your `.bashrc` file (or similar file).

> **Note**
>
> You may need to exit and enter again your terminal to make changes effective.

4. To test whether you have installed Apache Spark correctly or not, you can run the following command:

```
spark-shell
```

You should see output similar to the following one:

```
Welcome to
      ____              __
     / __/__  ___ _____/ /__
    _\ \/ _ \/ _ `/ __/  '_/
   /___/ .__/\_,_/_/ /_/\_\   version 3.1.2
      /_/

Using Scala version 2.12.15 (OpenJDK 64-Bit Server VM,
Java 1.8.0_322)
Type in expressions to have them evaluated.
Type :help for more information.

scala>
```

5. To exit the Apache Spark shell, you can use *Ctrl + C*.

Now that we have installed Apache Spark, we can move on to the final step, installing PySpark and Spark NLP.

Installing PySpark and Spark NLP

To install PySpark, you can run the following command:

```
pip install pyspark
```

To install Spark NLP, you can run the following command:

```
pip install spark-nlp
```

It may happen that some pretrained models of pipelines are available only for specific versions of Spark NLP. Therefore, you can create a virtual environment to contain a specific version of Spark NLP. You can proceed as follows:

1. Firstly, you create a new virtual environment:

    ```
    python3 -m venv /path/to/your/virtual/environment
    ```

2. Then, you activate it:

    ```
    cd /path/to/your/virtual/environment
    source bin/activate
    ```

3. Now, you can install the specific version of Spark NLP, as follows:

    ```
    pip install spark-nlp==x.y.z
    ```

> **Note**
>
> In the new virtual environment, you should also install all the required libraries to run your software. For example, if you use Jupyter to create your code, you should install and run it in your virtual environment.

Now that we have installed all the software required to run Spark NLP, let's move on to a practical example.

Using NLP, from project setup to report building

In this section, you will implement a practical example that performs sentiment analysis of some text extracted from Twitter. This example compares a pretrained model and a custom model, showing the results in Comet.

The full code of the example described in this section is available at the following link: `https://github.com/PacktPublishing/Comet-for-Data-Science/tree/main/09`.

We will focus on the following aspects:

*   Configuring the environment
*   Loading the dataset
*   Implementing a pretrained pipeline
*   Logging results in Comet
*   Using a custom pipeline
*   Building the final report

Let's start from the first point, configuring the environment.

Configuring the environment

Environment configuration involves the following two steps:

1. We initialize Spark NLP. We import and start the Spark NLP library as follows:

    ```
    import sparknlp
    spark = sparknlp.start()
    ```

> **Note**
>
> If you are writing you code in Jupyter, the previous command may fail because your kernel is not able to find the Apache Spark directory. To solve the problem, before calling `sparknlp.start()`, use the `findspark` Python library, as follows:
>
> ```
> import findspark
> findspark.init()
> ```

2. We initialize Comet. We import and initialize the Comet library, as follows:

    ```
    import comet_ml
    from sparknlp.logging.comet import CometLogger
    comet_ml.init()
    logger = CometLogger()
    logger.experiment.set_name('PretrainedModel')
    ```

    We should have a `.comet.config` file, which contains our Comet credentials and is located in the same directory as the working directory, as described in *Chapter 1, An Overview of Comet*.

Now that we have configured the environment, we can move to the next step, loading the dataset.

Loading the dataset

As a use case, you will use the `Disneyland Reviews` dataset, available on Kaggle at the following link, `https://www.kaggle.com/datasets/arushchillar/disneyland-reviews`, and released under the CC0: Public Domain license. The dataset contains 42,656 reviews of three Disneyland branches (Paris, California, and Hong Kong) posted by visitors on Tripadvisor. For each review, it also provides the rating, which ranges from 1 (totally unsatisfied) to 5 (satisfied). We group ratings into two categories: positive if the rating is greater than 2, and negative otherwise. For simplicity, we assume that a positive rating also corresponds to a positive sentiment within the text review, and a negative rating also corresponds to a negative sentiment within the text review.

The following table shows the first two rows of the dataset:

Review text	Rating
If you've ever been to Disneyland anywhere you'll find Disneyland Hong Kong very similar in the layout when you walk into main street! It has a very familiar feel. One of the rides its a Small World is absolutely fabulous and worth doing. The day we visited was fairly hot and relatively busy but the queues moved fairly well.	4
Its been a while since d last time we visit HK Disneyland .. Yet, this time we only stay in Tomorrowland .. AKA Marvel land!Now they have Iron Man Experience n d Newly open Ant Man n d Wasp!!Ironman .. Great feature n so Exciting, especially d whole scenery of HK (HK central area to Kowloon)!Antman .. Changed by previous Buzz lightyear! More or less d same, but I'm expecting to have something most!!However, my boys like it!!Space Mountain .. Turns into Star Wars!! This 1 is Great!!!For cast members (staffs) .. Felt bit MINUS point from before!!! Just dun feel like its a Disney brand!! Seems more local like Ocean Park or even worst!!They got no SMILING face, but just wanna u to enter n attraction n leave!!Hello this is supposed to be Happiest Place on Earth brand!! But, just really Dont feel it!!Bakery in Main Street now have more attractive delicacies n Disney theme sweets .. These are Good Points!!Last, they also have Starbucks now inside the theme park!!	4

Figure 9.6 – Extracts from the Disneyland Reviews dataset

The dataset has many columns; from them, we will use the following: `Review_Text`, which contains the text, and `Rating`, which contains the rating and which we will use as a sentiment.

We perform the following steps to load the dataset as a `pyspark` DataFrame:

1. Firstly, we load the dataset:

    ```
    from pyspark.sql.functions import when, col
    df=spark.read.format("csv").option("header", "true").
    load("source/DisneylandReviews.csv")
    ```

    We use the `format()` function provided by the Spark NLP library to load the package, as well as the option header, set to `true` to read the CSV header.

2. Then, we group rows as follows:

    ```
    df = df.withColumn("sentiment", when(col("Rating") > 2,
    "positive").otherwise("negative"))
    ```

    We mark rows with a `Rating` value greater than 2 as positive, and those with a `Rating` value fewer or equal to 2 as negative.

The previous operation is required by the pretrained model that we will use.

Now that we have loaded the dataset, we can move on to the next step, implementing a pretrained pipeline.

Implementing a pretrained pipeline

The first pipeline we will implement uses the pretrained model named `VieknSentimentModel`, provided by Spark NLP. You can find more information about the pretrained model and how to use it at the following link: `https://nlp.johnsnowlabs.com/2021/11/22/sentiment_vivekn_en.html`.

1. Firstly, we import all the required packages:

    ```
    from sparknlp.base import *
    from sparknlp.annotator import *
    from pyspark.ml import Pipeline
    ```

2. Now, we implement the Spark pipeline with the following five stages:

    ```
    document = DocumentAssembler() \
        .setInputCol("Review_Text") \
        .setOutputCol("document")

    token = Tokenizer() \
        .setInputCols(["document"]) \
        .setOutputCol("token")

    normalizer = Normalizer() \
        .setInputCols(["token"]) \
        .setOutputCol("normal")

    vivekn =  VieknSentimentModel.pretrained() \
        .setInputCols(["document", "normal"]) \
        .setOutputCol("result_sentiment") \

    finisher = Finisher() \
        .setInputCols(["result_sentiment"]) \
        .setOutputCols("final_sentiment")
    ```

    The pipeline stages include the document assembler, the tokenizer, the normalizer, the pretrained model of the `Vivekn` sentiment model, and the finisher.

3. We create the pipeline:

```
pipeline = Pipeline().setStages([document, token,
normalizer, vivekn, finisher])
```

4. Now, we fit the pipeline with the text:

```
X = df.select('Review_Text').toDF('Review_Text')
pipelineModel = pipeline.fit(X)
```

We have built an auxiliary variable called X, which contains only the text.

5. Finally, we calculate the results:

```
result = pipelineModel.transform(X)
```

Now that we have trained the pipeline, we are ready to log the results in Comet.

Logging results in Comet

In Comet, we can now log the evaluation results:

1. To perform this operation, we need to map the results produced by the pipeline with the original sentiment labels, as shown in the following piece of code:

```
df_tot = df.join(result, on=["Review_Text"])
```

We have built a new spark DataFrame, called df_tot, which has the following structure:

Review_Text	sentiment	final_ sentiment
If you've ever been to Disneyland anywhere you'll find Disneyland Hong Kong very similar in the layout when you walk into main street! It has a very familiar feel. One of the rides its a Small World is absolutely fabulous and worth doing. The day we visited was fairly hot and relatively busy but the queues moved fairly well.	positive	[positive]

Its been a while since d last time we visit HK Disneyland .. Yet, this time we only stay in Tomorrowland .. AKA Marvel land!Now they have Iron Man Experience n d Newly open Ant Man n d Wasp!!Ironman .. Great feature n so Exciting, especially d whole scenery of HK (HK central area to Kowloon)!Antman .. Changed by previous Buzz lightyear! More or less d same, but I'm expecting to have something most!!However, my boys like it!!Space Mountain .. Turns into Star Wars!! This 1 is Great!!!For cast members (staffs) .. Felt bit MINUS point from before!!! Just dun feel like its a Disney brand!! Seems more local like Ocean Park or even worst!!They got no SMILING face, but just wanna u to enter n attraction n leave!!Hello this is supposed to be Happiest Place on Earth brand!! But, just really Dont feel it!!Bakery in Main Street now have more attractive delicacies n Disney theme sweets .. These are Good Points!!Last, they also have Starbucks now inside the theme park!!	positive	[positive]

Figure 9.7 – The merged DataFrame, with both a true sentiment
(sentiment) and a predicted sentiment (final_sentiment)

2. To log the results in Comet, we need to convert the Spark DataFrame into a pandas DataFrame, as shown in the following piece of code:

```
pandas_df = df_tot.toPandas()
```

3. We need to convert to strings the lists contained in the final_sentiment columns:

```
pandas_df['predicted_sentiment'] = [','.join(map(str, l))
for l in pandas_df['final_sentiment']]
```

We have built a new column, called predicted_sentiment, which stores the cleaned version of final_sentiment. We will use predicted_sentiment for further analysis.

4. We use the classification_report() function provided by scikit-learn to calculate precision, recall, and the F1 score:

```
from sklearn.metrics import classification_report
report = classification_report(pandas_df['sentiment'],
pandas_df['predicted_sentiment'], output_dict=True,
labels=['positive', 'negative'])
```

5. We log the results in Comet:

```
for key, value in report.items():
    if key!='accuracy':
        logger.log_metrics(value,prefix=key)
    else:
        logger.log_metrics({"accuracy": value})
```

After running the experiment, you should see the results in Comet, as shown in the following figure:

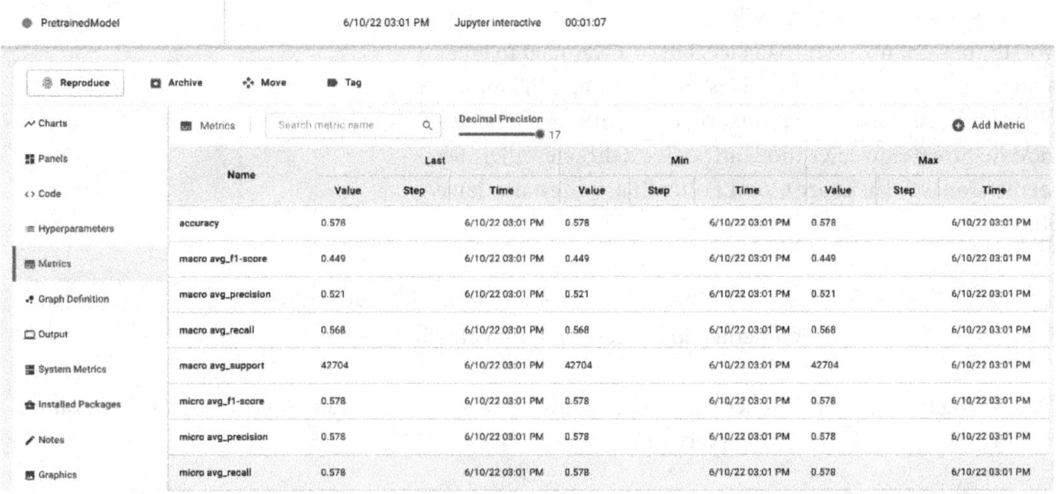

Figure 9.8 – The evaluation metrics of the pretrained model in Comet

Now that we have logged the results of the first pipeline in Comet, we can move on to the second pipeline.

Using a custom pipeline

The custom pipeline trains a custom model, based on the VieknSentimentApproach class. The pipeline contains the same stages as the previous examples, with the exception of the model:

```
vivekn_custom = VieknSentimentApproach() \
    .setInputCols(["document", "normal"]) \
    .setSentimentCol("sentiment") \
    .setOutputCol("result_sentiment")
```

While in the previous example the model was already trained, this time you need to train it.

To perform training, you can implement the following steps:

1. Firstly, we split the dataset into training and test sets, as described in the following piece of code:

```
(training_set, test_set) = df.randomSplit([0.8, 0.2])
X_train = training_set.select('Review_Text',
'sentiment').toDF('Review_Text', 'sentiment')
X_test = test_set.select('Review_Text', 'sentiment').
toDF('Review_Tex', 'sentiment')
```

   We have reserved 80% of the samples for the training set and the remaining 20% for the test set.

2. Then, we train the pipeline on the training set:

```
pipelineModel = pipeline.fit(X_train)
```

3. Finally, we calculate predictions on the test set:

```
result = pipelineModel.transform(X_test)
```

Once you have trained the model and calculated the predictions for the test set, you can log the results in Comet, as you did in the previous example. Since `VivekrSentimentApproach` does not implement the `setOutputLogsPath()` method, we cannot monitor the training process in Comet.

Now that we have built the custom model and evaluated its performance, we can proceed with the final step, building the final report.

Building the final report

Now, you are ready to build the final report. In this example, we will build a simple report with a comparison between the two models. As a further exercise, you can improve them by applying the concepts learned in *Chapter 5, Building a Narrative in Comet*. You will create a report with a section containing the comparison between the calculated performance metrics in the two models.

We perform the following two steps:

- Building the panels
- Building the report

Let's start from the first step, building the panels.

Building the panel

We implement the following panels:

- A **bar chart**, for the performance metrics at the macro level, and another for the performance metrics at the micro level

- A **parallel coordinates chart** to compare the different performance metrics, calculated for each target class (positive and negative sentiment)

- A **project summary**, which contains all the performance metrics for both the experiments in a tabular form

To build the first bar chart for metrics at the macro level, you can follow the following steps:

1. From the Comet project main dashboard, select the **Panels** tab, and then **Add | New Panel | BUILT-IN | Bar Chart | Add**.

2. In the **Y-Axis** input textbox, select the accuracy.

3. Click on the **Add field** button, and select **macro avg_f1_score**.

4. Repeat the previous two steps to also add **macro avg_precision** and **macro avg_recall**.

5. Click on the **Done** button. You should see a graph similar to the one shown in the following figure:

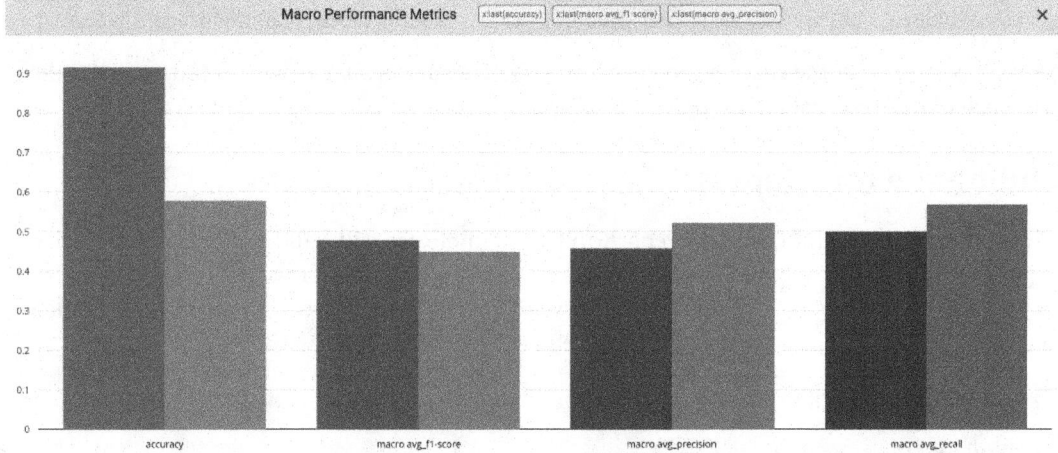

Figure 9.9 – The bar chart panel for the macro performance metrics

To build the parallel coordinates chart, you can perform the following steps:

1. From the Comet project main dashboard, select the **Panels** tab, and then **Add | New Panel | BUILT-IN | Parallel Coordinates Chart | Add**.

2. In the **Target Variable** input textbox, select the accuracy.

3. In the **Y-Axis** input textbox, you can add as many metrics as you want. In our example, we add the following metrics: **positive_precision**, **positive_recall**, **negative_precision**, and **negative_recall**.

4. Click on the **Done** button. You should see a graph similar to the one shown in the following figure:

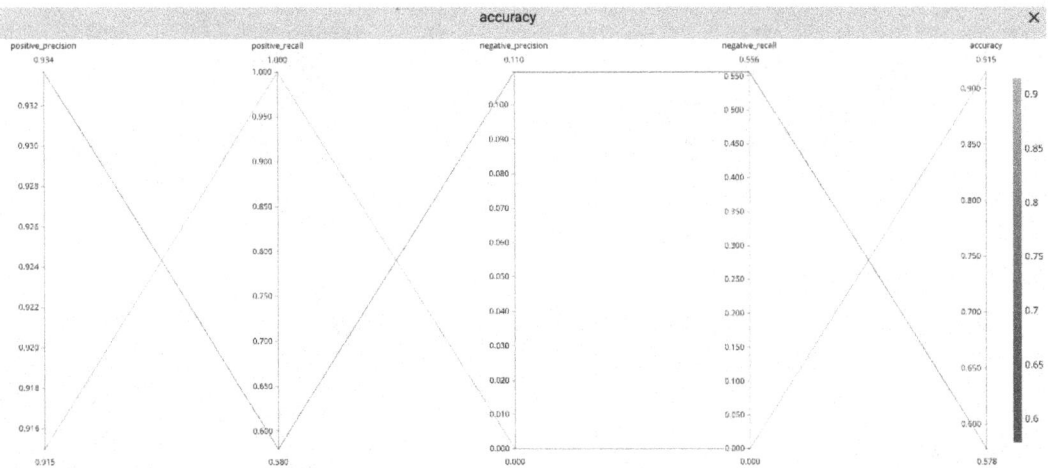

Figure 9.10 – The parallel coordinates chart

5. To build the **Project Summary** panel, select the **Panels** tab, and then **Add | New Panel | FEATURED | Project Summary | Add | Done**. The produced panel should be similar to the one shown in the following figure:

Metric	Count		Min	Max	Mean	Median
accuracy	2	Value	0.95	0.95	0.95	0.95
		Steps	1	1	1	1
macro avg_f1-score	2	Value	0.49	0.49	0.49	0.49
		Steps	1	1	1	1
macro avg_precision	2	Value	0.47	0.47	0.47	0.47
		Steps	1	1	1	1
macro avg_recall	2	Value	0.5	0.5	0.5	0.5
		Steps	1	1	1	1
macro avg_support	2	Value	205	205	205	205
		Steps	1	1	1	1

Figure 9.11 – The Project Summary panel

Now that we have built the panels, we can add them to the report. Thus, we can move on to the final step, building the report.

Building the report

To create the report, in the Comet dashboard, perform the following steps:

1. Click on the **Panels** tab, and then select **Add | Add to Report | New Report**.

2. A new report appears with the previously created panels.

3. You can customize the size of each chart in the panel by selecting the **Edit Layout** button and then adjusting the layout of each chart with the mouse.

4. You can customize your report with the title, descriptions, and so on.

5. Finally, click on the **Save** button.

    The following figure shows a snapshot of part of the final report:

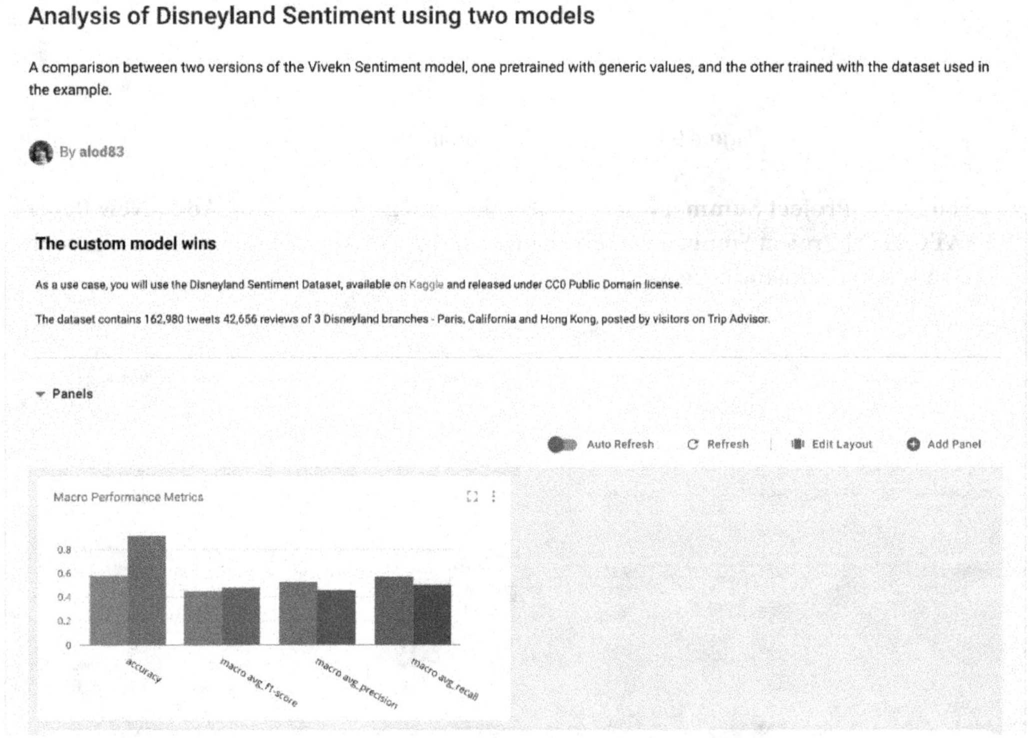

Figure 9.12 – A snapshot of the final report

You can see the full report at `https://www.comet.ml/packt/spark-nlp/reports/analysis-of-dysneyland-sentiment-using-two-models`, while the full project in Comet is available at the following link: `https://www.comet.ml/packt/spark-nlp/`.

Summary

We just completed the journey to build an NLP model in Spark NLP and track it in Comet!

Throughout this chapter, we described some general concepts regarding NLP, including the basic NLP workflow, how you can classify NLP tools, and the main NLP challenges. In addition, you have seen the main structure of the Spark NLP package and how to set up the environment to make it work. We also illustrated some important concepts, such as annotators and pipelines.

In the last part of the chapter, you implemented a practical use case that showed you how to track an NLP experiment in Comet, as well as how to build a report with the results of the experiment.

In the next chapter, we will review the basic concepts related to deep learning and how to perform it in Comet.

Further reading

- Kedia, A., and Rasu, M. (2020). *Hands-On Python Natural Language Processing: Explore tools and techniques to analyze and process text with a view to building real-world NLP applications.* Packt Publishing Ltd.

- Thomas, A. (2020). *Natural Language Processing with Spark NLP: Learning to Understand Text at Scale.* O'Reilly Media, Inc.

10
Comet for Deep Learning

Deep learning is a subfield of artificial intelligence that aims to extract knowledge from data through complex **neural networks**. You can imagine a neural network as a collection of nodes (or neurons), which are organized in different layers of processing. Similar to the human brain, which learns in incremental steps, in a neural network, the level of knowledge increases as you move from one layer to the next.

Compared to traditional machine learning algorithms, deep learning algorithms can automatically extract features from incoming data. Therefore, they are able to process even thousands of input features, which is unthinkable for a machine learning algorithm in which you should select the input features by hand.

Running deep learning algorithms is both computationally expensive and resource-consuming. However, over the last few years, their uptake has increased considerably thanks to the development of even more powerful machines as well as the spread of distributed systems for parallel calculations. Today, deep learning finds its major applications in the fields of image and speech recognition. In this chapter, you will review some basic concepts behind deep learning as well as the main types of deep learning networks.

In recent years, different open source tools and libraries have been implemented to perform deep learning, including TensorFlow, Keras, Caffe, PyTorch, and so on. In this chapter, you will review the TensorFlow library and how to integrate it with Comet.

In the last part of this chapter, you will implement a practical use case, which uses the TensorFlow library and tracks the result in Comet.

The chapter is organized as follows:

- Introducing basic deep learning concepts
- Exploring the TensorFlow package
- Using deep learning - from project setup to report building

Before we start to review basic deep learning concepts, let's install the required software needed to run the examples described in this chapter.

Technical requirements

We will run all of the experiments and code in this chapter using Python 3.8. You can download it from the official website at `https://www.python.org/downloads/`, choosing the 3.8 version.

The examples described in this chapter use the following Python packages:

- `comet-ml 3.23.0`
- `gradio 3.2.2`
- `matplotlib 3.2.2`
- `numpy 1.21.6`
- `pandas 1.3.4`
- `tensorflow 2.8.2`

We have already described the `comet-ml`, `matplotlib`, `numpy`, and `pandas` packages and how to install them in *Chapter 1, An Overview of Comet*. So, please refer back to that for further details on installation.

In this section, you will see how to install the other required packages.

gradio

`gradio` is a Python package that permits you to build fast demo apps that are fully integrated with your notebooks. You can install `gradio` as follows:

```
pip install gradio
```

For more details on `gradio`, you can read its official documentation, available at the following link: `https://gradio.app/`.

tensorFlow

`tensorFlow` is one of the most popular Python packages for deep learning. You can install it as follows:

```
pip install tensorflow
```

You can read more details about `TensorFlow` in its official documentation, available at the following link: `https://www.tensorflow.org/`.

Now that you have installed all of the software needed in this chapter, let's move on to how to use Comet for deep learning, starting with reviewing some basic concepts.

Introducing basic deep learning concepts

The following figure shows how deep learning fits into the field of artificial intelligence:

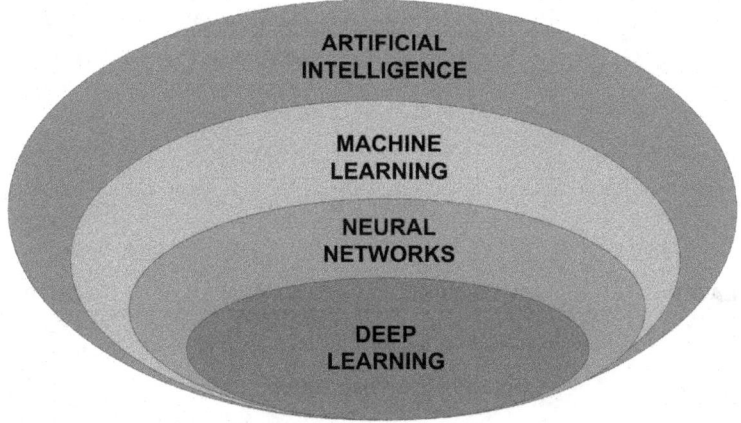

Figure 10.1 – How deep learning is related to other artificial intelligence fields

You can see that deep learning is a subfield of neural networks, which are a subfield of machine learning, which is a subfield of artificial intelligence. In this section, you will understand the difference between deep learning and neural networks as well as how you can classify deep learning networks.

You will learn some general concepts about deep learning. For more details, you can refer to the books contained in the *Further reading* section of this chapter.

The section is organized as follows:

- Introducing neural networks
- Exploring the difference between deep learning and neural networks
- Classifying deep learning networks

Let's start from the first point: introducing neural networks.

Introducing neural networks

The basic building block of a neural network is called a **neuron**. The objective of a neuron is to represent the behavior of a neuron in our brain through a mathematical model. The simplest neural network is called **perceptron**, and it is composed of N input features, one single neuron, and one output, as shown in the following figure:

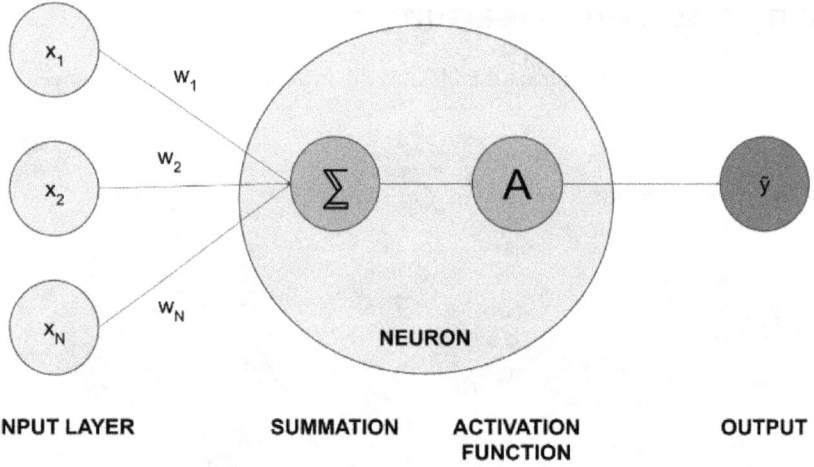

Figure 10.2 – A perceptron

The single neuron is composed of the following two main components:

- The **summation**, which calculates the summation of the product between the input features and the weights as shown in the following equation:

$$\sum = \sum w_i \times x_i$$

    Weights establish how much each input feature influences the output. Usually, the preceding formula should be corrected by introducing a new term called **bias**, as shown in the following formula:

$$\sum = \sum w_i \times x_i + b$$

- The **activation function**, which is a nonlinear function, calculates the final output. Common activation functions are the sigmoid function, the hyperbolic tangent, and the **rectified linear unit (ReLU)**.

This process, from input to output, is called **forward propagation**.

The objective of the learning algorithm for a neural network is to find the best values of weights and bias, which minimize a loss function. This process, which is also known as **backward propagation**, is achieved through an iterative process that computes the loss function and updates weights and bias at every iteration. As a loss function, usually, you can use **mean squared error** (**MSE**) for regression tasks and cross-entropy for classification tasks.

A perceptron is only the basic building block of a neural network. You can build more complex neural networks by adding many neurons and layers. For more details on neural networks, you can read the books listed in the *Further reading* section.

Now that you have learned the basic concepts behind neural networks, we can move on to the next point: exploring the difference between deep learning and neural networks.

Exploring the difference between deep learning and neural networks

A typical neural network is composed of an input layer, a single hidden layer, and an output layer. Each layer can have a variable number of nodes or neurons. The following figure shows an example of a neural network:

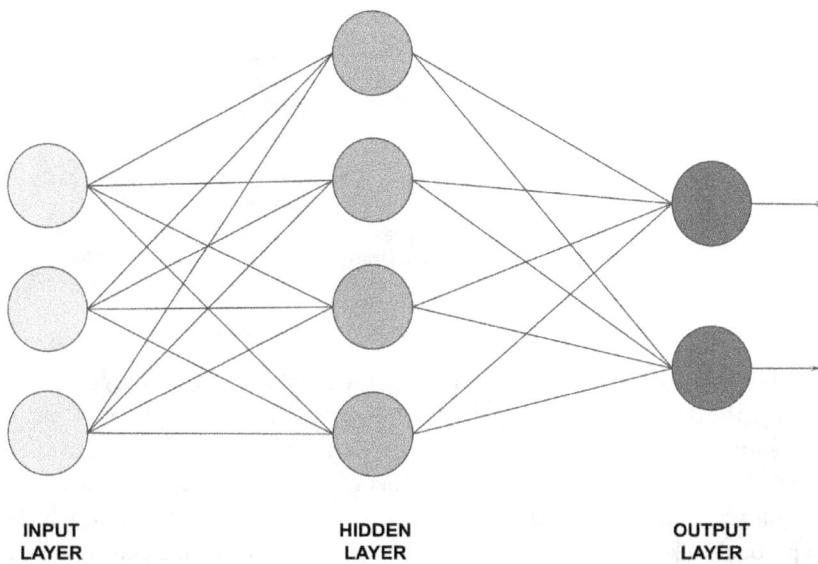

Figure 10.3 – A simple neural network

The neural network of the preceding figure is composed of the following three layers:

- The first layer (input layer) has three nodes.
- The second layer (hidden layers) has four nodes.
- The third layer (output layer) has two nodes.

It is worth noting that nodes in the same layer are not connected directly.

Deep learning is the name used to indicate a neural network that is composed of more than three layers, including the input and the output layers. The following figure shows an example of a deep learning network:

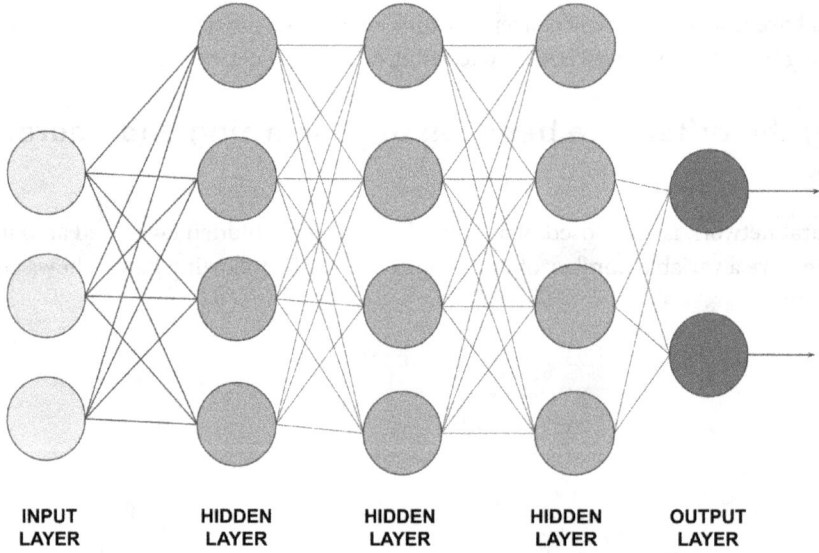

Figure 10.4 – A deep learning network

The preceding figure shows a deep learning network with three hidden layers.

In a deep learning network, each layer receives as input the set of features produced as output by the previous layer. Thus, the deeper the network, the more complex the features the layers can recognize. This is known as feature hierarchy. Thanks to feature hierarchy, deep learning networks can work with very large and high-dimensional data with a huge number of parameters that are learned automatically without human intervention. The output of a deep learning network is a **classifier**, which assigns a probability to a particular class or label. Deep learning is quite resource-consuming because it requires high-performance GPUs and large amounts of storage to train models.

Now that you have learned the difference between deep learning and neural networks, we can move on to the next point: classifying deep learning networks.

Classifying deep learning networks

You can classify deep learning networks into the following three main categories:

- **Artificial neural networks (ANN)**
- **Recurrent neural networks (RNN)**
- **Convolutional neural networks (CNN)**

Let's investigate each category separately, starting with the first: artificial neural networks.

Artificial neural networks

An **artificial neural network (ANN)**, also known as a **feedforward neural network**, is the simplest type of neural network with many neurons in each layer. An ANN processes inputs only in the forward direction, as shown in *Figure 10.3*.

An ANN has no memory because information moves only in one direction: from the input to the output. This means that an ANN is not able to remember what happened in the past. An ANN makes a decision only based on the current input.

Typically, you use ANNs to solve problems related to tabular data, images, and texts.

Recurrent neural networks

A **recurrent neural network (RNN)** is a type of neural network that makes decisions based on the current input as well as the inputs received previously. This is achieved by adding a recurrent connection to the hidden layers, as shown in the following figure:

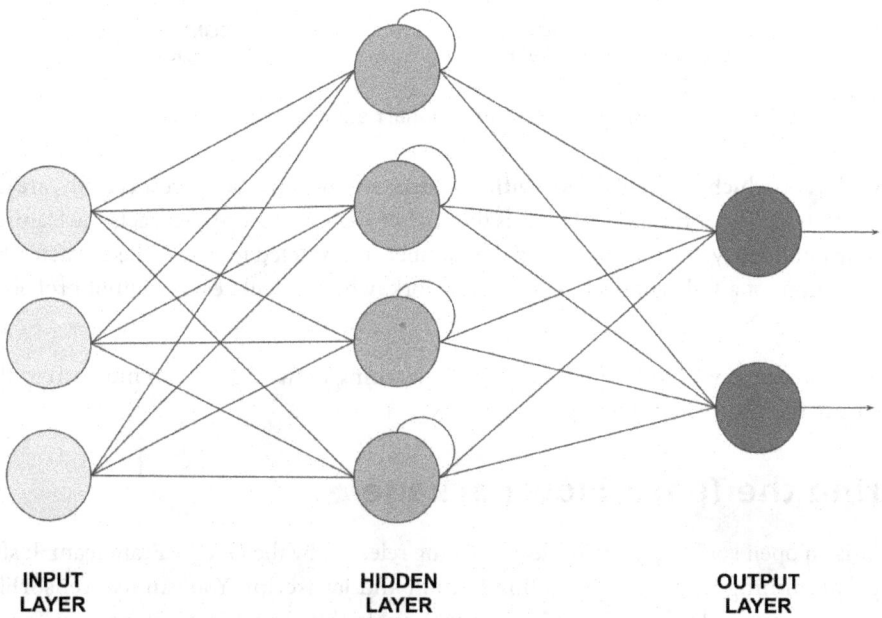

INPUT LAYER   **HIDDEN LAYER**   **OUTPUT LAYER**

Figure 10.5 – A recurrent neural network

In the preceding figure, the neurons in the hidden layer contain a recurrent connection, which permits the neurons to remember past inputs. In other words, RNNs are networks with memory. Typically, you use RNNs to solve problems related to time series, texts, and audio.

Convolutional neural networks

A **convolutional neural network** (**CNN**) is the most popular type of deep learning network, and it is especially used for image recognition and classification, object detection, and recognition of faces. The following figure shows an example of a CNN:

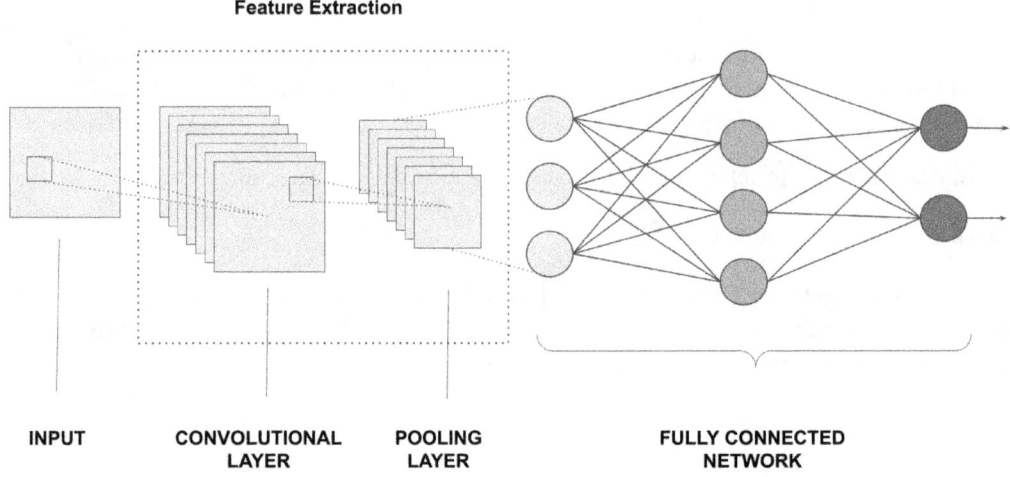

Feature Extraction

| INPUT | CONVOLUTIONAL LAYER | POOLING LAYER | FULLY CONNECTED NETWORK |

Figure 10.6 – A convolutional neural network

The first two layers, which are called **convolutional layer** and **pooling layer**, respectively, are devoted to feature extraction. The convolutional layer is the core of a CNN because it extracts the features from the input. The pooling layer aims at reducing the number of extracted features. The extracted features constitute the input of a fully connected neural network, which calculates the output probability for a given task.

Now that you have reviewed the main types of deep learning networks, we can move on to the next topic: exploring the TensorFlow package.

Exploring the TensorFlow package

TensorFlow is an open source library for deep learning released by the Google Brain team. It supports different programming languages, including Python and Javascript. You can use TensorFlow for different purposes, especially for audio and image analysis. In this chapter, we will focus on TensorFlow 2.x. Since training a model in TensorFlow could be time and resource-consuming, TensorFlow also provides many pre-trained models, stored in the **TensorFlow Hub**, available at the following link: `https://www.tensorflow.org/hub`.

Running TensorFlow on your local machine could be computationally expensive and resource-consuming, thus you use **Google Colab**, a collaborative framework provided by Google, to train your models. In fact, Google Colab provides you with free access to GPU and powerful machines. Google Colab is a valid alternative to Jupyter Notebook and is compatible with it. You can run your first Google Colab notebook at the following link: `https://colab.research.google.com/`. You can use it as you usually do with Jupyter Notebook.

In this section, we will cover the following aspects:

- Introducing the TensorFlow package
- Integrating TensorFlow with Comet

Let's start from the first point: introducing the TensorFlow package.

Introducing the TensorFlow package

The following figure shows the main tasks and the related sub-packages provided by the core TensorFlow library:

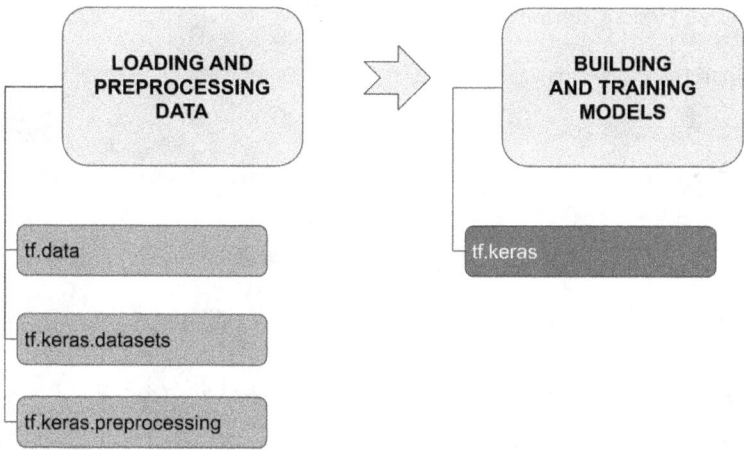

Figure 10.7 – The main tasks in TensorFlow

There are two main tasks as follows:

- Loading and preprocessing data
- Building and training models

The full list of implemented tasks is available in the TensorFlow official documentation, which is available at the following link: `https://www.tensorflow.org/api_docs/python/tf`.

TensorFlow includes many additional extensions that permit you, for example, to preprocess text, optimize models, and so on. You can find the list of all of the extensions at the following link: `https://www.tensorflow.org/resources/libraries-extensions`.

Let's investigate the two described tasks separately, starting with the first one: loading and preprocessing data.

Loading and preprocessing data

To work with TensorFlow, you need to load your dataset as an object belonging to the `tf.data.Dataset` class. You can do it by hand or by using the API provided by the `tf.keras` sub-package. To get familiar with TensorFlow datasets, you can use the toy datasets provided by the `tf.keras.datasets` module, which are ready for use. In addition, you can use the public TensorFlow datasets provided by the TensorFlow extension library, available at the following link: `https://github.com/tensorflow/datasets`.

In this section, we will explore two simple ways to load your datasets from CSV or images.

Loading a TensorFlow dataset from CSV

The following steps will load a TensorFlow dataset from CSV:

1. First, you load the dataset as a `pandas` DataFrame as follows:

    ```
    import pandas as pd
    dataset_path = 'path/to/csv/file'
    df = pd.read_csv(dataset_path)
    ```

2. Then, you convert it into a dictionary as follows:

    ```
    dict_df = dict(df)
    ```

3. Finally, you load the dictionary as a TensorFlow dataset as follows:

    ```
    import tensorflow as tf
    data = tf.data.Dataset.from_tensor_slices(dict_df)
    ```

Now you can review how to build a TensorFlow dataset from a set of images.

Loading a TensorFlow dataset from a set of images

Next, let's try images with the following steps:

1. First, you need to organize your set of images. To do so, you need to create a folder for each class label, and each folder must contain the images associated with that class, as shown in the following figure:

Figure 10.8 – An example of the structure of an images folder

In the preceding figure, there are two directories, cat and dog, which correspond to two classes. Each directory contains three images.

2. Then, you can extract the training set from your images, as shown in the following code:

```
train_ds = tf.keras.utils.image_dataset_from_directory(
    data_dir,
    validation_split=0.2,
    subset="training",
    seed=123,
    image_size=(img_height, img_width),
    batch_size=batch_size)
```

The `data_dir` parameter specifies the directory where the images are located.

3. You can then extract the validation set as follows:

```
validation_ds = tf.keras.utils.image_dataset_from_
directory(
    data_dir,
    validation_split=0.2,
    subset="validation",
    seed=123,
    image_size=(img_height, img_width),
    batch_size=batch_size)
```

Before using your dataset as an input to a TensorFlow model, you can preprocess it. The `tf.keras.preprocessing` module provides many functions to preprocess images, texts, and sequence data. For further details, you can refer to the TensorFlow official documentation, available at the following link: `https://www.tensorflow.org/api_docs/python/tf/keras/preprocessing`.

Now that you have learned how to load and preprocess a dataset in TensorFlow, we can move on to the next point: building and training models.

Building and training models

The simplest way to build a TensorFlow model is to use the classes and functions provided by the `tf.keras` package as follows:

1. First, you define the model as follows:

    ```
    from tensorflow import keras
    model = keras.models.Sequential([
        tf.keras.layers.Conv2D(),
        tf.keras.layers.MaxPooling2D(),
        tf.keras.layers.Flatten(),
        tf.keras.layers.Dense(),
    ])
    ```

    The preceding example implements a `Sequential` model, which is a network where each layer has one input tensor and one output tensor. Within the `Sequential` model, you should specify the list of layers. In the preceding example, the first layer is a convolutional layer for 2D spatial data (`Conv2D`), followed by a pooling layer for 2D spatial data (`MaxPooling2D`). Then, there is a flatten (`Flatten`) layer, and finally, a dense layer (`Dense`). For simplicity, the preceding code does not show the configuration parameters for each layer. You can read the TensorFlow official documentation at `tensorflow.org` to learn how to configure the parameters of each layer.

2. Then, you compile the model as follows:

    ```
    model.compile(
        optimizer='adam',
        loss='sparse_categorical_crossentropy',
        metrics=['accuracy']
    )
    ```

    The `compile()` method receives as input `optimizer`, the `loss` function, and the list of `metrics` to evaluate during the training process.

3. Finally, you train the model as follows:

    ```
    model.fit(
        train_features,
        train_labels,
        validation_data=(test_features, test_labels),
    )
    ```

The `fit()` method receives as input the training set and the validation set. In addition, you can specify other parameters, including the number of batches and epochs. You can read the TensorFlow official documentation at `tensorflow.org` for further details.

Now that you have learned how to build and train a model in TensorFlow, let's move on to the next point: integrating TensorFlow with Comet.

Integrating TensorFlow with Comet

To make Comet log your TensorFlow models, you need to import the `comet_ml` library before TensorFlow, as shown in the following piece of code:

```
from comet_ml import Experiment
import tensorflow as tf
```

> **Note**
> You should always import the `comet_ml` library before any machine learning library.

If you are using a Keras model included in the TensorFlow library, Comet will automatically log most of the metrics and parameters. In addition, you can configure some specific configuration parameters to log histograms, graphs, hyperparameters, and so on. You can find the list of configuration parameters in the Comet documentation, available at the following link: `https://www.comet.ml/docs/v2/integrations/ml-frameworks/keras/`.

You can configure the configuration parameters in the following three different (alternative) ways:

- By adding a new line for each configuration parameter in the `.comet.config` file, as shown in the following piece of code:

```
[comet]
api_key=YOUR_COMET_API_KEY
workspace=YOUR_WORKSPACE
project_name=YOUR_PROJECT_NAME
[comet_auto_log]
histogram_weights=True
histogram_gradients=True
histogram_activations=True
```

In the preceding example, in addition to the classical `[comet]` section, we have added the `[comet_auto_log]` section and specified that we want to plot histograms related to weights, gradients, and activations.

- By passing the configuration parameters as input to the `Experiment` class, as shown in the following piece of code:

```
experiment = Experiment(auto_histogram_weight_
logging=True,auto_histogram_gradient_logging=True,auto_
histogram_activation_logging=True)
```

Similar to the case in the earlier example, we have specified that we want to plot histograms related to weights, gradients, and activations.

- By using environment variables, as shown in the following piece of code:

```
export COMET_AUTO_LOG_HISTOGRAM_WEIGHTS=True
export COMET_AUTO_LOG_HISTOGRAM_ACTIVATIONS=True
export COMET_AUTO_LOG_HISTOGRAM_ACTIVATIONS=True
```

Also, in this case, we have specified to plot histograms related to weights, gradients, and activations. In Python, we can also set the environment variables directly in our script, as shown in the following piece of code:

```
import os
os.environ['COMET_AUTO_LOG_HISTOGRAM_WEIGHTS'] = True
```

We have set the `environ` variable provided by the `os` library.

Now that you have learned how to integrate Comet with TensorFlow, we can move on to the final topic: using deep learning- from project setup to report building.

Using deep learning- from project setup to report building

In this section, you will implement a practical example that performs an image classification task. The objective of this example is to build a TensorFlow model that predicts the type of dress represented in an image. The model is fitted with some images representing clothes, and then it is used to predict the type of dress. You will track the model in Comet, and you will build a simple demo interface using Gradio to test the model performance interactively.

The full code of the example described in this section is available at the following link: `https://github.com/PacktPublishing/Comet-for-Data-Science/tree/main/10`.

You will focus on the following aspects:

- Introducing Gradio
- Loading the dataset
- Implementing a basic model
- Exploring results in Comet

- Building a prediction interface
- Building the final report

Let's start from the first point: introducing Gradio.

Introducing Gradio

Gradio is a Python library that permits you to build demos and quick web interfaces for testing purposes. Conceptually, a Gradio interface is composed of the following three components:

- **Input**, which can be either a single element or a list of elements, including textboxes, checkboxes, radio buttons, and many more.

- **Function**, which is a Python function that receives the input component as input, performs some computation, and returns the output.

- **Output**, which can be either a single element or a list of elements, including labels, images, plots, and many more.

For more details on the Gradio library, you can read its official documentation, available at the following link: `https://gradio.app/docs/`. You will see a practical example on how to use Gradio and combine it with Comet in the following sections.

Now that you have learned some basic concepts of Gradio, we can move on to the next step: loading the dataset.

Loading the dataset

As a use case, you will use the `Fashion-MNIST` set released by Zalando Research and available on GitHub at `https://github.com/zalandoresearch/fashion-mnist`, released under the MIT license. The dataset contains a training set of 60,000 examples and a test set of 10,000 examples. Each example consists of a 28x28 grayscale image associated with a label from one of the following 10 classes:

Label	Class
0	T-shirt/top
1	Trouser
2	Pullover
3	Dress
4	Coat
5	Sandal
6	Shirt
7	Sneaker
8	Bag
9	Ankle boot

Figure 10.9 – The mapping between label and class in the Fashion-MNIST dataset

You can load the dataset through TensorFlow as follows:

1. First, we set all of the environment variables as follows:

```
import os
os.environ['COMET_KERAS_HISTOGRAM_ACTIVATION_INDEX_LIST']
= "1,2"
```

The preceding variable indicates that we want to build the activation histograms for layers 1 and 2.

2. Then, we import all of the required libraries as follows:

```
from comet_ml import Experiment
import tensorflow as tf
from tensorflow import keras
```

> **Note**
> To make Comet log a TensorFlow model, you should import the comet_ml library before the tensorflow one.

3. Then, we use the load_data() method to load the dataset directly from keras as follows:

```
fashion_mnist = keras.datasets.fashion_mnist
(train_images, train_labels), (test_images, test_labels)
= \
    fashion_mnist.load_data()
```

We use the keras.dataset.fashion_mnist object provided by keras, and then the load_data() method to retrieve both the training and test sets.

4. We can plot the first 25 examples extracted from the training set, as shown in the following piece of code:

```
n_row = 5
n_col = 5
_, axs = plt.subplots(n_row, n_col, figsize=(12, 12))
axs = axs.flatten()
for img, ax in zip(train_images, axs):
    ax.imshow(img)
    ax.axis('off')
plt.show()
```

We have defined a grid of size 5x5, and then we have plotted an image in each cell of the grid through the `imshow()` method. We have also hidden both axes through the `ax.axis('off')` statement. The following figure shows the first 25 examples extracted from the training set:

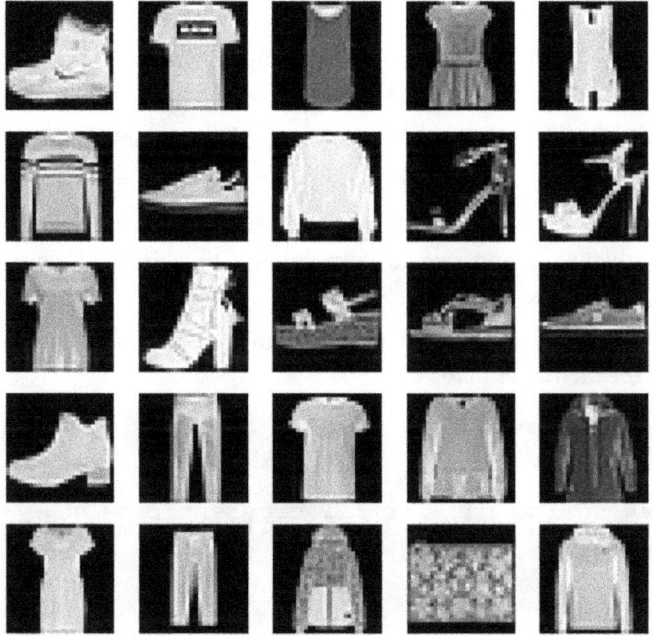

Figure 10.10 – The first 25 examples extracted from the training set

To the human eye, it is quite easy to recognize the class to which each image belongs.

Now that you have loaded the dataset, we can move on to the next step: implementing a simple model.

Implementing a basic model

It is time to build the model with the following steps:

1. Since TensorFlow is automatically integrated with Comet, we will first create a Comet experiment where we specify the additional parameters to also log histograms as follows:

    ```
    experiment = Experiment(auto_histogram_weight_
    logging=True,auto_histogram_gradient_logging=True,auto_
    histogram_activation_logging=True)
    ```

    You need to configure the `.comet.config` file as explained in *Chapter 1, An Overview of Comet*. Alternatively, to pass the additional parameters to the Comet experiment, you can include them in the `.comet.config` file.

2. Now, we implement a Sequential model, which is the simplest available model. A Sequential model contains exactly one input tensor and one output tensor for each layer. In our case, we implement a Sequential model with three layers as follows:

```
model = keras.models.Sequential([
    keras.layers.Flatten(input_shape=(28, 28)),
    keras.layers.Dense(32, activation='relu'),
    keras.layers.Dense(10, activation='softmax')
])
```

The first layer is a Flatten layer, which receives the image as input. The second layer is a Dense layer with 32 units and a relu activation function. The third layer contains 10 units and a softmax activation function.

3. Once you have created the model, we need to compile it by defining the loss function, the optimizer, and the metrics as follows:

```
model.compile(
    optimizer='adam',
    loss='sparse_categorical_crossentropy',
    metrics=['accuracy']
)
```

In our case, we have set the optimizer to adam, the loss function to sparse_categorical_cross_entropy, and the list of metrics to accuracy.

4. The model is ready to be fitted as follows:

```
model.fit(
    train_images,
    train_labels,
    epochs=8,
    validation_data=(test_images, test_labels),
)
```

We use the fit() method, which receives as input the training images and their associated labels as well as the number of epochs and data used for validation.

5. After fitting, we can plot a model summary as follows:

```
print(model.summary())
```

The summary() method returns the output shown in the following figure:

```
Model: "sequential"
_____
 Layer (type)                Output Shape              Param #
=================================================================
 flatten (Flatten)           (None, 784)               0

 dense (Dense)               (None, 32)                25120

 dense_1 (Dense)             (None, 10)                330

=================================================================
Total params: 25,450
Trainable params: 25,450
Non-trainable params: 0
_____

None
```

Figure 10.11 – The output of the summary() method

It is worth noting how the number of parameters and the output shape are calculated. The Flatten layer does not have any input channel. Since we have set the input shape to (28, 28), you can calculate the output shape as follows: 28 x 28 = 784. The output shape of the Flatten layer is given as input to the first Dense layer. In general, you can calculate the parameters of a Dense layer as the product between the input shape plus one (784 + 1) and the output shape plus one (32). In our case, (784 + 1) * 32 = 25,120. The output shape of the first Dense layer is given as input to the second Dense layer. Thus, you can calculate the number of parameters as follows: (32+1) * 10 = 330.

6. You can plot the model graph as follows:

```
tf.keras.utils.plot_model(model, expand_nested=True)
```

The following figure shows the produced graph:

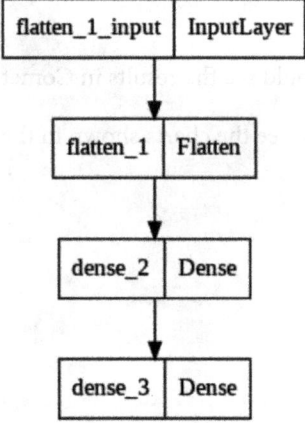

Figure 10.12 – The implemented model graph

7. Comet will automatically log the training phase as well as the defined metrics. In addition, you can log the confusion matrix as follows:

```
preds = model.predict(test_images)
experiment.log_confusion_matrix(test_labels, preds)
```

First, we calculate predictions through the `predict()` method, then we log the confusion matrix directly in Comet.

You can also log some sample images in Comet through the `summary` object provided by the TensorFlow library. We log five images for the training set and five images for the test set as follows:

1. We prepare the images for logging by reshaping them as follows:

```
train_img = np.reshape(train_images[0:5], (-1, 28, 28, 1))
test_img = np.reshape(test_images[0:5], (-1, 28, 28, 1))
```

2. We create a `file_writer` object, which will store the logged images as follows:

```
LOG_DIR = 'logs'
file_writer = tf.summary.create_file_writer(LOG_DIR)
```

3. We log the images:

```
with file_writer.as_default():
    tf.summary.image("Training data", train_img, step=0)
    tf.summary.image("Test data", test_img, step=0)
```

Now that you have built and fitted the basic model, we are ready to see the results in Comet.

Exploring results in Comet

After running the experiment, you should see the results in Comet as follows:

1. Under the **Chart** menu, you can see the charts shown in the following figure:

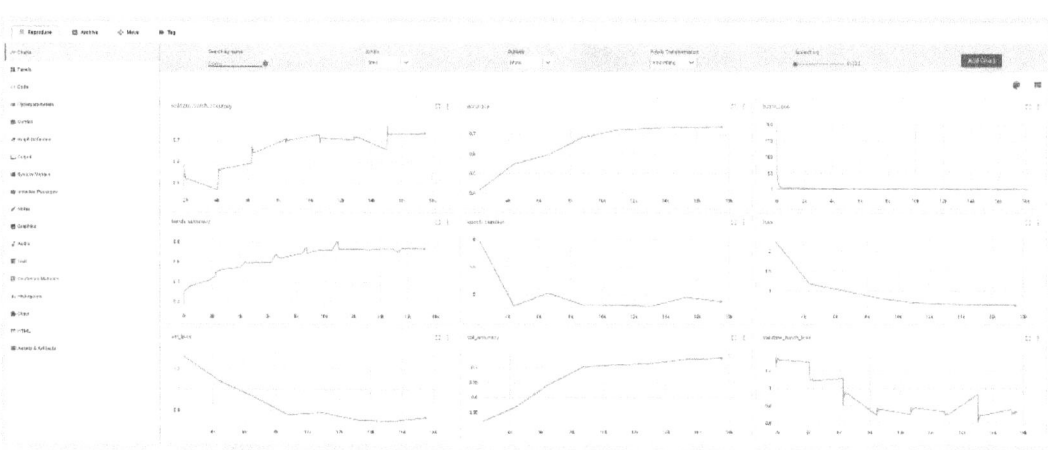

Figure 10.13 – The charts automatically produced by Comet

You can access the produced charts at the following link: `https://www.comet.ml/packt/deep-learning`. They include the following:

- Accuracy

- Loss

- Validate batch accuracy

- Validate batch loss

- Batch loss

- Batch accuracy

- Epoch duration

- Val loss

- Val accuracy

Each graph shows the metrics versus the number of steps. Considering a default batch size of 32 and a training set of 60,000 samples, the number of steps per batch size is 60,000/32 = 1875. Since we have set the total number of epochs to 8, we can calculate the total number of steps for the training set as follows: 1875 x 8 = 15,000. For the test set (which has 10,000 samples), the number of steps per batch size is 10,000 x 32 = 312.5, which is about 313. Considering a number of epochs equal to 8, we can calculate the total number of steps for the test set as follows: 313 x 8 = 2,504. The total number of steps shown on the x-axis is given by the sum of the number of steps for the training and test phases, as follows: 15,000 + 2,504 = 17,504.

2. Under the **Metrics** menu, you can find all of the logged metrics, as shown in the following figure:

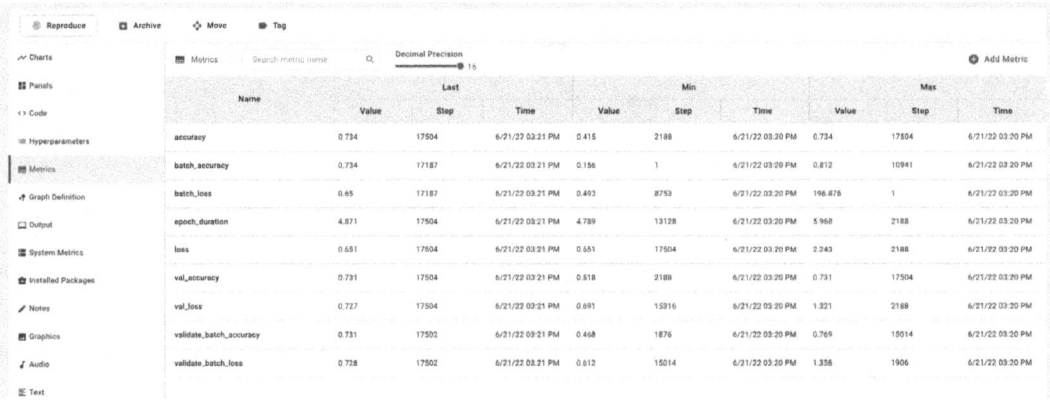

Figure 10.14 – Logged metrics in Comet

3. Under the **Graphics** menu, you can find the examples produced through the `tf.summary` object, as shown in the following figure:

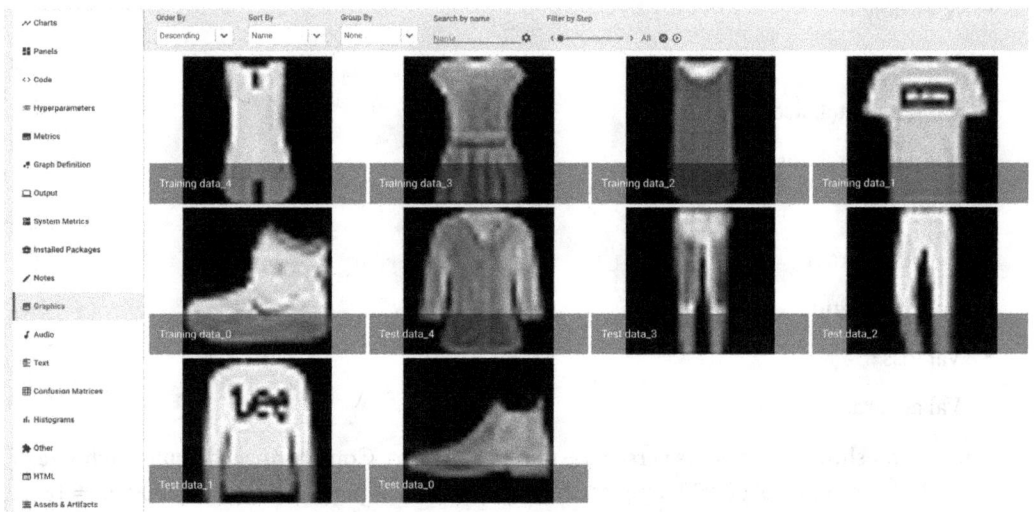

Figure 10.15 – Some images saved in Comet

The prefix *Training data* or *Test data* specified whether the image belongs to the training or test set, respectively.

4. Under the **Confusion Matrices** menu, you can view the confusion matrix, as shown in the following figure:

Figure 10.16 – The confusion matrix in Comet

If you move the mouse over a cell of the matrix, you will be able to see some details on the total number of values, the real ones, and the predicted ones. In addition, if you click on the cell, you will also see some examples.

5. Under the **Histograms** menu, you can see the produced histograms, as shown in the following figure:

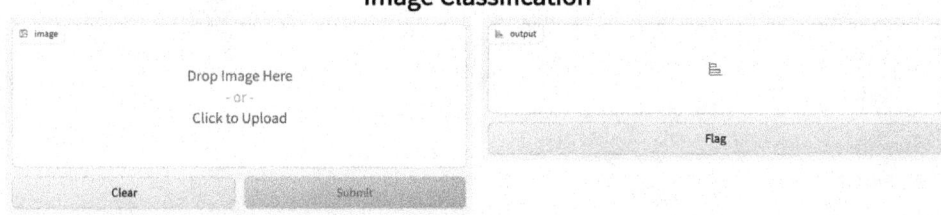

Figure 10.17 – The histograms in Comet

The section contains six histograms: bias, activation, and kernel for layers 2 and 3. You can see the complete graphs directly in Comet by clicking on the following link: `https://www.comet.ml/packt/deep-learning`, in the Histograms menu.

Now that you have explored the results in Comet, we can move on to the next step: building a prediction interface.

Building a prediction interface

You can use a prediction interface to calculate the class for an image dynamically provided as input. To build the prediction interface, we can use Gradio, a Python library, that permits you to quickly build simple interfaces for tests. In our case, the prediction interface should include an input form, where you upload an image, and an output box, which shows the predicted class. The following figure shows a possible prediction interface built in Gradio:

Image Classification

image

Drop Image Here
- or -
Click to Upload

output

Flag

Clear   Submit

Figure 10.18 – A possible prediction interface in Gradio

Comet is fully integrated with Gradio, so you can run a Gradio interface as a Comet panel. To build the prediction interface and add it to Comet, you can perform the following steps:

1. First, you need to import the Gradio library and define the Gradio interface before importing the Comet library. You can define an auxiliary function, named `predict()`, that calculates the predicted class for the image provided as input as follows:

    ```
    import gradio as gr
    model = None
    def predict(image):
       image = image.reshape(-1, 28, 28, 1)
       prediction = model.predict(image).flatten()
       return {class_names[i]: float(prediction[i]) for i in
    range(len(class_names))}
    ```

    Since we have not defined the model yet, we initialize a dummy variable, named `model`, which will be set to the actual model at runtime. Here, the problem is that you must import Gradio before Comet, and TensorFlow after Comet, thus you cannot train the TensorFlow model before importing Gradio.

2. Now, you can build the interface in Gradio as follows:

    ```
    image = gr.inputs.Image(shape=(28, 28))
    label = gr.outputs.Label()
    io = gr.Interface(fn = predict,inputs = image,outputs =
    label, title="Image Classification")
    io.launch(inline=False)
    ```

    We have defined the input and the output as well as the Gradio interface. Then, we launched the interface through the `launch()` method, without showing it (`inline = False`), since we will use it in Comet.

3. We create the Comet experiment as we usually do and save the experiment key because we will use it later as follows:

    ```
    experiment = Experiment()
    experiment_key = experiment.get_key()
    ```

    We have used the `get_key()` method to retrieve the experiment key.

4. Now, we integrate the Gradio interface in Comet as follows:

    ```
    io.integrate(comet_ml=experiment)
    ```

5.   `integrate()` must be called immediately after the creation of the experiment. Unfortunately, this method also terminates the experiment, thus making it not possible to continue building the TensorFlow model. To overcome this issue, we reopen the experiment by creating an `ExistingExperiment` object as follows:

```
from comet_ml import ExistingExperiment
experiment = ExistingExperiment(previous_
experiment=experiment_key)
```

To create an `ExistingExperiment` object, you need to pass the key of the preceding experiment. The `ExistingExperiment` object continues the preceding experiment.

6.   After reopening the experiment, you can continue working on it by creating and training your TensorFlow model as described in the preceding section.

Once you have run the experiment, you can access the Comet dashboard and create a new panel with the Gradio interface. To create the Gradio panel, you can perform the following steps:

1.   From the Comet project main dashboard, select the **Panels** tab, then **Add** | **New Panel** | **PUBLIC** | **Gradio Panel** | **Add** | **Done**.

2.   You should see a new panel in your Comet dashboard as shown in the following figure:

Figure 10.19 – The Gradio interface

In the preceding example, we uploaded an image representing a shirt (shown on the left), and our predictor returned the **T-shirt/top** with a probability of 38% and the **Shirt** class with a probability of 24%.

Now that you have built the Gradio interface, you can move on to the final step: building the final report.

Building the final report

Now you are ready to build the final report. In this example, we build a simple report with the model results. As a further exercise, you could improve them by applying the concepts learned in *Chapter 5, Building a Narrative in Comet.*

To create the report, in the Comet dashboard you can click on the **Panels** tab, then select **Add | Add to Report | New Report**.

You will create a report with the following three sections:

- Data
- Model evaluation
- Real-time prediction

The report automatically loads the model evaluation report; thus, we will describe only how to build the first and the third sections. Let's start with the first section: data.

Data

Let's begin by performing the following steps:

1. First, we need to create a new section by clicking on the **Add section here** button, which is available when you hover the mouse at the top of the previous section.

2. Then, we insert the logged sample images. We use the Comet Viewer Panel, which you can add by clicking on **Add | New Panel | PUBLIC | Comet Viewer Panel | Add | Done**.

    The following figure shows the produced Comet Viewer Panel:

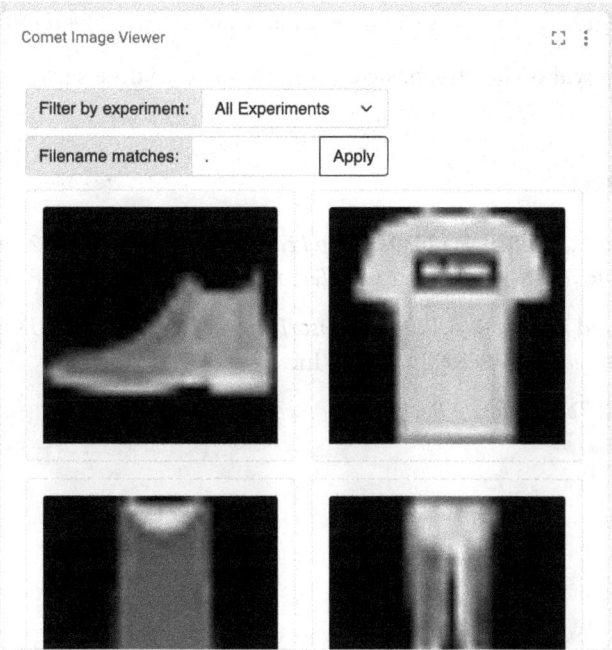

Figure 10.20 – The Comet Viewer Panel

The panel permits you to filter images by experiments. In our case, we have just one experiment, thus the panel automatically shows it.

Real-time prediction

In this section, we add the Gradio panel as follows:

1. First, we create a new section by clicking the **Add section here** button, which is available when you hover the mouse at the bottom of the Model Evaluation section.

2. Then, we add a new panel, as described in the *Building a prediction interface* section of this chapter.

Your report is ready! You can view the final result directly in Comet at the following link: `https://www.comet.ml/packt/deep-learning/reports/clothes-classification`.

Summary

We have just completed the journey to build a deep learning model in TensorFlow and track it in Comet!

Throughout this chapter, we described some general concepts regarding deep learning as well as the main structure of the TensorFlow package, and some related concepts, including how to load a dataset and build and train a model in TensorFlow.

In the last part of the chapter, we implemented a practical use case that showed you how to track a deep learning experiment in Comet as well as how to build a report with the results of the experiment.

In the next chapter, we will review the basic concepts related to time series analysis and how to perform it in Comet.

Further reading

- Gulli, A., Kapoor, A., and Pal, S. (2019). *Deep Learning with TensorFlow 2 and Keras: Regression, ConvNets, GANs, RNNs, NLP, and More with TensorFlow 2 and the Keras API*. Packt Publishing Ltd.

- Ramsundar, B., and Zadeh, R. B. (2018). *TensorFlow for Deep Learning: From Linear Regression to Reinforcement Learning*. O'Reilly Media, Inc.

- Tung, KC (2021). *TensorFlow 2 Pocket Reference*. O'Reilly Media, Inc.

11
Comet for Time Series Analysis

Time series analysis is the study of the evolution of phenomena over time, in order to predict their future trends. It finds its application in various sectors, such as the price trend of a given product, tourist flows to a given location, and the performance of a product on the stock exchange.

Generally speaking, a time series can be represented by different components, such as the trend, which can be increasing, stable or decreasing; the seasonality – that is, the repetition over time; and the presence of breaking points, due to external events, which interrupt its normal trend.

In this chapter, you will review some basic concepts behind time series analysis, including the concept of stationarity, time series components, and how to check for the presence of breakpoints in a time series.

Over the last few years, different open source tools and libraries have been implemented to perform time series, including Prophet, statsmodels, and Kats. In this chapter, we will review the `Prophet` library and how to integrate it with Comet.

In the last part of this chapter, you will implement a practical use case, which uses the Prophet library, and tracks the result in Comet.

The chapter is organized as follows:

- Introducing basic concepts related to time series analysis
- Exploring the Prophet package
- Using time series analysis from project setup to report building

Before starting to review the basic concepts related to time series analysis, let's install the required software needed to run the examples described in this chapter.

Technical requirements

We will run all the experiments and codes in this chapter using Python 3.8. You can download it from the official website, https://www.python.org/downloads/, choosing the 3.8 version.

The examples described in this chapter use the following Python packages:

- comet-ml 3.23.0
- matplotlib 3.2.2
- numpy 1.21.6
- pandas 1.3.4
- prophet 1.1
- scikit-learn 1.0
- statsmodels 0.13.2

We have already described the comet-ml, matplotlib, NumPy, pandas, and scikit-learn packages and how to install them in *Chapter 1, An Overview of Comet*, so please refer to that for further details on installation.

In this section, you will see how to install the other required packages.

Prophet

Prophet is an open source Python package for time series analysis. You can install it as follows:

```
pip install prophet
```

For more details about Prophet installation, you can read its official documentation, available at the following link: https://facebook.github.io/prophet/docs/installation.html.

statsmodels

statsmodels is a Python library for statistical analysis. You can install it as follows:

```
pip install statsmodels
```

For more details about statsmodels installation, you can read its official documentation, available at the following link: https://www.statsmodels.org/stable/install.html.

Now that you have installed all the software needed in this chapter, let's move on to how to use Comet for time series analysis, starting with reviewing some basic concepts.

Introducing basic concepts related to time series analysis

A time series is an ordered sequence of values over time, representing the variation of a certain phenomenon. Examples of time series include the trend of the prices of a certain product, and the trend of rainfall in a given region over time. The following figure shows an example of time series representing the natural gas price from 2000 to 2020:

Figure 11.1 – The natural gas price time series

Data was extracted from the DataHub website and is available at `https://datahub.io/core/natural-gas` under the public domain and the use of **Energy Information Administration (EIA)** content license.

Time series analysis, also known as **time series forecasting**, is the study of the past values of a time series, with the purpose of building a model that predicts its future values.

In this section, you will learn the following basic concepts and aspects related to time series:

- Loading a time series in Python
- Checking whether a time series is stationary
- Exploring time series components
- Identifying breakpoints in a time series

Let's start with the first aspect, loading a time series in Python.

Loading a time series in Python

To load a dataset as a time series in Python, you can proceed as follows:

1. Firstly, you load the dataset as a `pandas` DataFrame:

   ```
   import pandas as pd
   df = pd.read_csv('https://datahub.io/core/natural-gas/r/
   monthly.csv', parse_dates=['Month'])
   ```

   You should make sure that the column related to dates is parsed as a datetime (`parse_dates=['Month']`). As an example, we have used the natural gas prices dataset, as described previously.

2. Then, we set the index of the DataFrame to the column containing dates:

   ```
   df = df.set_index('Month')
   ```

3. Finally, we assign to the time series the column containing values:

   ```
   ts = df['Price']
   ```

Now that you have learned how to load a time series in Python, you are ready to check whether a time series is stationary.

Checking whether a time series is stationary

Stationarity is a time series property, meaning that the statistical properties of the process that generates the time series do not change over time. This property does not mean that the time series is constant over time, just that the way it changes does not change over time. For example, the time series of *Figure 11.1* is not stationary because you cannot recognize a constant generating process for that time series.

The section is organized as follows:

- The stationarity test
- Dealing with non-stationary time series

Let's start from the first point, the stationarity test.

The stationarity test

To check whether a time series is stationary, you can use different methods, including the **Augmented Dickey-Fuller (ADF)** test and the **Kwiatkowski-Phillips-Schmidt-Shin (KPSS)** test. In this chapter, we will focus on the ADF test. For more details, you can refer to the books contained in the *Further reading* section.

The ADF test is a statistical test, which is conducted with the following assumptions:

- **Null Hypothesis** (H0): The time series is non-stationary.
- **Alternate Hypothesis** (HA): The time series is stationary.

If the test fails to reject the null hypothesis, the series is non-stationary. In the ADF test, there are two conditions to reject the null hypothesis:

- If test statistic < critical value
- If p-value < alpha

The test statistic, the p-value, and the critical value are variables returned by the test. Conversely, the alpha variable is usually set to 0.05. If both the conditions are satisfied, you can conclude that the time series is stationary. The statsmodels Python package provides a function to perform the ADF test. You will explore it through a practical example in the *Using time series analysis from project setup to report building* section.

If a time series is stationary, you can build a prediction model, which, in theory, could be very accurate. Instead, if the time series is not stationary, you can even build a prediction model, but the results of the prediction could be unreliable. Thus, it would be better if your time series were stationary.

What should you do with a non-stationary time series? Let's investigate it in the next section.

Dealing with non-stationary time series

If a time series is not stationary, you can perform a transformation to make the time series stationary. Examples of transformations include differentiating the current value from the previous one, and a logarithmic transformation.

To apply a differencing transformation to the time series, you can proceed as follows:

1. Firstly, you calculate the difference between the values in the time series:

    ```
    ts2 = ts.diff()
    ```

    This diff() method calculates the difference between the current value and the previous one of the time series. Obviously, this operation cannot be done for the first value of the time series, which is set to NaN by the diff() method.

2. Then, we drop all the NaN values in the new time series, as follows:

    ```
    ts2.dropna(inplace=True)
    ```

The following figure shows the effects of the differencing transformation on the time series of *Figure 11.1*:

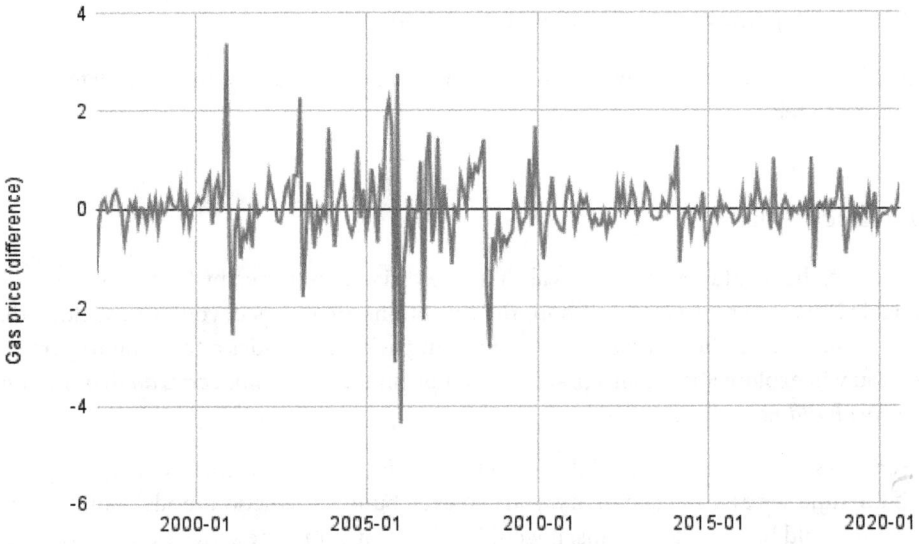

Figure 11.2 – The effects of the differencing transformation

In the example, the differencing transformation makes the time series stationary. Thus, you can use the differentiated time series to build the model. However, if you want to reconstruct the original time series, you need to perform the inverse operation. In the case of differencing, you should calculate the cumulative function:

```
ts3 = ts2.cumsum()
```

The new time series, ts3, is equal to the original time series. However, in ts3, the first value is missing, since it is also missing in ts2. Now that we have learned some basic concepts regarding stationarity, we can move on to the next aspect, exploring the time series components.

Exploring the time series components

You can think of a time series as being composed of the following three components:

- **Trend**, which represents the long-term direction. It can be either increasing or decreasing.

- **Seasonality**, which is a cycle that repeats regularly over time (for example, every day, month, or year). You can identify seasonality by finding regularly spaced peaks with approximately the same magnitude for every period of time.

- **Residuals**, which are short-term fluctuations.

You can use different techniques to decompose a time series in the previous three components and how to deal with seasonality. In this section, you will see the following:

- Decomposing a time series in Python
- Dealing with seasonal time series

Let's start from the first point, decomposing a time series in Python.

Decomposing a time series in Python

In Python, you can decompose a time series into its three components through the seasonal_decompose() function, provided by the statsmodels package. Let's suppose that you want to decompose the time series shown in the following figure:

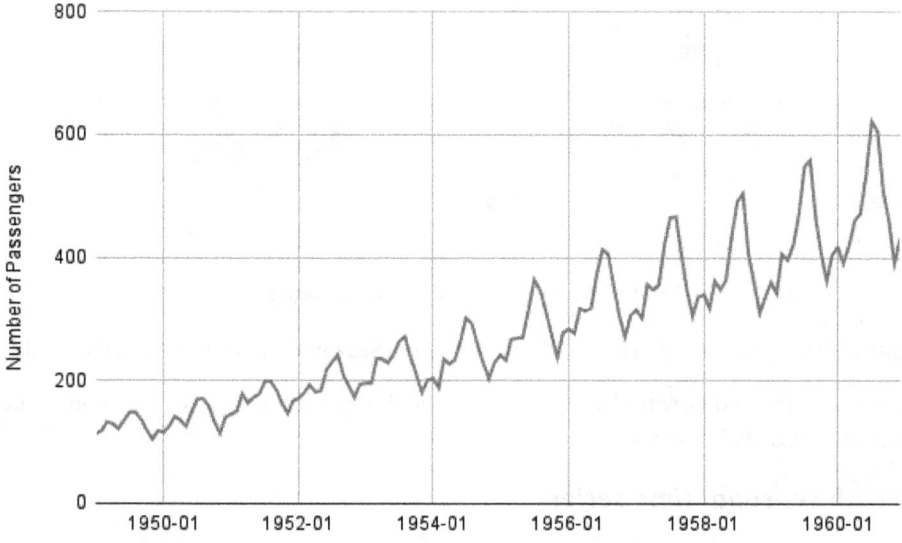

Figure 11.3 – The air passengers time series

The time series represents the number of air passengers per month. The dataset is available on Kaggle at https://www.kaggle.com/datasets/rakannimer/air-passengers under the Open Data Commons license.

The following piece of code shows how to decompose the time series through the seasonal_decompose() function:

```
from statsmodels.tsa.seasonal import seasonal_decompose
result = seasonal_decompose(ts, model='additive', period=12)
```

The function receives the time series as input, the model used to decompose (either multiplicative or additive), and the period, in months.

You can also plot the results of decomposition through the `result.plot()` method:

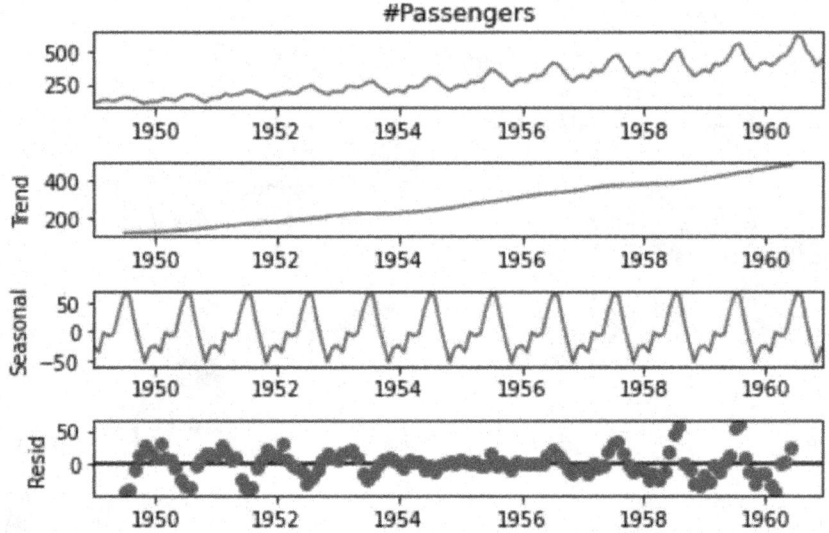

Figure 11.4 – The decomposed time series

You can easily identify the trend (**Trend**), the seasonality (**Seasonal**), and the residuals (**Resid**).

Now that you have learned how to decompose a time series in Python, we can move on to the next step, dealing with seasonal time series.

Dealing with seasonal time series

When you build a model for time series forecasting, you should take into account whether your time series has a seasonal component or not. In fact, some models perform better if the time series does not present seasonality, and others if they do.

One strategy to use any model regardless of the presence of seasonality is to remove seasonality from the time series and predict future values for the seasonally adjusted time series. Then, you can add the seasonality to the predicted values, to obtain the correct predictions.

Follow these detailed steps:

1. First, you can decompose the time series through the `seasonal_decompose()` function, as described in the previous section.

2. Then, you can compute the seasonally adjusted time series, as follows:

```
seasonality = result.seasonal
adjusted_ts = ts - seasonality
```

Since we have used an additive model, we can remove seasonality simply by calculating the difference between the original time series and the seasonality. The following figure shows the original time series and the seasonally adjusted time series:

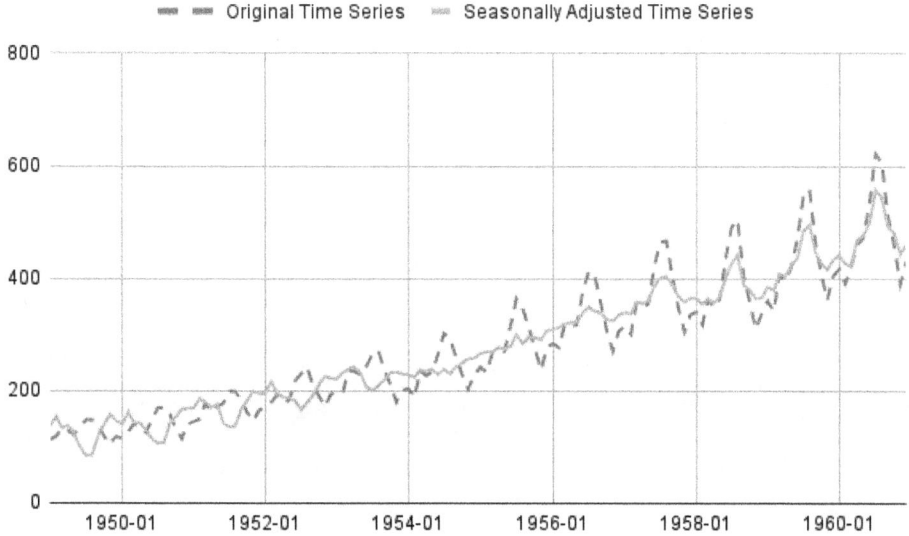

Figure 11.5 – The original time series and the seasonally adjusted time series

3. Now, you can use any model you prefer to predict the future values of the seasonally adjusted time series.

4. Finally, you add the seasonality of the last time period to the predicted values, to obtain the predictions for the original time series.

Now that we have learned about time series components, we can discuss how to identify breakpoints in a time series.

Identifying breakpoints in a time series

A breakpoint is a structural change in a time series, such as an anomaly or an expected event. The following figure shows an example of time series with two breakpoints:

Figure 11.6 – A time series with at least two breakpoints

The figure shows the time series related to the number of new daily cases of infection of COVID-19 in Italy, extracted from the Humanitarian Data Exchange, available at the https://data.humdata.org/dataset/coronavirus-covid-19-cases-and-deaths under the Creative Commons attribution for the Intergovernmental Organizations license.

You can use the following two main techniques to identify breakpoints:

- **Detection**, which aims to identify breakpoints automatically. In Python, many libraries exist to detect breakpoints, such as ruptures (https://pypi.org/project/ruptures/) and jenkspy (https://pypi.org/project/jenkspy/).

- **Test**, which tests whether a point is a breakpoint or not. An example of this category of techniques is the Chow test (https://pypi.org/project/chowtest/).

Now that we have learned the basic concepts of breakpoints in a time series, we can move on to the next step, exploring the Prophet package.

Exploring the Prophet package

Prophet is an algorithm for time series analysis, released by Facebook's Core Data Science team. Other algorithms exist for time series analysis, including **Autoregressive Integrated Moving Average (ARIMA)** and **Seasonal Autoregressive Integrated Moving Average (SARIMA)**. You can refer to the books in the *Further reading* section if you are interested in them.

In this section, you will investigate Prophet, with a focus on the following aspects:

- Introducing the Prophet package
- Integrating Prophet with Comet

Let's start with the first point, introducing the Prophet package.

Introducing the Prophet package

To build a model using `Prophet`, you can proceed as follows:

1. Firstly, you import the `Prophet` library:

    ```
    from prophet import Prophet
    ```

2. Then, you build a `Prophet()` object:

    ```
    model = Prophet()
    ```

3. You train the model:

    ```
    model.fit(df)
    ```

4. You build a dataset with future dates:

    ```
    future = model.make_future_dataframe(periods=N)
    ```

    The N variable indicates the number of months.

5. Finally, you use the model to predict future values:

    ```
    forecast = model.predict(future)
    ```

You can customize the `Prophet` model with different parameters, which permit you to deal with the following main aspects:

- **Breakpoints**: `Prophet` automatically detects breakpoints. However, you can set them manually for more precise control, as follows:

    ```
    model = Prophet(changepoints=['2020-05-06', '2022-07-06'])
    ```

We use the `changepoints` parameter to set the list of changepoints.

- **Seasonality**: `Prophet` automatically searches for weekly and yearly seasonality, and for time series with more than two cycles in length. However, you can set a custom seasonality as follows:

```
model = Prophet(weekly_seasonality=False, yearly_
seasonality=False)
model.add_seasonality(name='monthly', period=30.5,
fourier_order=3)
```

We disable the weekly and yearly seasonality (if not needed), and we use the `add_seasonality()` method to add a custom seasonality. This method receives as input the associated name, the period of seasonality, and the Fourier order. For more details on the Fourier order, you can read the official Prophet documentation, available at the following link: `https://facebook.github.io/prophet/docs/seasonality,_holiday_effects,_and_regressors.html`.

- **Cross-validation**: You can use this diagnostic feature to calculate the performance error of your model. This is achieved by means of cross-validation, where you need to specify the period and the length of prediction, as shown in the following piece of code:

```
from prophet.diagnostics import cross_validation
df_cv = cross_validation(model, initial='365 days',
period='30 days', horizon = '100 days')
fig3 = plot_cross_validation_metric(df_cv, "mse")
```

You should pass to the `cross_validation()` function the following additional three parameters:

- `Initial`: the size of the training set used to fit the model. The training set is extracted from the first values of the dataset.

- `Period`: how much data the algorithm must add to the training set in every iteration of cross-validation.

- `Horizon`: the length of prediction.

The `cross_validation()` function returns a DataFrame that contains the predicted values and the actual values, which you can use to calculate the classical performance metrics, such as MSE, RMSE, and the **Mean Absolute Percentage Error** (**MAPE**).

For more details on the Prophet package, you can refer to its official documentation, available at the following link: `https://facebook.github.io/prophet/`.

Now that you have learned some basic concepts of the Prophet package, we can move on to the next step, integrating Prophet with Comet.

Integrating Prophet with Comet

Comet is fully integrated with Prophet, so whenever you build a Prophet model, Comet will log the following elements automatically:

- Hyperparameters
- The model
- All the `matplotlib` figures

You can control which elements you want to log by setting the corresponding parameter, as described in the Comet official documentation, available at the following link: `https://www.comet.com/docs/python-sdk/prophet/`.

To make automatic logging work, you should make sure that the `comet_ml` library is imported before the `prophet` one:

```
from comet_ml import Experiment
from prophet import Prophet
```

It may happen that you have a version of `Prophet` that is not supported by Comet. In this case, we can use the configuration parameters to make Comet log elements. For example, we can set the `COMET_AUTO_LOG_FIGURES` environment variable to 1 to make Comet log figures. We can set this variable in many ways. For example, we can use the `os` library, as follows:

```
import os
os.environ['COMET_AUTO_LOG_FIGURES'] = '1'
```

Alternatively, you should log the elements manually, as shown in the following piece of code:

```
experiment.log_figure(figure_name='forecast', figure=fig1)
```

The code shows how to plot a `matplotlib` figure named `fig1`.

Now that we have learned how to integrate Comet with Prophet, we can implement a practical example, which starts from project setup to report building.

Using time series analysis from project setup to report building

In this section, you will implement a practical example that builds two models to predict the future trend of a time series describing arrivals at tourist accommodation establishments. The dataset shows the trend from 1990 to 2022; thus, it contains a breakpoint in correspondence in April 2020, when the COVID-19 pandemic began.

In the example, you will build two models, one which considers the breakpoint at the beginning of the COVID-19 pandemic and another which does not. You will compare the two models in Comet to establish which one performs better.

The full code of the example described in this section is available at the following link: `https://github.com/PacktPublishing/Comet-for-Data-Science/tree/main/11`.

You can write the code using the editor or the notebook you prefer. In this example, you will use Deepnote, a popular online notebook, which is fully integrated with Comet.

You will focus on the following aspects:

- Configuring the Deepnote environment
- Loading and preparing the dataset
- Checking stationarity in data
- Building the models
- Exploring results in Comet
- Building the final report

Let's start from the first point, configuring the Deepnote environment. If you do not want to use Deepnote for your project, you can skip the next section and jump directly to the *Loading and preparing the dataset* section.

Configuring the Deepnote environment

Deepnote is a platform that permits you to create notebooks over the web. The entry point to the Deepnote website is available at the following link: `https://deepnote.com/`.

With respect to the classical Jupyter notebooks, you can share notebooks created in Deepnote with your colleagues in real time. In addition, Deepnote is compatible with Jupyter notebook, so you can easily import notebooks implemented in Jupyter directly into Deepnote, and vice versa. Deepnote also provides you with a virtual machine that you can configure according to your needs. For example, you can configure the Python version, as well as the Python libraries used by a specific project.

Comet is fully integrated with Deepnote, so you can run your Comet experiments directly in Deepnote, and then you can display the Comet dashboard directly in Deepnote.

Let's investigate how to configure Deepnote to work with Comet:

1. To use Deepnote, firstly you should create an account at the following link: `https://deepnote.com/sign-up`. The automatic procedure will ask you to write the name of your workspace, as well as how you are using Deepnote (for example, as a student or a teacher).

2. Once you have created your account, you can log in to the Deepnote platform. You will see a dashboard similar to the one shown in the following figure:

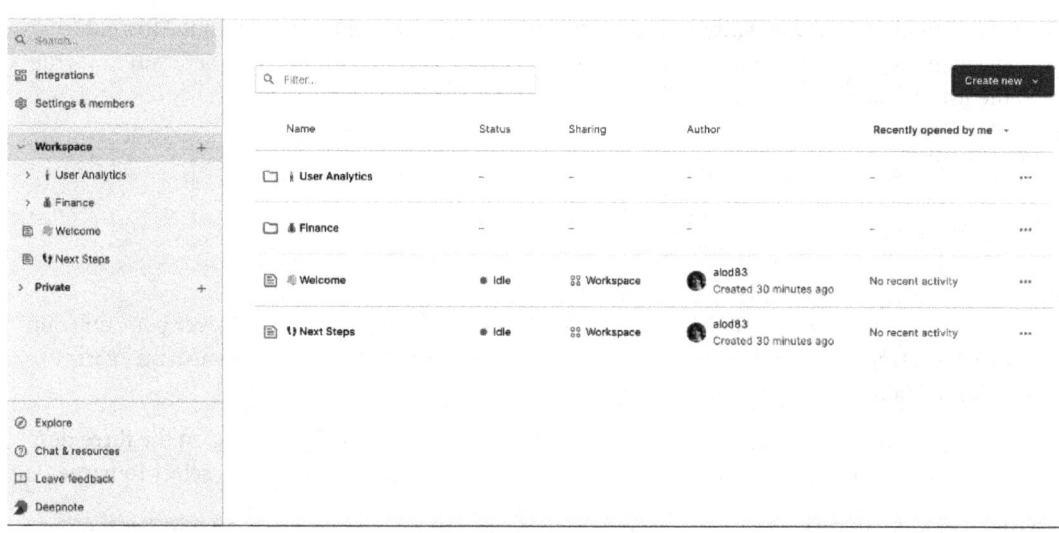

Figure 11.7 – The Deepnote dashboard

3. Now you can create a new project by clicking on **Create new** | **New Project**. A new notebook opens, where you can start writing your code. You can also import an existing project from a Jupyter notebook, GitHub, or Google Drive. In the right part of the notebook, you can see a section that allows you to configure the environment, including the required hardware and the Python package, add new files, and so on.

4. To install the required libraries for the project, you should create a new file, called `requirements.txt`, containing the list of libraries to install. You can create a new file by clicking on the + symbol, located in the **Project** tab of the section on the right, as shown in the following figure:

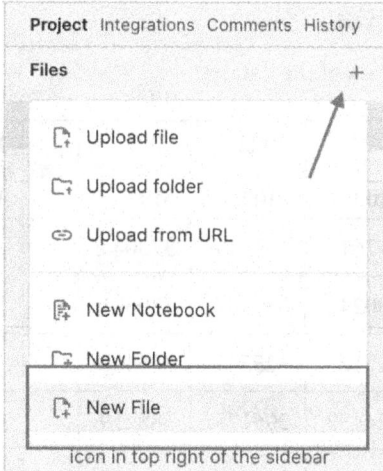

Figure 11.8 – The menu to create a new object in a Deepnote project

5. After clicking on the + symbol, you can select **New File** from the popup menu. After entering the filename (`requirements.txt` in this example), a text editor opens, where you can write the list of required libraries, as shown here:

```
pandas
comet_ml
statsmodels
prophet
```

You do not need to save the file because Deepnote will do it for you. Whenever you run your project, Deepnote will install all the software contained in the `requirements.txt` file automatically.

6. Now, you can come back to the previous notebook and rename it, by clicking on the three dots near the file, located in the **Project** tab of the section on the right, and then select **Rename**.

We have just configured Deepnote to host our code, so we can move on to the next step, loading and preparing the dataset.

Loading and preparing the dataset

As a use case, you will use the arrivals at tourist accommodation establishments, released by Eurostat and available at `https://ec.europa.eu/eurostat/web/tourism/data/database` and released under the Creative Commons CC BY 4.0 License. The dataset contains the tourist arrivals related to 42 European countries from January 1990 to April 2022. For each country, the dataset provides details related to the type of tourism (foreign, domestic, or both), the type of accommodation, and the unit used to measure the tourist arrivals. In this example, we will select Italian tourism, with a focus on hotels and similar accommodation (code `I551`). In addition, we will select absolute numbers as the unit to measure the tourist arrivals.

The following figure shows an extract of the dataset:

c_resid, unit, nace_r2, and geo\time	2022 M04	2022 M03	2022 M02	2022 M01	...	1990 M03	1990 M02	1990 M01
DOM, NR, I551, AL	:	31761 e	35551 e	37304 e		:	:	:
DOM, NR, I551, AT	:	680242	649287	539601		357811	330121	266230
DOM, NR, I551, BE	:	397127	325543	284400		:	:	:
DOM, NR, I551, BG	:	224482	206135	188719		:	:	:
DOM, NR, I551, CH	:	831017	751113	614597		426512	337820	306973

DOM, NR, I551, CY	:	23821	30126	30594		:	:	:
DOM, NR, I551, CZ	:	630907	594435	428509		:	:	:
DOM, NR, I551, DE	:	6073604	3677290	3677290		3354058	2785396	2530603
DOM, NR, I551, DK	:	306650	227656	147052		:	:	:
DOM, NR, I551, EA	:	22616176 e	18255402 e	15035003 e		:	:	:
DOM, NR, I551, EE	:	98792	91520	86076		:	:	:
DOM, NR, I551, EL	:	342585 e	276111 e	240491 e		:	:	:
DOM, NR, I551, ES	:	3437064	3167837	2328516		1597756	1299985	1190501

Figure 11.9 – An extract of the dataset used in the example

The dataset contains 2,358 rows and 389 columns. Each row contains data related to a country, while columns contain values related to different years. Before you can use the dataset, we should prepare it by extracting a subset containing only the Italian time series.

Let's proceed to extraction:

1. Firstly, we load the dataset as a pandas DataFrame:

    ```
    import pandas as pd
    df = pd.read_csv('source/tour_occ_arm.tsv', sep='\t', na_
    values=': ')
    ```

    As additional parameters for the read_csv() function, we pass the sep argument to specify the separator character, as well as the na_values parameter to make read_csv() recognize the ': ' character as an additional NaN value. The following figure shows an extract of the loaded dataset:

c_resid,unit,nace_r2,geo\time	2022M04	2022M03	2022M02	2022M01	2021M12	2021M11	2021M10	2021M09	2021M08	...
0	DOM,NR,I551,AL	NaN	31761 e	35551 e	37304 e	41636 e	32897 e	33916 e	47912 e	91353 e	...
1	DOM,NR,I551,AT	NaN	680242	649287	539601	318146	415244	989519	1108301	1325158	...
2	DOM,NR,I551,BE	NaN	397127	325543	284400	338310	372601	443372	439419	528126	...
3	DOM,NR,I551,BG	NaN	224482	206135	188719	205099	139072	192263	334894	576847	...
4	DOM,NR,I551,CH	NaN	831017	751113	614597	688729	584084	973563	1069163	1234215	...

Figure 11.10 – The tourist arrivals dataset loaded as a pandas DataFrame

The first column of the dataset is not loaded properly, since columns are not separated by the tab character (\t) but by the comma character (', ').

2. To split the first column into four columns, we write the following code:

```
df[["c_resid", "unit", "nace_r2", "geo_time"]] = df["c_
resid,unit,nace_r2,geo\\time"].str.split(",",expand=True)
```

We access the single cell of the selected column through the `str` accessor. Then, we use the `split()` function to extract the tokens in the string and assign each of them to a new column.

3. Now, we select only the rows related to Italy:

```
df = df[df['geo_time'] == 'IT']
```

The country code is contained in the `geo_time` column.

4. For `unit`, we select numbers:

```
df = df[df['unit'] == 'NR']
```

The unit is contained in the `unit` column.

5. We also select the total number of tourist arrivals:

```
df = df[df['c_resid'] == 'TOTAL']
```

The type of tourists is contained in the `c_resid` column.

6. Finally, we select the type of accommodation associated with code `I551`:

```
df = df[df['nace_r2'] == 'I551']
```

The type of accommodation is contained in the `nace_r2` column.

The following figure shows the dataset produced by applying the previous operations:

	c_resid,unit,nace_r2,geo\time	2022M04	2022M03	2022M02	2022M01	2021M12	2021M11	2021M10	2021M09	2021M08	...
1669	TOTAL,NR,I551,IT	NaN	3892999	3464820	3034874	4258208	3159371	5389662	7413684	10119537	...

Figure 11.11 – The filtered dataset

Now, we should convert the previous dataset into a time series, where each row represents a different date. You can do it by transposing the DataFrame. However, before doing it, we should remove all the columns that do not refer to dates, as well as columns containing NaN values.

7. We drop all the columns that are not dates:

```
df = df.drop(['c_resid,unit,nace_r2,geo\\time', 'c_
resid', 'unit', 'nace_r2', 'geo_time'], axis=1)
```

8. We drop all the NaN values:

```
df = df.dropna(axis=1)
```

9. We transpose the dataset:

```
df = df.T
```

10. We reset the index of the resulting dataset:

```
df = df.reset_index()
```

11. We format the dates according to a common format:

```
df['index'] = df['index'].str.replace('M', '-').str.
strip() + '-01'
```

12. We rename the columns:

```
df = df.rename(columns={'index' : 'ds', 1669 : 'y'})
```

13. We convert columns to the correct type:

```
df['ds'] = pd.to_datetime(df['ds'])
df['y'] = pd.to_numeric(df['y'])
```

14. The dataset contains the last date as the first row, so we need to reverse it:

```
df = df.iloc[::-1]
df.reset_index(inplace=True)
df.drop(['index'], axis=1,inplace=True)
```

The following figure shows the first five rows of the resulting dataset:

	ds	y
0	1990-01-01	2543920.0
1	1990-02-01	2871632.0
2	1990-03-01	3774702.0
3	1990-04-01	5107712.0
4	1990-05-01	4738376.0

Figure 11.12 – The extracted time series

15. You can quickly plot the extracted time series, as follows:

```
import matplotlib.pyplot as plt
plt.plot(df['ds'], df['y'])
plt.show()
```

The following figure shows the produced plot:

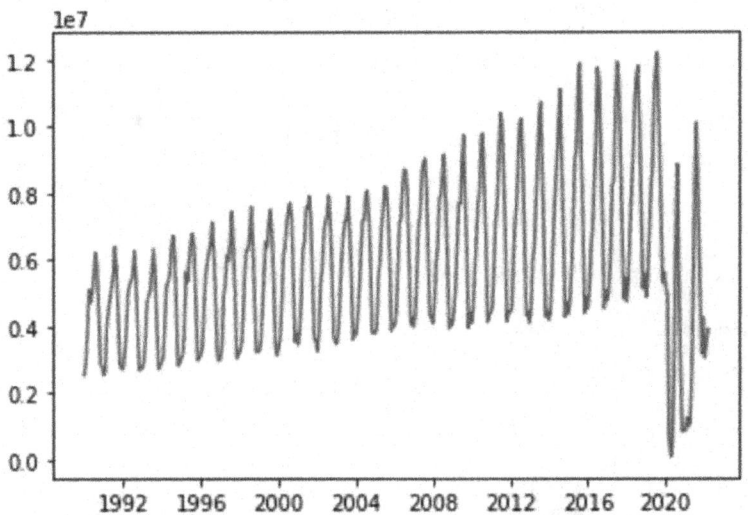

Figure 11.13 – The time series representing the total number of
arrivals at Italian accommodation establishments

The figure clearly shows the breakpoint in correspondence during April 2020, when the lockdown produced by the COVID-19 pandemic began.

Now that we have loaded and prepared the dataset, we can move on to the next step, checking stationarity in data.

Checking stationarity in data

To check stationarity in data, we use the adfuller() function, provided by the statsmodels library. This function performs the ADF test. We proceed as follows:

1. Firstly, we define a function, which returns True if the time series passed as an argument is stationary and False otherwise:

```
from statsmodels.tsa.stattools import adfuller
def is_stationary(df):
```

```
df2 = df.set_index('ds')
ts = df['y']
dftest = adfuller(ts)
adf = dftest[0]
pvalue = dftest[1]
critical_value = dftest[4]['5%']
if (pvalue < 0.05) and (adf < critical_value):
    return True
else:
    return False
```

The `adfuller()` function returns a tuple containing the test statistic (`dftest[0]`), the p-value (`dftest[1]`), and other values, including the critical ones.

2. We call the defined function as follows:

```
test_result = is_stationary(df)
if test_result == True:
    print('The series is stationary')
else:
    print('The series is NOT stationary')
```

In our case, the series is stationary, so do not need to transform it to make it stationary.

We are ready to build the two prediction models, so let's proceed to the next step, building the models.

Building the models

We build two different models; the first one does not consider the COVID-19 breakpoint, and the second one does. In both cases, we perform the following operations:

- Building the Comet experiment
- Building the Prophet model
- Logging model results in Comet
- Logging performance metrics in Comet

In the description that follows, we will analyze the model without breakpoints. The procedure adopted for the model with breakpoints is very similar, with only one difference when creating the model.

Let's start from the first step, building the Comet experiment.

Building the Comet experiment

To build the Comet experiment, you can proceed as follows:

1. Firstly, we import the libraries:

    ```
    from comet_ml import Experiment
    from prophet import Prophet
    ```

    You should make sure that you import the comet_ml library before the Prophet one.

2. Finally, we create the experiment and set its name:

    ```
    experiment = Experiment()
    experiment.set_name('WithoutChangePoints')
    ```

To make the preceding code work, you should configure the .comet.config file, as described in *Chapter 1, An Overview of Comet*.

We are ready to build the model, so let's proceed.

Building the Prophet model

We split the dataset into training and test sets. The training set contains the first rows, up to the date of 2020-12-01, and the test set contains the remaining rows. We have chosen to keep in the training set some rows related to the effects of the COVID-19 pandemic (from April 2020 to December 2020), to give the model the possibility to learn the presence of the breakpoint:

1. To split the dataset into training and test sets, we write the following code:

    ```
    index = df.index[df['ds'] == '2021-01-01'].tolist()[0]
    n = df.shape[0] - index
    df_train = df.head(index)
    df_test = df.tail(n)
    ```

    We extract the index used to separate the training and test sets, and then we assign the first index rows to the training set and the remaining row to the test set.

2. We create the model and train it with the training set, as follows:

    ```
    m = Prophet()
    m.fit(df_train)
    ```

    In the case of the model that also considers the breakpoint, we create a different model, as shown here:

    ```
    m = Prophet(changepoints=['2020-03-01'])
    ```

3. We are ready to use the model to make predictions:

```
future = m.make_future_dataframe(periods=n, freq='MS')
forecast = m.predict(future)
```

We predict n values, with a monthly frequency (freq='MS'). The forecast variable is a DataFrame, which contains all the information related to the predicted values. The following figure shows the columns of the forecast DataFrame:

```
Index(['ds', 'trend', 'yhat_lower', 'yhat_upper', 'trend_lower', 'trend_upper',
       'additive_terms', 'additive_terms_lower', 'additive_terms_upper',
       'yearly', 'yearly_lower', 'yearly_upper', 'multiplicative_terms',
       'multiplicative_terms_lower', 'multiplicative_terms_upper', 'yhat'],
      dtype='object')
```

Figure 11.14 – The list of columns of the forecast DataFrame

After building the model, we are ready to log model results in Comet, so let's proceed.

Logging model results in Comet

We log three graphs in Comet:

1. Firstly, we log the forecast graph, as follows:

```
fig1 = m.plot(forecast)
```

The following figure shows the produced plot:

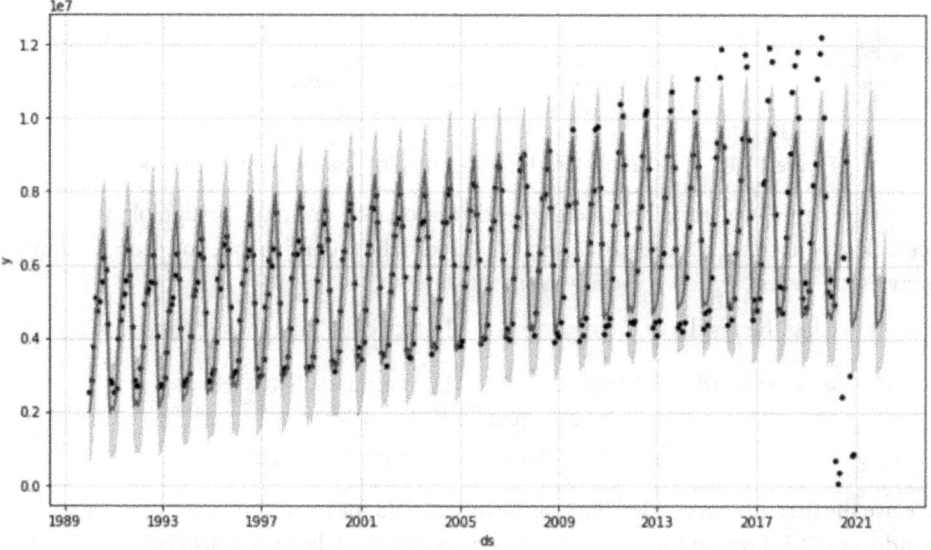

Figure 11.15 – The graph produced by the plot() method

The points in the figure indicate the data points used to train the model, the bold line indicates the prediction, and the light area indicates the uncertainty intervals. Note that the model is not able to recognize the breakpoint in correspondence at the beginning of the lockdown caused by the COVID-19 pandemic.

2. Then, we log the `forecast` components, as follows:

```
fig2 = m.plot_components(forecast)
```

Similar to the `plot()` method, `plot_components()` should log the figure in Comet automatically. The following figure shows the produced graphs:

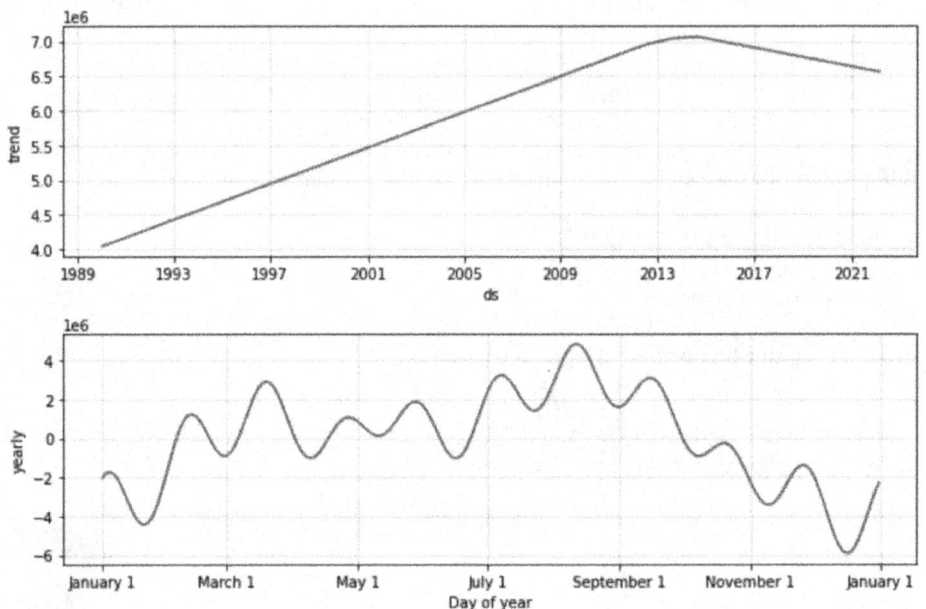

Figure 11.16 – The graphs produced by the plot_components() method

The graph shows the trend line and the yearly seasonality. From the first graph, you can clearly see a change in trend between 2013 and 2017. From the second graph, you see a peak in tourist arrivals in August, which is the hottest month in Italy.

3. Finally, we log the results of the `cross_evaluation()` diagnostic function:

```
df_cv = cross_validation(m, initial="7300 days",
period="365 days", horizon="730 days")
fig3 = plot_cross_validation_metric(df_cv, "rmse")
```

We use the first 20 years (7,300 days) as historical data for the training set, then we use a cutoff window of 365 days, and we predict the next 2 years (730 days). We also calculate RMSE. The following figure shows the plotted figure:

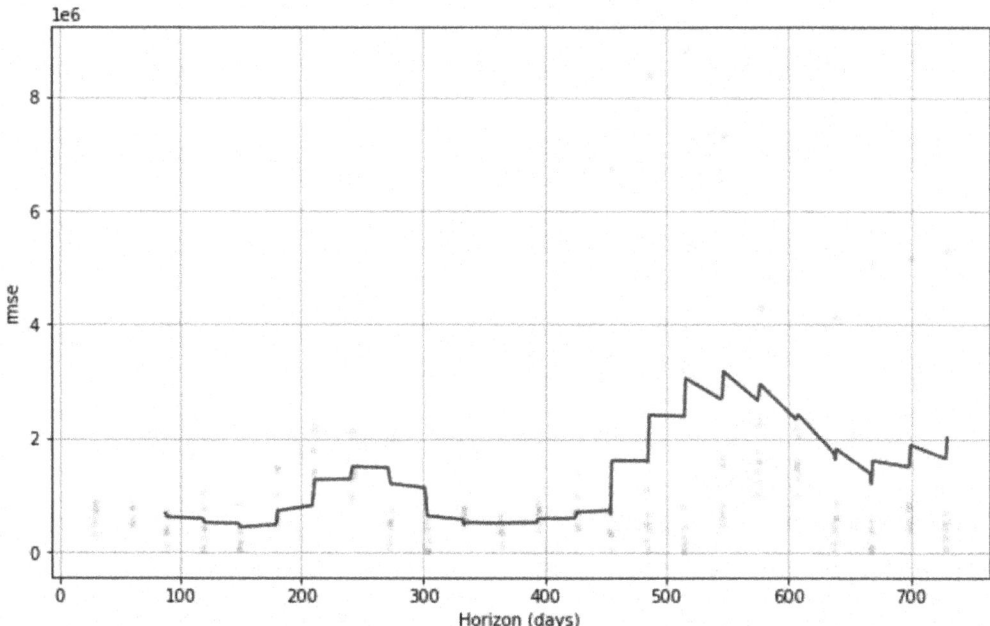

Figure 11.17 – RMSE over Horizon, as produced by the plot_cross_validation_metric() function

Now that we have logged the model results in Comet, we can move on to the next step, logging performance metrics in Comet.

Logging performance metrics in Comet

We log three performance metrics, MAE, MAPE, and RMSE, as follows:

1. We use the functions provided by the `scikit-learn` package to calculate the metrics, so we first import them:

    ```
    from sklearn.metrics import mean_absolute_error
    from sklearn.metrics import mean_absolute_percentage_
    error
    from sklearn.metrics import mean_squared_error
    ```

2. Then, we define a function, which receives the original DataFrame and the forecast as input; then, it calculates the metrics and, finally, logs the results in Comet:

    ```
    def log_metrics(ds, forecast, experiment):
        df_merge = pd.merge(df[['ds', 'y']],
    forecast[['ds','yhat']],on='ds')
    ```

```
y_true = df_merge['y'].values
y_pred = df_merge['yhat'].values
metrics = {}
metrics['mae'] = mean_absolute_error(y_true, y_pred)
metrics['mape'] = mean_absolute_percentage_error(y_
true, y_pred)
metrics['rmse'] = mean_squared_error(y_true, y_pred,
squared=False)
experiment.log_metrics(metrics)
```

3. Finally, we call the function with the actual parameters, as follows:

```
log_metrics(df,forecast, experiment)
```

You should remember that, at the end of your code, you should call the `experiment.end()` method, since you are using Deepnote.

You can repeat the previous steps for the second `Prophet` model, which also considers the COVID-19 breakpoint.

Now that we have learned how to log performance metrics in Comet, we can move on to the next step, exploring results in Comet.

Exploring results in Comet

After running the experiments, you should see the results in Comet:

1. For each experiment, in the **Chart** menu, you can see the charts shown in the following figure:

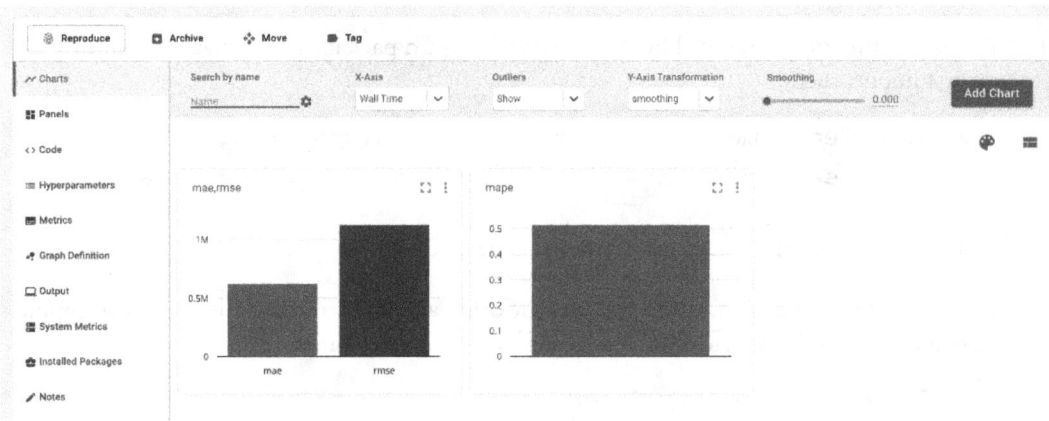

Figure 11.18 – The charts produced automatically by Comet

2. In the **Hyperparameters** menu, you can see the hyperparameters logged by the model, as shown in the following figure:

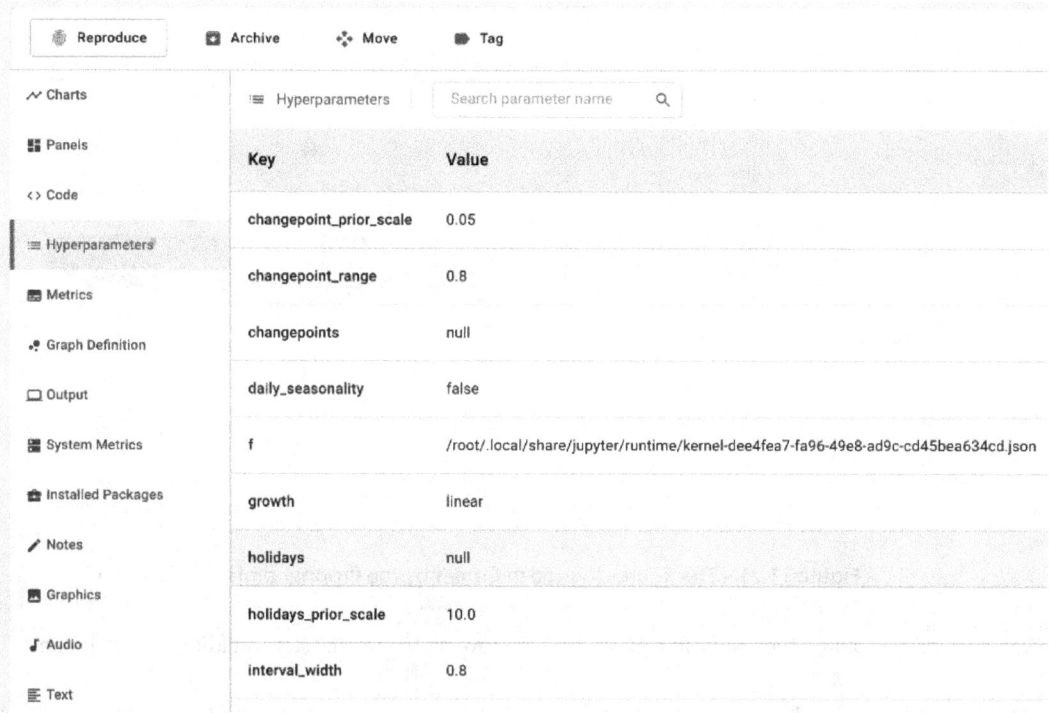

Figure 11.19 – The hyperparameters logged automatically by Comet

3. In the **Metrics** menu, you can see all the logged metrics, as shown in the following figure:

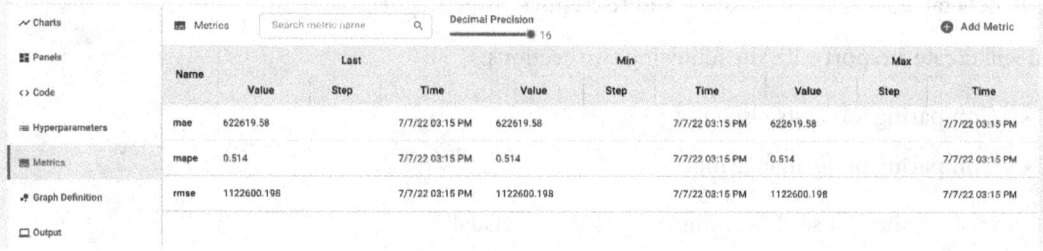

Figure 11.20 – The logged metrics in Comet

The table in the preceding screenshot shows the last (**Last**), minimum (**Min**), and maximum (**Max**) values for each metric.

4. In the **Graphics** menu, you can find the figures logged by Prophet, as shown in the following figure:

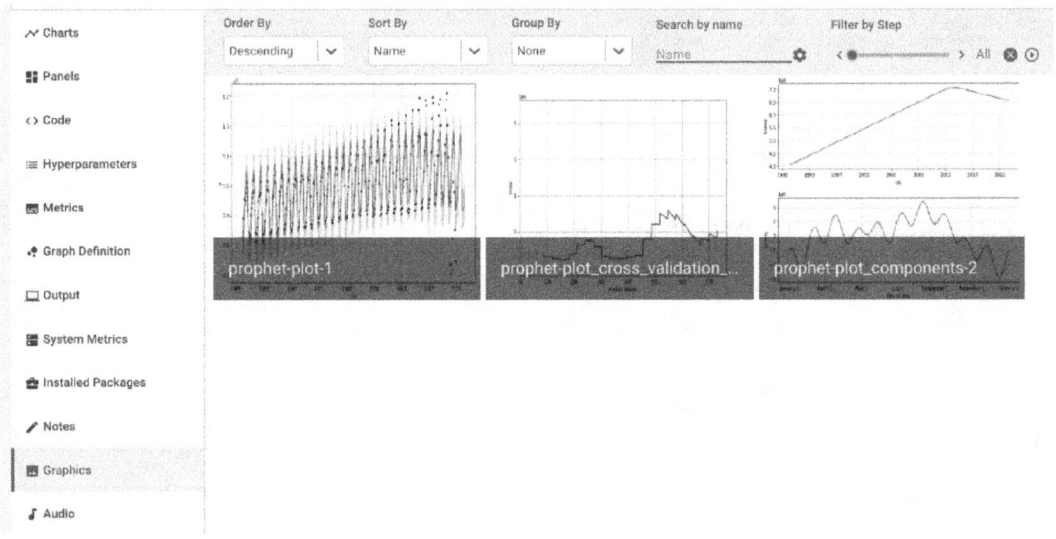

Figure 11.21 – The figures logged in Comet by the Prophet plots

Now that we have explored the results in Comet, we can move on to the final step, building the final report.

Building the final report

Now you are ready to build the final report. In this example, we will build a simple report with the results of both models. As a further exercise, you could improve them by applying the concepts learned in *Chapter 5*, *Building a Narrative in Comet*. To create the report, in the Comet dashboard, click on the **Panels** tab and then select **Add | Add to Report | New Report**.

You will create a report with the following two sections:

- Comparing forecasts visually
- Comparing performance metrics

Let's start with the first section, comparing forecasts visually.

Comparing forecasts visually

In this section, we add all the graphics available under the **Graphics** menu:

1. Firstly, we click on the **Add your fist panel** area, within your section in the report.
2. A popup window opens. Select **FEATURED | Show Images | Add | Done**.

3. A new panel is added to your section, but the graphs are small. You can adjust the panel size by clicking on the **Edit Layout** button, located at the top-right part of the section, and drag and drop the panel to adjust its size.

4. Within the panel, you can select one of the two experiments and show the related graphs.

Comparing performance metrics

In this section, we add two panels related to performance metrics:

- The first panel contains a bar chart showing MAE and RMSE for the two experiments.

- The second panel contains a bar chart showing MAPE for the two experiments. We use a different panel for MAPE because it has a different order of magnitude with respect to MAE and RMSE.

To add the two panels, you can proceed as follows:

1. Firstly, we click on the **Add your fist panel** area, within our section in the report.

2. A popup window opens. Select the **BUILT-IN** tab | **Bar Chart** | **Add**. Select **MAE** in the **Y-Axis** textbox.

3. You can click on the **Add field** button under the **Y-Axis** textbox and then select RMSE.

4. Then, click **Done**. A new panel is added to your section.

You can repeat *steps 1, 2, and 3* to add the **MAPE** panel.

Your report is ready! You can view the final result directly in Comet at the following link: `https://www.comet.com/packt/time-series-analysis-deepnote/reports/time-series-forecasting`.

Summary

We have just built a time series analysis model in `Prophet` and tracked it in Comet!

Throughout this chapter, we described some general concepts regarding time series analysis, including stationarity, seasonality, and breakpoints. In addition, you have learned the main concepts behind the Prophet package and how to combine it with Comet.

In the last part of the chapter, you implemented a practical use case that showed you how to track and compare two time series analysis experiments in Comet, as well as how to build a report with the results of the experiments.

The world of data science is very promising and challenging. Both research and industry sectors are constantly trying to improve current knowledge with new algorithms, frameworks, and tools. Throughout this book, you have investigated Comet, one of the promising platforms for experiment tracking and monitoring.

I hope that all the concepts you learned in this book will help you to increase your knowledge to build better models, track them with a valid tool, and eventually, become a better data scientist and be able to increase your overall knowledge for the future.

Further reading

- Atwan, T.A. (2022). *Time Series Analysis with Python Cookbook*. Packt Publishing Ltd.

- Nielsen, A. (2019). *Practical Time Series Analysis: Prediction with Statistics and Machine Learning*. O'Reilly Media.

- Pinheiro, C.A.R. and Patetta, M. (2021). *Introduction to Statistical and Machine Learning Methods for Data Science*. Cary, NC: SAS Institute Inc.

Index

Packt.com

Subscribe to our online digital library for full access to over 7,000 books and videos, as well as industry leading tools to help you plan your personal development and advance your career. For more information, please visit our website.

Why subscribe?

- Spend less time learning and more time coding with practical eBooks and Videos from over 4,000 industry professionals

- Improve your learning with Skill Plans built especially for you

- Get a free eBook or video every month

- Fully searchable for easy access to vital information

- Copy and paste, print, and bookmark content

Did you know that Packt offers eBook versions of every book published, with PDF and ePub files available? You can upgrade to the eBook version at packt.com and as a print book customer, you are entitled to a discount on the eBook copy. Get in touch with us at customercare@packtpub.com for more details.

At www.packt.com, you can also read a collection of free technical articles, sign up for a range of free newsletters, and receive exclusive discounts and offers on Packt books and eBooks.

Other Books You May Enjoy

If you enjoyed this book, you may be interested in these other books by Packt:

Applied Machine Learning Explainability Techniques

Aditya Bhattacharya

ISBN: 9781803246154

- Explore various explanation methods and their evaluation criteria
- Learn model explanation methods for structured and unstructured data
- Apply data-centric XAI for practical problem-solving
- Hands-on exposure to LIME, SHAP, TCAV, DALEX, ALIBI, DiCE, and others
- Discover industrial best practices for explainable ML systems
- Use user-centric XAI to bring AI closer to non-technical end users
- Address open challenges in XAI using the recommended guidelines

Practical Deep Learning at Scale with MLflow

Yong Liu

ISBN: 9781803241333

- Understand MLOps and deep learning life cycle development
- Track deep learning models, code, data, parameters, and metrics
- Build, deploy, and run deep learning model pipelines anywhere
- Run hyperparameter optimization at scale to tune deep learning models
- Build production-grade multi-step deep learning inference pipelines
- Implement scalable deep learning explainability as a service
- Deploy deep learning batch and streaming inference services
- Ship practical NLP solutions from experimentation to production

Packt is searching for authors like you

If you're interested in becoming an author for Packt, please visit `authors.packtpub.com` and apply today. We have worked with thousands of developers and tech professionals, just like you, to help them share their insight with the global tech community. You can make a general application, apply for a specific hot topic that we are recruiting an author for, or submit your own idea.

Share Your Thoughts

Now you've finished *Comet for Data Science*, we'd love to hear your thoughts! Scan the QR code below to go straight to the Amazon review page for this book and share your feedback or leave a review on the site that you purchased it from.

`https://packt.link/r/1-801-81443-0`

Your review is important to us and the tech community and will help us make sure we're delivering excellent quality content.